INTERNATIONAL TELECOMMUNICATION MANAGEMENT

For a complete list of the *Artech House Telecommunications Library,*
turn to the back of this book . . .

INTERNATIONAL TELECOMMUNICATION MANAGEMENT

Bruce R. Elbert

Artech House
Boston • London

Library of Congress Cataloging-in-Publication Data

Elbert, Bruce R.,
  International telecommunication management / Bruce R. Elbert
      p.   cm.
  Includes bibliographical references.
  ISBN 0-89006-383-4
  1. Telecommunication--Handbooks, manuals, etc.
  2. Telecommunication systems--Handbooks, manuals, etc. I. Title.
  HE7621.E43   1990   90-30130
  384--dc20   CIP

British Library Cataloguing in Publication Data

Elbert, Bruce R.
  International telecommunication management.
  1. International telecommunication systems
  I. Title
  384

  ISBN 0-89006-383-4

International Standard Book Number: 0-89006-383-4
Library of Congress Catalog Card Number: 90-30130

10   9   8   7   6   5   4   3   2   1

This book is dedicated to communication,
the medium of progressive change throughout the world.

Contents

PREFACE

International Telecommunication Management is the third in a series of books by this author on telecommunication systems and applications in the commercial world. The book is written at an introductory level, but considerable detail on national environments and digital telecommunication systems is included. For the reader to have some understanding of commercial telecommunication is desirable, either by personal experience in business or through study and reading of some of the trade press; this is not essential, however, insofar as necessary technical material is included. This can be skipped by those who already are familiar with the technologies involved, or whose main focus is management and operational issues.

We take the view of the strategic unit, which is either a business operation or a government agency. Examples of strategic units include commercial ventures, like airlines, retail store chains, automobile manufacturers and distributors, and financial institutions; electronic media organizations, like television and radio networks, news organizations, cable television services, video production companies and syndicators of programming; and governmental groups, like foreign services and embassies, public information agencies, and international regulatory bodies and trade associations. The basic philosophy is that a strategic unit requires telecommunication and information technology to accomplish its designated mission.

Many strategic units find that their activities are becoming international in scope and their telecommunication and information systems are not easily adaptable to the multiplicity of country environments. In addition to the strategic unit, those organizations which provide equipment and services for telecommunication and information processing would also benefit from the information contained in this book.

There are other books on the international environment for telecommunication, but we believe that a completely new approach needs to be taken. Users and providers of services are confronted with a diverse array of technologies and regulatory environments in different countries and in different geopolitical regions.

The older references are out of date, due to the rapid changes in the composition of and procedures for using telecommunication systems throughout the world.

The audience for *International Telecommunication Management* cuts across many lines in the fields of telecommunication, information technology, television and radio broadcasting, international business, and government relations. Our primary focus is on the role of and challenges faced by the telecommunication or network managers of strategic units as they enter the international environment. The individual who holds the title of Chief Information Officer (CIO) on the executive staff can benefit greatly from exposure to this material, although other specialists are directly served by the detailed discussions included within. We summarize the key points at the beginning of each chapter to aid the reader interested in an overview.

Television and radio professionals often require international hookups to transmit and distribute programming around the world. The viewer and listener have become accustomed to programs, news, and sporting events coming from overseas, and broadcasters find that they must deal in different countries to satisfy these needs. News directors will find the information helpful as they often require a variety of telecommunication services, such as special telephone access lines and electronic messaging.

Many readers will approach the international telecommunication challenge from the perspective of the engineer, marketing professional, and business manager working for equipment manufacturers, service providers, and system integrators. These readers will find that the book is organized to give them a good introduction to the various elements of international networking for voice, data, imagery, and video. Importantly, we discuss the key players both internationally and within the major regions of North America, Europe, and Asia. Obviously, you cannot design and implement systems without having the "lay of the land" reasonably well understood. This is certainly the case for marketeers who need current information on potential partners and competitors, which we cover in some detail. Also of interest are examples of user applications which have become popular among strategic units.

An important audience consists of students, teachers, and consultants seeking reliable information on international telecommunication availability and regulation. Many universities offer degree programs in telecommunication management, for which this book would provide background reading and source material for the preparation of case studies.

The two other books by this author are *Introduction to Satellite Communication* [Elbert, 1987] and *Private Telecommunication Networks* [Elbert, 1989]. As stated previously, the three books complement each other and are useful as references for the practitioner in telecommunication management. The first book is a complete review of commercial satellite systems, which, as readers are aware, are used for a variety of video, audio, and data communication services. The book provides a good underpinning in the technology and applications for satellite networks, particularly those for businesses. The second book broadens the scope in private telecommunication to include terrestrial networks and services within a common country, such as the United States. Finally, *International Telecommunication Management* extends the framework internationally, where the technology often takes a secondary role to the complexity of dealing in different countries. Regulatory issues, standards concerns, and competition among newly privatized service providers are of great interest, and consequently they are described in detail.

The book is organized into ten chapters that develop each topic in a progressive way. Chapter 1 is an overview of the book and can be read as an executive "summary." Most of the basic points are mentioned, informing the reader of many of the issues that confront the telecommunication or network manager as he or she embarks on an international network project.

The specific arrangement of the worldwide physical plant is provided in Chapter 2, The International Network Environment. We emphasize modern digital fiber optic and satellite facilities, as these support high-quality voice, data, image, and video transmission. The domestic telecommunication authority (TA) — also known as the post, telephone, and telegraph (PTT) agency in many countries — is introduced and various profiles are presented. This is important because some TAs operate like commercial telephone companies, whereas others behave like typical governmental agencies. TA profiles are provided for countries in key regions, including North America, Western Europe, Eastern Europe, Latin America, the Pacific Basin, and the Far East. The degree to which competition is allowed is also discussed in terms of its effect on domestic services. The major fiber optic and satellite networks of the world are reviewed because they provide the international backbone for all voice, data, image, and video services. This material is up to date as of 1990 and includes capabilities that will be important well into the decade.

Chapter 3, Circuit Switched and Private Line Services, covers the primary services offered by the TAs and international carriers providing for many conventional needs of strategic units. For the most part, these services are implemented over the international fiber optic and satellite backbone networks operated by the traditional carriers. We present applications in switched telephone service for voice and low-speed data using modems. Strategic units also use the backbone for fulltime connections called *private lines;* these are applied for voice communication and dedicated data networks. The dedicated channels are provided via fiber optic cables and satellite links, notably INTELSAT Business Service (IBS). Other facilities must be added by the strategic unit, as discussed in Chapters 5 and 6.

Chapter 4, Packet-Switched Data Network Services, is a complete review of the options for obtaining information networking on an international basis from third-party suppliers. Data communication and information technology are particularly important for strategic units in today's competitive world. Although building international private networks is a challenge, we may be able to deal fairly effectively by using public data networks that employ computer messaging and packet switching. These are also denoted *value-added networks,* where the service provider takes care of many of the ticklish interfacing problems and often provides information services as well. These services include basic packet-switched transport using X.25 and SNA, electronic mail (E-mail), electronic data interchange (EDI), and information or "videotex" services. Providers such as Infonet, GE Information Systems, and IBM Information Network have made extensive global networks and computing capabilities available to users, solving many of the data communication problems that may be encountered. Public network operators including AT&T, British Telecom, France Telecom, and Singapore Telecom are also very active in this field.

We mentioned in the context of Chapter 2 that strategic units often acquire dedicated capacity and equipment to implement a private network capability. In Chapters 5 and 6, we describe the types of telecommunication equipment that can be purchased and operated as a part of an overall system, primarily via private lines. The basic structure of private networks, treated in more detail in [Elbert, 1989], is reviewed in Chapter 5, Private Ownership of Networks and Customer Premises Equipment. These include local area networks (LANs), wide area networks (WANs), T1 backbones using high-level multiplexers, and specialized systems for bandwidth compression. The chapter reviews customer premises equipment that users connect to public and private networks, for voice, data, facsimile, and video teleconferencing applications.

Chapter 6, Telecommunication Node Equipment in Private Networks, reviews the nodal devices that serve user application systems and route traffic over the backbone. Examples include packet switches, private branch exchanges (PBXs), digital cross-connects (DACSs), and statistical multiplexers. Fast packet switching and very small aperture terminal (VSAT) satellite networks are covered, as these are some of the more interesting of the emerging technologies. All told, the devices for user access and transmission represent "tools of the trade" for the telecommunication or network manager.

Facilities and services for the carriage of international television and audio are presented in Chapter 7. This information is directed primarily at broadcasting professionals responsible for transporting television and radio programming material between countries. In addition, telecommunication managers and audiovisual professionals in strategic units would find Chapter 7 helpful with regard to the setting up of international teleconferences. We review the three color television systems in use around the world and describe the role of digital television for transmission over fiber optic cables. Other key technologies like video scrambling and high definition television (HDTV) are reviewed, particularly as they relate to international video transmission. Because satellite communication provides the bulk of current international television transmission, we review the procedures for employing INTELSAT for video services in some detail. Organizations such as the National Association of Broadcasters (NAB) and regional broadcasting unions are discussed, as these can provide meaningful support.

The process for managing an international network project is covered in Chapter 8, Management and Logistics. We present workable strategies for dealing in foreign countries, where cultural and political differences can make even a simple task seem impossible. We emphasize that an in-country presence will greatly aid in overcoming the problems that will also arise (and are unpredictable, even for the most experienced manager). Network management is presented as one of the technologies and processes that will prove important.

International Telecommunication Regulation and Data Flows, Chapter 9, delineates the structure that currently governs country-to-country communication. The International Telecommunication Union (ITU), a specialized agency of the United Nations, is the major force in international regulation, as it allocates radio frequencies and recommends standards for telecommunication services and equipment interfaces. Most countries of the world follow these recommendations, although there is still considerable variation from one country to another. The organization and results of ITU conferences, particularly the World Administrative Radio Conference (WARC) and the World Administrative Telegraph and Telephone Conference (WATCC), are discussed. Depending on the particular circumstances, the telecommunication plans of a strategic unit need to consider the ITU as well as other organizations, such as the International Organization for Standardization (ISO), INTELSAT, and regional bodies such as the European Telecommunication Standards Institute. Rules with regard to content of communication can hamper the transfer of data across borders, which is the reason that transborder data flows are considered in this chapter.

Chapter 10, Trends in International Telecommunication, looks ahead toward the year 2000, when many of the recent political and technical changes will have taken effect. We discuss the importance of the opening of frontiers in Western Europe as a result of the passage of the Single European Act by the European Community (EC). Another important factor is the increased role of information technology in business and the global economy. We anticipate that telecommunication networks of the future will be more effective due to full digitization, wide bandwidth, and the use of consistent standards such as ISDN (integrated services digital network).

Reflecting on the development of this book, the author recognizes that he could not have covered so much ground without the help of many able people. To read about global networking and business is not enough; you have to experience it. The author's personal experiences in telecommunication management as part of the U.S. Army Signal Corps in the United States and Southeast Asia were key to understanding the delivery of services under difficult conditions. While working for COMSAT, much help was provided by the worldly experiences of Pier

Bargellini and Emeric Podraczky. Later, at Hughes Aircraft, Lloyd Harrison and Bill Grayer provided many opportunities to learn about international telecommunication regulation as we built the Indonesian satellite network, Palapa A. Mr. Wikanto of Perumtel also provided valuable lessons in how to practice telecommunication network development in a developing country. Philip Van der Veen, the European representative of Hughes, had on many occasions introduced this author to Europe and the complexity of doing business there, particularly with the ITU, the PTTs, and European manufacturers. Likewise important is to recognize the information transfer from Takaji Kuroda of NEC Corporation, a man of many technical talents (including art forms such as music and kite making).

Particular help on the book came from Naoki Endo, Leslie Taylor, David Allen, Peter Neuman, Kathy Hansel, Mat Fujita, Pat Dinneen, Wes Hanemayer, Malek Asghar, Pierre Conruyt, Gary Epstein, Jim McEachern, Francisco Linares Merino, John Milman, Bill Page, Dan Miller, Steve Carroll, Joanne Tanner, Elizabeth O'Beirne, Jan Kaechele, Ed Miller, Walter Morgan, Norm Wienhouse, Derek Thorpe, Luc-Pierre Terral, Kas Kalba, Joaquin Patron, Gerry Ragsdale, Nobuhiko Shimasaki, Rebel Brown, and Chris Schram. Organizations to be recognized are COMSAT, INTELSAT, France Telecom, BrightStar, Infonet, IBM Information Network, British Telecom, MIT, CBS, Turner Broadcasting, NHK, the ITU, Teknekron, KJH Communications, Sprint International, COMSYS, Logica, KDD, C. Itoh, Latham and Watkins, and Hughes Communications. Also appreciated is the support and friendship of Professor Norman Abramson of the University of Hawaii.

Overall organization and focus were helped by the many useful suggestions of Vint Cerf, the series editor, and by Mark Walsh and Julie Burnham of Artech House. Dennis Ricci, always the professional, did his able part in bringing this book to press.

The final recognition goes to my wife, Cathy. She is my greatest supporter in the writing of these books, and those of you who appreciate this work should understand how important she is to it.

Chapter 1
INTRODUCTION

Many organizations deal in an international context, and communication conse-quently becomes a much greater challenge than that which is experienced at home. Through information technology in general and telecommunication in particular, businesses and government agencies are better able to cope with the distances, time spans, and cultural differences that we encounter around the globe. That voice patterns, data, and images can be carried on radio waves and via cables is well recognized. These classical forms of information travel the telecommunication highways within and between countries.

Television and radio broadcasters are likewise concerned about how they will bring signals of high quality from event locations halfway around the world. During the student demonstrations in Beijing in May 1989, viewers were rewarded with live pictures and reports from the scene, thanks to the efforts of CBS News and Cable News Network. Broadcasts from the opening of the Berlin Wall in November of that same year were equally dramatic. Event coverage, news broadcasts, and even entertainment are being transmitted between countries with much greater frequency than ever. International television and audio, both of which have de-manding technical standards, have become an important medium to broadcasters, as they continually adjust to the tastes and interests of viewers and listeners. Consequently, broadcasters require reliable international links, which can be set up in a matter of days or less.

International telecommunication management, the theme of this book, deals more with the organization, selection, and operation of these highways than with the internal workings of the devices that compose the networks of the world. Therefore, we do not examine international networks as would an engineer, but rather as one who would use and apply services to solve business problems. The chief information officer or telecommunication manager of a corporation or gov-ernment agency (termed a "strategic unit") is usually very familiar with the concepts which underlie the private telecommunication network. In an earlier book [Elbert, 1989], the purpose and architecture of private telecommunication networks are developed in detail for domestic uses within regions such as North America, Europe, or Japan. This type of information can be valuable to a strategic unit that carries on operations within the borders of an industrialized country with a well developed telecommunication infrastructure. Strategic units are now familiar with the use of PBXs, virtual private networks, T1 and packet-switched networks, VSATs, distributed relational databases, optical disk storage and image processing, *et cetera*. Extending these concepts globally, although theoretically possible, can be extremely difficult, even as we approach the year 2000. The political, cultural, and economic barriers can be frustrating to the business person if adequate patience and knowledge are not applied.

The bulk of the information in this book is designed to help the CIO, tele-communication manager, or news director develop his or her own process to build a private network that is international in extent. An important consideration is how one deals with service providers throughout the world, notably the PTT agency with monopoly control of domestic telecommunication. In a survey of U.S. companies involved in international business conducted by the National Telecommunications and Information Agency (NTIA), the consistently most serious problem was the difficulty of working with the local PTT [NTIA, 1985b]. This subject receives considerable attention throughout this book. In this introductory chapter, we explain why international telecommunication management is even necessary in the first place.

1.1 OVERVIEW OF NEED

The previous twenty or thirty years have brought incredible changes in the ways that business is conducted. Even in the United States, all major corporations conduct a certain amount of business with or in foreign countries. This considers the flow of goods, services, and particularly capital, because many of the dollars that have left the country are returning in the form of foreign investment. U.S. companies that are successfully finding foreign markets need good communication to be competitive with distant rivals. After all, the competitive edge once enjoyed by the United States has certainly been dulled by the diffusion of technology and the growth of foreign capital wealth. Many U.S. corporations maintain a competitive edge due to a superior technology base. The U.S. market, however, is viewed as perhaps the most lucrative in the world; hence, foreign corporations find that they can successfully compete on U.S. soil. To do so, they also must develop excellent communications.

1.1.1 Globalization of Business

The term "multinational corporation" was coined several years ago in reference to a business that operates in a number of countries. One of the prime examples is ITT, which once had substantial manufacturing operations in Europe, Latin America, and the United States. A key point was that the multinational corporation tended to allow the individual foreign siblings to operate more or less on their own, with management personnel often recruited from the local economy. In the days of mediocre communication, this was almost an imperative. The situation today appears to be much different because modern telecommunication and information technology tend to permit companies to extend control to distant points. The headquarters in a foreign country can now manage distant operations almost as effectively as a local rival.

1.1.2 Europe in 1992

The European Community has begun an historic program to remove many of the administrative barriers that exist between member countries. In the key year of 1992, people and trade will begin to move freely across borders between countries in Western Europe. Work is underway to facilitate the flow of information as well. Liberalization in Eastern Europe is also having a major effect on the thinking and realization of the integration of economies on the continent. There is the strong belief that a new Europe will emerge, wherein business can be conducted almost as easily as in the United States.

Another important trend in Europe is the consolidation of the largest business ventures into multinational corporations. These companies can aggregate their activities within the more open Europe and potentially compete more effectively in markets outside of Europe. For such firms, international telecommunication will play an extremely vital role. Conversely, competitors in the United States and Japan could seize the opportunity to set up more unified operations in Europe. This would be attractive because the European market should become more homogeneous.

1.1.3 Linkages in National Relations

Telecommunication is a common language among the practitioners of the art in every country of the world. This author was particularly impressed upon meeting the staff telecommunication engineers in Indonesia in 1974. As was apparent, they had received world class training on the same technology that we employed in the United States. One young engineer even produced a copy of the standard text on telecommunication engineering [Schwartz, 1980]. During the ensuing months of contact between these Indonesian engineers and their American counterparts, the common language of telecommunication clearly facilitated a successful project.

When President Richard M. Nixon made his historic visit to the People's Republic of China in 1971, the first concrete step taken was to install a satellite communication terminal. This permitted the U.S. government staff as well as the press and broadcasters to maintain contact with its home base more than a hemisphere away. Every major visit between East and West has involved an advance party for the purpose of establishing reliable communication.

Once established, telecommunication provides a linkage between entities in different countries. The Washington–Moscow hotline is one of the earliest examples. It was installed in 1969 to allow the President of the United States to make immediate contact with his Soviet counterpart. A balance was struck by implementing two dedicated circuits, one over the public INTELSAT system and one over the Soviet Molnya satellites. This is the highest level of linkage for national relations. The same principle, however, applies to such dedicated circuits used by lower levels of government and, of course, businesses.

1.1.4 Importance of Information Technology

Telecommunication and data processing constitute approximately 10% of the cost of doing business. This is because most organizations now rely on information technology to enhance or maintain a competitive position in their respective activities. Modern computer systems are increasingly being used to run a business. In the past we referred to the "back office" of a company where records were kept, orders processed, and all other administrative tasks were performed. The rapid growth of computer automation has allowed companies to replace the back office with a data processing environment, and to shift the focus from administration to other, more important areas such as customer service and marketing.

In some books, [Keen, 1988] and [Porter, 1985], the manner in which information technology is employed by leading companies is emphasized. Many companies have become "time-based competitors," employing information technology to reduce the time required to develop a new product or service. The apparel industry is characterized by rapid shifts in consumer desires. One particular company, Benetton, operates out of its home country of Italy, where it manufactures all of its clothing. A real-time global telecommunication network, however, allows Benetton to couple manufacturing decisions to the instant buying habits of its retail customers. This same concept is being tried by Japanese automobile companies to allow a customer order to be sent directly to the manufacturing cycle and produce a specific car (with options) in a matter of weeks.

1.2 HIERARCHY OF NETWORKS

Virtually all telecommunication networks are arranged in a hierarchical structure to achieve efficient loading of transmission and switching facilities. This simply means that calls and messages are routed from low-level terminal devices through increasingly higher levels of aggregation; once the highest level of the network is reached, the process is reversed through to the distant terminal device. The domestic telephone network in most countries is arranged in this manner. In international telecommunication, we encounter a different kind of hierarchy, as there is no true international system. This hierarchy consists of the different transmission media that can be employed to complete a temporary or permanent connection between countries. Global networks mainly operate to provide major paths between continents, typically spanning oceans. These networks have been designed to be universal, much like the postal system, so that any country can connect and thereby provide international telecommunication services.

Complementing the global networks are regional networks and links available to countries on the same continent. Such countries usually are tied economically or even politically; hence, the telecommunication infrastructure may already be designed to the same standards. Final connection will require the facilities and services of domestic telecommunication networks where the ultimate subscribers or users are located. The long-haul links are provided by international satellites, transoceanic cables, and transborder microwave, cable systems, and domestic satellite links. Theoretically, any one of these media could be employed universally; as a practical matter, however, an international telecommunication network must employ an appropriate mix. This mix tends to minimize cost and provide reliable service (through such capabilities as diverse routing where it is economically feasible).

1.2.1 Global Networks Topology

The primary role of global telecommunication networks is to provide transmission capacity between countries. When international communication proliferated in the last century, it was in the form of direct physical connections between neighbors by using telegraph poles and wire. This was expanded with undersea cables and eventually high frequency radio. Microwave line-of-sight radio links were installed in the 1950s and 1960s to reach across land masses and between countries, providing high transmission bandwidth and excellent reliability. This was not possible with high frequency radio because of the limitations of ionospheric "skip" propagation. Over-the-horizon microwave links of up to 200 miles are possible by using tropospheric scatter propagation, but the installations are expensive, particularly considering their low bandwidth capability. These factors indicate the superiority of satellite communication because a repeater located 22,300 miles above the equator can relay between points in different hemispheres. International satellite links are attractive because they provide reliable communication without transiting the territory of a third country.

The latest evolutionary trend is the creation of worldwide fiber optic networks. Cables are being laid under vast bodies of water, giving rise to the opportunity for telecommunication managers to have high-quality, digital, fiber-based services on a global basis. The first links to be installed have been across the Atlantic and Pacific, and they have had more than their share of implementation problems. As discussed in the next chapter, however, many more links are being installed, which will result in ample bandwidth and high reliability through the provision of alternative routes. These undersea cables are supplemented by the land-based fiber networks already in operation in North America, Japan, and Western Europe.

An overview of these systems is presented in Figure 1.1. Indicated are the primary global networking systems available today. These are identified below and discussed in more detail in Chapter 2.

1.2.1.1 Satellite Communication

Satellite communication is a reliable means of establishing international telecommunication links between multiple points on the ground. In this book, we focus on communication among fixed points in different countries. Considerable detail on the design, use, and economics of satellite communication systems can be found in our earlier book [Elbert, 1987]. Satellite communication is also practical for domestic and regional communication, employing VSAT technology, particularly for business data and video communication. The North American and Japanese markets are the best developed in the world, but VSATs have gained a foothold in Western Europe due to liberalization of telecommunication.

The principal provider of international satellite capacity is the International Telecommunication Satellite Organization (INTELSAT), one of the most successful joint telecommunication ventures in the world. Formed under U.S. leadership in the 1960s, INTELSAT today provides the true international "backbone" for voice, data, and television transmission services. Total traffic on the system has continued to increase rapidly, even in the age of fiber optic cables. Some 114 countries are represented, including every major industrialized nation. INTELSAT owns and operates the satellites (totaling 18 in 1990) while the member nations and affiliates provide the earth segment. Membership is open to all nations, and a number of Eastern European nations provide access to the system. INTELSAT has been a driving force in the development of commercial satellite communication, and many of the innovations in digital communication systems resulted from INTELSAT experimentation and implementation projects.

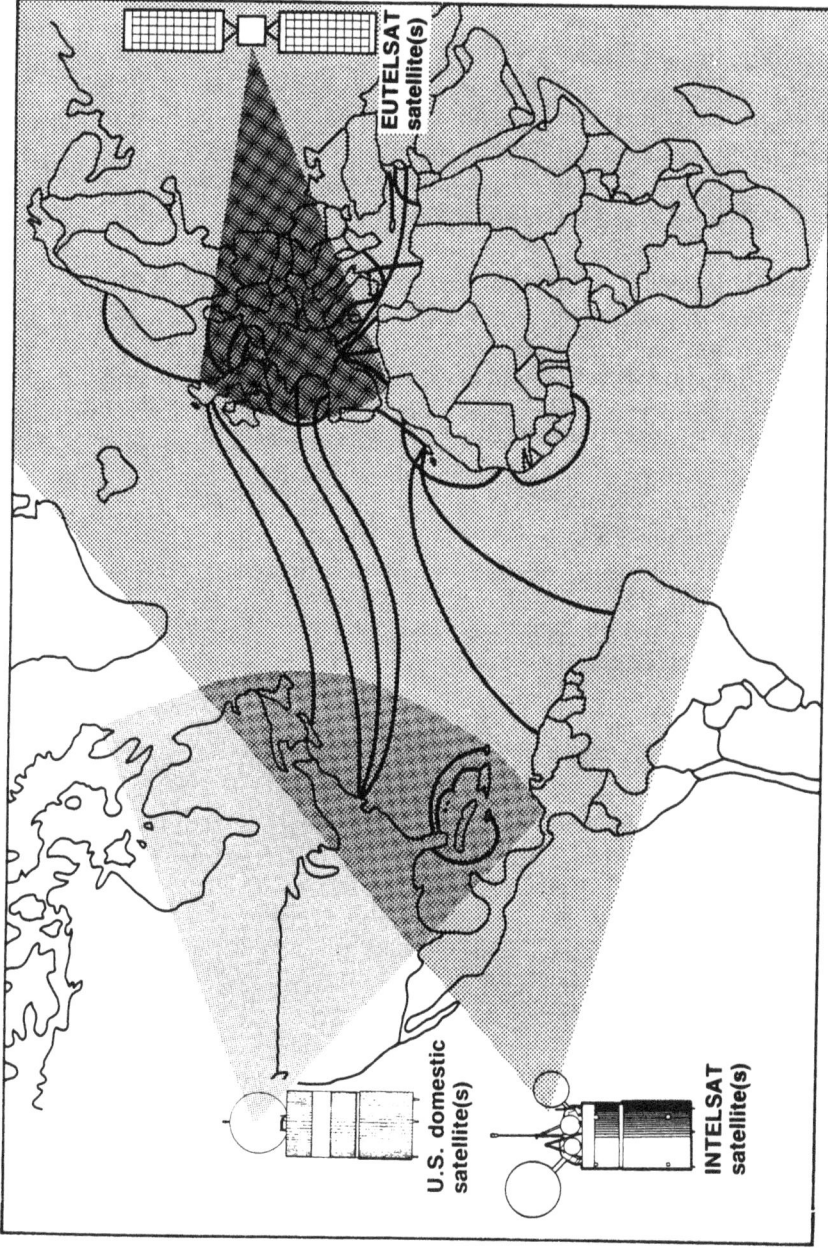

Figure 1.1 Major undersea and oversea links for Atlantic and Mediterranean regions are augmented by satellite links through domestic and international satellite systems (courtesy of Hughes Communications, Inc.).

The Eastern European nations, under the leadership of the Soviet Union, have implemented their own international satellite system, called INTERSPUT-NIK. This system began by employing the Soviet Molnya series of nongeosynchronous satellites. Operating in a 12-hour inclined orbit, the Molnya satellites must be tracked with elaborate earth station antennas, which tended to make the system less attractive than the INTELSAT system. Since then, INTERSPUTNIK has launched geosynchronous satellites so that use of the system has been greatly facilitated. This system is generally regarded as a means to communicate within the countries of Eastern Europe, however, and is therefore not a true international backbone, as is INTELSAT.

Traditionally, access to INTELSAT satellites has been limited to the member countries and their respective *telecommunication authorities* (TAs). As will be discussed later, these authorities are typically governmental agencies. Costs of using the system are competitive with other means of international communication; the charges made to users, however, tend to be high, relative to similar services on domestic satellite networks in the United States. Furthermore, users must gain access to the satellites through major earth stations operated by the local TA.

Application of the INTELSAT system to private telecommunication networks has been facilitated by the INTELSAT Business Service (IBS), a digital transmission service which permits the use of relatively small earth stations. This capability, which first appeared in 1985, effectively revolutionized private international communication. Because of its importance to international telecommunication management, IBS is treated in detail in Chapter 3.

1.2.1.2 Transoceanic Cables (Coaxial and Fiber)

Transoceanic cables were the first means of providing international telegraph and telephone links across bodies of water. Developments in high frequency (HF) radio and satellite communication temporarily eclipsed the use of cables, but improvements in bandwidth and reliability have generated a rebirth. Many of the cables in the Atlantic and Mediterranean regions are shown in Figure 1.1. On links using copper wire, bandwidth is severely limited. This limitation precluded the use of early cables for television transmission; hence, this led to the motivation for satellites. In December 1988, the TAT-8 fiber optic cable was completed across the Atlantic Ocean. TAT-8 starts in New Jersey, making its way across the Atlantic to a point off the Irish coast, where it splits into legs to the United Kingdom and France. The Atlantic link was complemented in the Pacific with the introduction of the first transpacific fiber optic cable system in April 1989. Called TPC-3, the Pacific cable connects from California to Hawaii, then continuing to Japan and Guam. An all-fiber telephone circuit can now be established from Japan to Europe, by transiting one of several land-based fiber networks in the United States. The inherent bandwidth of fiber optic cables permits the transmission of greater quantities of telephone channels as well as high-speed data and even television. For the time being, however, these cables will continue to be used for international telephone, telex, and data transmission.

The TAs of major countries find investment in fiber optic cables that connect their territories to be attractive. This is, in fact, how TAs obtain a commitment to portions of the capacity. AT&T, British Telecom (BT), and France Telecom, among others, are partners in the TAT-8 cable. By working together, they share the investment and operating expenses, and are in a position to obtain substantial revenues from traffic that others place on the cable system. The TAs, in turn, benefit from the growth of international telecommunication services.

The fact that cables and satellites are complementary is borne out by the need for satellites to provide temporary replacement capacity to restore service when a high capacity cable is broken. Services on cables must also transit countries along the path, whereas satellite links can go directly from city to city. Another synergy is obtained when a satellite link hops over a land mass and cable extends the circuit across a body of water. This restricts the end-to-end circuit to a single satellite hop, which is consistent with acceptable quality for most applications. A double hop (using two satellites) is usually unacceptable for interactive services such as voice and data communication. For the foreseeable future, satellites will continue to be the main way to transmit international video.

1.2.1.3 Transborder Links

A transborder link is simply a telecommunication link that crosses the border of two neighboring countries. Where the telecommunication infrastructure is well developed, transborder links can operate very efficiently. This is particularly the case on the continent of Europe, which for many years did not see the introduction of regional satellite communications. Transborder links are joint undertakings of the neighboring countries and their respective telecommunication authorities. Occasionally, the links are implemented by the user organization. The use of radio paths, particularly microwave, is subject to international regulation because the transmissions can interfere with the operation of systems across the border. This "spillover" is the basis for the importance of the International Telecommunication Union and the International Frequency Registration Board, which regulate the international use of the radio frequency spectrum. This topic is covered in more detail in Chapter 9.

The radio spillover phenomenon can be exploited for transborder services where domestic satellites are employed. Illustrated in Figure 1.1 is a U.S. domestic satellite which radiates its beam over portions of Canadian and Mexican territory. Domestic satellites serving those respective countries have the potential for transborder service into the United States. A significant number of U.S. domestic satellite networks have been extended to locations in Canada and Mexico through spillover coverage, permitting earth stations in these areas to be part of a larger network. Companies in the automobile manufacturing business find this particularly attractive because they operate factories and sales offices throughout North America.

1.2.2 Regional Telecommunication

Regionalism exists today in telecommunication as it does in other areas of human activity. Its origins are easy to understand — communication was difficult before the advent of modern electrical systems. Consequently, human activity was limited to smaller regions, such as a country or continent. The introduction of radio and cable links across continents and oceans, however, has tended to make telecommunication regions much larger. Perhaps in the not-too-distant future, today's regions will cease to serve any useful purpose.

At the present time, the local telecommunication authorities have tended to operate in concert with their neighbors. Regionalism is strong in North America, Western Europe, and Eastern Europe, with South America and Africa also tending to follow common courses. The ITU recognizes three regions for the purpose of regulating the use of radio frequencies. Region 1 consists of Europe and Africa, and tends to represent the interests of the "old world." The "new world," consisting of North and South America, is contained in Region 2. The rest of the world is designated as Region 3, consisting primarily of the Pacific region and Asia. To observe the regionalism that develops at the ITU World Administrative Radio Conferences is interesting, particularly when Region 1 pursues a policy which Region 2 sees as not in its interest. In such instances, the United States and Cuba (renowned adversaries) may oppose friends of the United States, such as the United Kingdom and West Germany.

1.2.2.1 North America

The United States, Canada, and the Caribbean Basin constitute the North American region. The telecommunication networks of the region interconnect in numerous places along the borders, supplemented with links through transborder coverage of the respective domestic satellite systems. The U.S. telecommunication infrastructure is the most diverse and dynamic in the world because it is almost completely open to competition. The breakup of the Bell System into seven Regional Bell Holding Companies (RBHCs) and AT&T (which retains the manufacturing operations of Western Electric, now called AT&T Technologies, and the dominant long-distance carrier) in 1983 was an action that decentralized the telecommunication service business. Today, there are three major long-distance carriers: AT&T, MCI Telecommunications, and US Sprint (a subsidiary of United

Telecommunications, Inc.). In addition to the RBHCs, local telephone service in the United States is also provided by independent telephone companies, the largest being GTE Corporation, Centel, Contel, United Telecommunications, and Pacific Telecom. Local telephone service is one of the last protected monopolies in the United States, under the regulation of state public utility commissions (PUCs) and the Federal Communications Commission (FCC). The manner in which the RBHCs, independent telephone companies, and long-distance companies provide service and compete is reviewed in our previous book [Elbert, 1989].

Supplementing the public network facilities of these carriers are various transmission systems owned and operated by investors. These include regional fiber optic networks such as Telecom®USA (formed by the merger of Teleconnect Co. and SouthernNet, Inc.), Lightnet, and Williams Telecommunications (which is a subsidiary of a major gas pipeline company). The domestic satellite operators, including Hughes Communications, Inc. (a GM subsidiary), GE American Communications, AT&T, and GTE Spacenet, all provide transponder capacity that is employed for domestic telecommunication. As mentioned previously, these domestic satellites provide some coverage of Canada, Mexico, and the Caribbean Ocean, and are utilized for various transborder services on a somewhat limited basis.

The terrestrial and satellite network facilities are heavily used for voice services, data communication, television, and audio, and as conduits for value-added services of public and private data networks. The latter include SprintNet (formerly Telenet), BT Tymnet, MCI Mail, and a wide variety of other offerings. The quantity and variety of public data network (PDN) offerings is changing almost daily, in response to the desire of domestic and foreign business in the United States to employ information technology.

Public telecommunication in Canada can be characterized in two ways. Owing to the smaller size of the population and economy, the telecommunication infrastructure is much smaller than that of the United States. This is in spite of the fact that Canada is comparable in land mass, including vast distances over which services must be extended. The population centers in Canada are clustered along the border with the United States, facilitating interconnection of the two national networks. Common telecommunication standards are applied for the most part, a fact which has allowed the major Canadian telephone manufacturing company, Northern Telecom, Inc., to gain a major share of the U.S. market. The split between local and long-distance service in Canada is in some ways similar to, and in others different from, the United States. There are several large, private telephone companies providing local service. Long-distance service is available from two national networks: the Trans-Canada Telecommunication System (TCTS), which is jointly owned by the local telephone companies, and CN/CP, which is owned by the national railway.

There is only one satellite system operator in Canada, Telesat Canada, which is protected from competition by the Canadian government. Unlike U.S. domestic satellite operators, which depend mainly on the satellites themselves to generate revenue, Telesat is a fully integrated telecommunication service provider. The commitment to satellite communication is necessary because the population in Northern Canada can only be reached by satellite links.

1.2.2.2 Europe

Europe represents the classic government-dominated telecommunication environment, although this is changing as a result of strong political pressure for change. Much will be said later about the PTT agencies that control and operate the domestic and international telecommunication services of Western European countries. With only a few exceptions (as discussed in Chapter 2), each PTT is a monopoly within its own borders, enjoying a protected market and government support. Typically, the government owns the networks and hence can use them as an important source of revenue. Anyone who has been in Europe knows that use of the public telephone network is quite expensive as compared to the United States.

That Europe is a powerful and dynamic region of the world is undeniable. The region has a collective economy similar to that of the United States. One of the most significant problems in dealing across borders is the lack of a common language. Without a common language, technical interchange is more complex. Any meeting must be set up like a UN conference, with continuous translation of two or more languages available. From the standpoint of television and radio broadcasting, the need for separate languages increases the cost of doing business. In addition, the cultural differences require that programming be tailored to the specific tastes of each country, thus fragmenting the market.

Within Europe, the PTTs tend to coordinate the activities relating to transborder communications, a process which is essential to the passage of trade and people around the continent. Networks are connected primarily with terrestrial links across the borders. This is facilitated by a common set of technical interface standards and rules for tariff adjustments. The Western Europeans are particularly adept at this coordination process as it relates to conventional long-distance telephone service. The availability of more modern services like those provided in the United States is generally restricted to certain countries in Europe. Leaders such as the United Kingdom, France, Germany, Belgium, and the Netherlands have created advanced digital public networks with the ability to provide the widest array of advanced service offerings. These services, however, often are not available between countries.

Supplementing the domestic network and transborder links is a regional satellite system operated by the European Telecommunication Satellite Organization (EUTELSAT). The system is primarily used for the distribution of television and radio programming; VSAT networks are also gaining in importance. Acting as a cooperative owned by the PTTs, EUTELSAT purchases and launches commercial communication satellites from the European industrial base. Access to EUTELSAT is also controlled by the PTTs, which are the exclusive sales agents. This situation is being debated in Europe as end users are demanding the right to deal directly with EUTELSAT, a position also being taken by the Commission of the European Communities (CEC).

Several European nations have implemented their own domestic satellite networks. These include: France (TELECOM 1); West Germany (Kopernikus); Sweden (Tele-X); Italy (Italsat); the United Kingdom (Olympus — a joint project with other nations); Luxembourg (SES-ASTRA); Spain (Hispasat); and Ireland (Atlantic Satellites). There are also direct broadcasting satellites in operation in France and the United Kingdom. As an example of the application of these satellites, all cable television (CATV) programming in West Germany is carried over the Kopernikus satellite; and ASTRA is employed to relay video channels to several European countries for either direct-to-home (DTH) or CATV service. The Hispasat project began in Spain in 1989, and will combine DTH, CATV, and telecommunication services on the same system. Links to Latin America are also contemplated.

Europe is moving rapidly toward a single economy under the banner of the European Community. Begun as a means to compete with the United States and Japan, the EC now promises to convert the continent into a more unified economic and political power. The conversion is expected to be completed by the end of 1992. The extent of this conversion remains to be seen, but the major enterprises and governments in Europe are treating this very seriously, investing a great deal of time and effort into establishing a more effective framework.

One of the outgrowths of this continental integration is the development of integrated services digital networks in Europe. As discussed in [Elbert, 1989] and elsewhere, ISDN is an evolving set of interfacing standards for high-speed digital networks. The ITU is particularly active in this field; consequently, ISDN is really a worldwide movement. In the United States, ISDN is more of a curiosity for telecommunication managers and government regulators; the wide variety of existing network offerings would appear to satisfy most current requirements. The Europeans, however, are more in need of a modern networking structure such as ISDN due to the wide variation of development across country borders.

1.2.2.3 Southeast Asia and the Pacific

Southeast Asia and the Pacific, also referred to as the Pacific Rim, includes large continental areas, vast stretches of ocean, and literally thousands of islands. Limited numbers of undersea cables are stretched between the most prosperous regions and cities, bypassing much of the underdeveloped areas. Consequently, the international telecommunication mainstay of the region is the satellite system operated by INTELSAT. A detailed discussion of INTELSAT may be found in Chapter 2 of [Elbert, 1987].

Telecommunication capabilities available within the countries of the Pacific Rim vary from the advanced all-digital networks of Japan, to the modern infrastructures of Hong Kong, Singapore, and the Republic of Korea, to the more elemental facilities of developing countries like Thailand, the Philippines, and Indonesia. Clearly absent is the kind of uniformity that exists in North America or even that of Europe. Much of the focus of this book is on international telecommunication management in the more advanced economies. Pacific Rim countries that have substantial manufacturing bases and are trading partners with the major industrialized powers usually have good telecommunication networks.

Discussion of the Pacific Rim usually begins with Japan, the Asian country which has established itself as a leading economic power. Recent legislative action has caused the privatization of the domestic telecommunication monopoly, Nippon Telegraph and Telephone (NTT). The high price of the stock of this company has made it the most valuable private firm in the world, not because of its telecommunication plant, but due to the real estate that it owns. In any case, NTT has invested heavily over recent years in an all-digital network with the capability of incorporating ISDN well ahead of the United States and Europe. Unknown to most people outside of Japan, NTT also provides data processing and data communication services for the major banks of the country, precisely the type of business that the court has restricted AT&T from entering in the United States.

The Japanese Ministry of Post and Telecommunication (MPT) still oversees all domestic and international telecommunication for Japan, exercising much more control than the U.S. Federal Communication Commission. The Japanese government now also allows competition with NTT in the areas of long-distance, private data networks, and value-added services. Resellers may also acquire bulk transmission capacity from NTT and offer it to a market segment.

MPT has authorized a number of interesting ventures in competition with NTT. In April 1989, Japan Communications Satellite (JC Sat) went into operation, providing Ku-band satellite capacity to the Japanese islands. As a joint venture of C. Itoh, Mitsui and Company, and Hughes Communications, Inc. (of the U.S.), JC Sat has already garnered a major share of the anticipated market for domestic satellite services. A second operator, Satellite Communications Corporation (SCC), is also actively in the market with the successful launch of its first satellite. SCC is controlled by the powerful Mitsubishi Group with the satellites having been built by Ford Aerospace. In the terrestrial arena, the major railroads have been authorized to construct long-haul fiber optic and microwave transmission systems and to offer the capacity to commercial companies for private network applications.

International services from Japan had been the domain of the overseas carrier, Kokusai Denshin Denwa (KDD), which literally means "international telegraph and telephone." Limited competition has been introduced by the creation of two new overseas carriers. One of them, International Digital Communications (IDC), is a joint venture of the Japanese trading company, C. Itoh, and other industrial and banking organizations. The United Kingdom's Cable and Wireless PLC and a medium-sized U.S. telecommunications holding company, Pacific Telecom, Inc., are cooperating with IDC in the North Pacific Cable, a fiber optic line that connects Japan and the United States. By combining the financial, technical, and marketing know-how of these companies, IDC plans to offer transpacific fiber-optic-quality digital services to the Japanese telecommunication market. More information on the North Pacific Cable is provided in Chapter 2.

The Asian city-states of Hong Kong and Singapore are similarly expanding their horizons in telecommunication. Therefore, telecommunication managers who need to interconnect with these commercial centers will have a relatively easy time of doing so. Telecommunication services in Hong Kong are provided directly or indirectly by Cable and Wireless, making this very much a commercial business. Hong Kong's telecommunication infrastructure is relatively modern, with excellent connections to foreign locations by cable and INTELSAT satellites. The situation in Singapore is comparable, the main difference being that the service provider, Singapore Telecom, is still totally government-owned. You only need to talk to the local management, however, to realize quickly that telecommunication is very much a business.

The major countries of Australia and New Zealand offer telecommunication networks comparable to the most advanced in the world. Obviously, these Westernized countries have the technical and financial resources to keep pace. The telecommunication systems are still owned and operated by government agencies. In addition to the Australian terrestrial services provider, Telecom Australia, there is a Ku-band national satellite system operated by a domestic satellite carrier called Aussat.

In the so-called "Little Tiger" nations of Taiwan and the Republic of Korea, telecommunication receives priority from the governmental telecommunication authority. The networks are relatively modern, and upgrades have been made by using digital technology. During the Seoul Olympics, for example, the Korean Telecommunication Authority (KTA) provided the quality and variety of facilities that broadcasters expected. KTA made a considerable effort to plan for the Olympics, installing local fiber, digital exchanges, and broadcasting center facilities for the international news and sports organizations' use. A domestic satellite system, possibly with Korean-built space hardware, is planned for operation some time in the 1990s.

The upper tier of developing nations along the Pacific Rim, particularly Thailand, Malaysia, Indonesia, and the Philippines, are still using much antiquated technology, including switchboards in outlying regions. The bigger cities in these countries have received attention over the past decade and now employ modern digital exchanges in many cases. Indonesia has led the region by developing the

Palapa System, a network of satellites and earth stations throughout this vast nation of over 10,000 islands. The Palapa System has done much to tie the nation together through the availability of telephone and television channels of reasonably high quality. Even with such advances, installing a simple telephone still can be very difficult. Often, the new telephone will not receive a dial tone for a considerable time due to overloading. More advanced digital services and ISDN are very much in the future.

Deserving of mention are the developments in the People's Republic of China. With the largest population on the planet, China has always had an allure for the foreign business person. The reader should recall that contact between China and the West opened with the visit of President Nixon in 1972. Along with the presidential entourage, a satellite communication earth station was set up to provide the first live television coverage from that country. The Chinese government has pursued a policy of internal development, and has opened the borders to allow foreign technology and, to some extent, capital to enter the country. The telecommunication infrastructure is still one of the most backward in the world, but through the human resources available, the problem is being attacked with some speed. China relies on terrestrial communication systems, including wire and microwave radio. In a recent development, a regional satellite system, called Asiasat, is being partially implemented with the support of an agency of the Chinese government. Asiasat will serve China and the surrounding region, offering satellite capacity on a condominium basis. Availability of wide area coverage from such a satellite should greatly improve the availability of telecommunication services throughout the country.

Another country that has worked hard to maintain a pace in modern telecommunication is India. With one of the few domestic satellite systems in the region, India has made contributions to the technology. For Indian scientists and engineers to have actually designed the equipment and systems employed in the domestic network is not uncommon. As is typical, all telecommunication networks are owned and operated by the government, including the satellite system.

As is evident from the preceding discussion, international telecommunication around the Pacific Rim tends to be fragmented. No common structure exists, except for links over the pervasive INTELSAT system. Nevertheless, the region has great potential for economic development, and it enjoys an incredibly interesting diversity of cultures and geography. There are efforts to rationalize the networks of the region. The Pacific Telecommunications Council, headquartered in Honolulu, Hawaii, hosts an annual conference to provide the type of forum needed to allow the interchange of thoughts and solutions. The program typically includes such current topics as ISDN, advances in artificial intelligence (AI), very small aperture terminal (VSAT) satellite networks, and electronic data interchange. There are also other activities within subregions, such as for the Association of Southeast Asian Nations (ASEAN) and for Australia, New Zealand, and Papua New Guinea.

Japan has increased its involvement on the international scene by actively participating in international conferences and providing technical and financial aid to the lesser developed countries along the Pacific Rim.

1.2.2.4 Latin America

The Latin American Region includes Mexico, Central America, and South America. This region has a common heritage derived from its colonization by Spain and Portugal. More economically advanced countries in the region, like Argentina, Brazil, Colombia, Costa Rica, and Venezuela, have public telecommunication networks employing modern stored-program exchanges and digital transmission links. Services on the local loop are still analog in nature and a considerable time will pass before ISDN services will appear. A microwave radio backbone operates through Central America from Mexico to Colombia. There are also cable and tropospheric scatter (over-the-horizon microwave) links between several of the Caribbean islands and to the United States.

Domestic satellite systems have been implemented in Brazil and Mexico, and in 1989 both of these countries began the process of procuring replacement satellites. Satellite communication is very popular in most South American nations which do not own their own satellites. This type of service is provided by INTELSAT through the leasing of transponders. Domestic communication services are then provided with earth stations which are typically owned and operated by the local PTT. Plans are being laid by the PTTs of nations in the Andean region to implement a regional satellite system called Condor. This system would presumably operate like EUTELSAT. A private company, Pan American Satellite, has launched and placed into service a private international satellite. Capability is provided for domestic and regional telecommunication in Latin America. Already, some South American nations are allowing commercial companies to gain access to PanAmSat.

1.2.2.5 Africa

Composed of various subregions and underdeveloped nations, the African continent represents a considerable challenge to the telecommunication manager. Ranging from Middle Eastern sheikdoms to nearly Westernized democracies, there is a wide variety of environments. Domestic telecommunication services in the modern economies of Israel and South Africa are on a par with what is available in Western Europe. International calling to these countries is typically carried over the INTELSAT system. In fact, the INTELSAT system is so pervasive in Africa that it provides the only practical means to call between the lesser developed countries, even those that share borders. These general circumstances will force any telecommunication manager to do his or her own legwork, including the installation of earth stations and local links.

1.2.2.6 Eastern Europe

Dramatic political change has occurred, yet the provision of telecommunication services in the region is both basic and limited. Telecommunication technology has always had solid support in the Soviet Union and its satellite nations in Eastern Europe. Both from a theoretical and practical standpoint, the Soviets have available to them any capability that those in the West employ. The main limitations, however, have been due to constraints on capital. The Soviets are active in the International Telecommunication Union, being strong supporters of practical means of radio communication and interfaces between networks. They operate the INTER-SPUTNIK system and are connected into the INTELSAT system.

Many of the Eastern European nations had established telecommunication manufacturing industries prior to World War II. Following the war, these nations have had a difficult time rebuilding their infrastructure. In recent years, industry in Hungary and Czechoslovakia, for example, developed a reasonably advanced capability in the manufacture of telecommunication hardware. The domestic networks of these countries produced services on a par with what was available in the West as recently as a decade ago.

Let us first consider Czechoslovakia, where the domestic telecommunication network is basically analog in nature, using cable and radio transmission along with conventional switching exchanges. Automation is being introduced in the form of computer control of the switches. It will still be several years, however, before digital services will be available on an end-to-end basis. Currently, digital links using pulse code modulation (PCM) have been implemented between exchanges, a process done in the United States approximately thirty years ago.

The public telecommunication network in Hungary has evolved along similar lines. In this case, the capital city, Budapest, is the economic and telecommunication center of the country. The Hungarian PTT, Magyar Posta, is proceeding with digital transmission and data communication services somewhat faster than many of the other Eastern European nations, reflecting its more advanced economy. Hungary, Czechoslovakia, and Poland have pursued membership in EUTELSAT.

INTELSAT earth stations are common in Eastern Europe because they can provide international access to the domestic networks. This is particularly important as these nations open more economic ties to the West. On numerous occasions, the Western press and broadcast networks are allowed to cover news and sporting events in areas where press coverage is state-controlled and highly censored. The case of President Nixon's visit to China was cited as one in which live coverage was provided by a transportable earth station brought in by American companies. Telephone circuits could be tied from Eastern Europe to the U.S. domestic network. All of this is technically possible, and the opening of the East is allowing what is possible actually to happen.

Otherwise, to interconnect between these nations and the West will be difficult. The relative austerity of Eastern Europe raises barriers to effective use of telecommunication and information technology. With the new freedoms that Eastern Europe will experience, an expansion of telecommunication seems logical. Concurrently, access to the domestic network from the outside should also be facilitated. Probably, the telecommunication manager who will be successful in the region is the one who makes the first move.

1.3 PREDOMINANT APPLICATIONS

Telecommunication managers are accustomed to having a wide variety of services and applications available at their fingertips. We are relatively complacent in North America, Europe, and Japan because the service providers are highly motivated to meet our expanding needs. The purpose of this section is to review the common denominator of services that businesses and government agencies employ in the pursuit of their strategic goals. We review an appropriate subset of the full range of possibilities; readers who wish to consider the entire spectrum may refer to the detailed discussions in our previous book [Elbert, 1989]. Likewise, television and radio broadcasters are in the business of covering events virtually anywhere on the globe. They require a number of service capabilities, some of which will require special construction for a relatively short period of operation. Subsequent chapters expand upon these brief introductions.

1.3.1 Business Communications

The role of telecommunication in business has increased dramatically in the past decade. Owing to the availability of new service options and sophisticated data and graphic networking technology, business managers are in an excellent position to achieve efficiencies never before possible. We have already cited the case of Benetton, an Italian clothing manufacturer with hundreds of retail stores in Europe and North America. The range of possible network options is illustrated in Figure 1.2, which envisages the private telecommunication network as a pervasive cloud. This simplified model nevertheless illustrates how a modern network is to function, permitting user information to be routed over a flexible medium. Digital communication networks via fiber optic cables, microwave links, and satellites are quite capable of providing this kind of connectivity among diverse points in a domestic or global network.

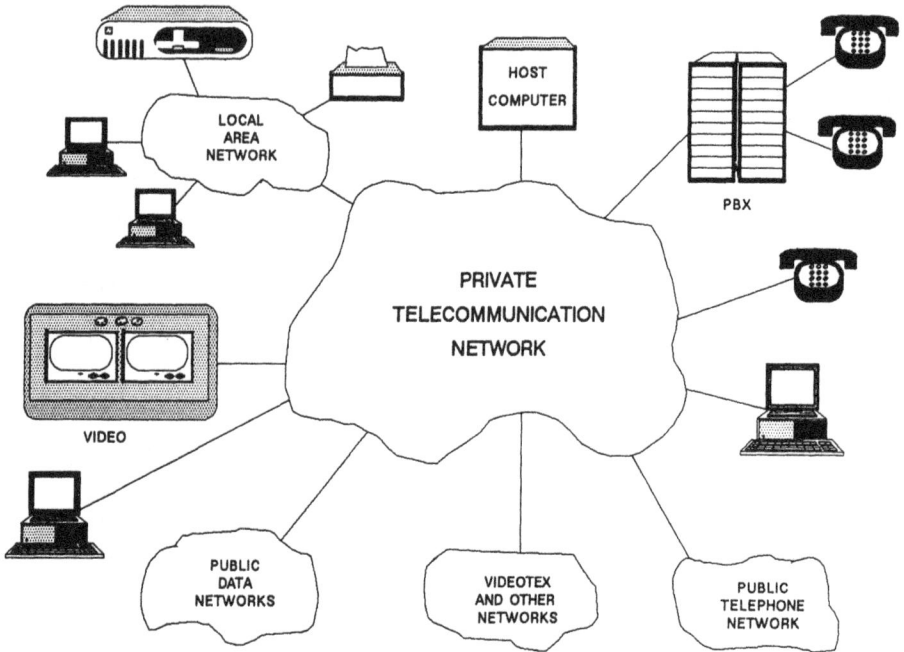

Figure 1.2 The private telecommunication network, where multiple services are integrated for convenience of access and economy of scale.

Integrated services digital networks have been mentioned previously and will be discussed at various points in this book. Initial reaction to ISDN has generally been very cautious because users still find it difficult to exploit its capabilities. Nevertheless, ISDN is an important innovation in telephone networks because of the common set of interface and operational specifications that it encompasses. The real promise of ISDN in the international context is that the domestic networks of the world can be tied together more efficiently than ever before. Users may connect terminal devices to ISDN facilities in any country, at least in principle. Digital services that are circuit-switched and packet-switched can be carried between countries; at the same time, PTTs can cooperate in the provision of value-added services, such as those that are now only available on the most advanced domestic networks.

1.3.1.1 Telephone

Basic telephone service, also known as "plain, old telephone service" (POTS), is the common denominator of telecommunication. The conventional telephone instrument is nearly standardized around the world, and essentially every world citizen knows how to operate it. Domestic telephone networks are arranged to allow any subscriber to reach any other through a hierarchy of telephone switching centers, called *exchanges*. To gain access to foreign countries, telephone calls are routed to international *gateway* exchanges so that they can connect to links to other countries. The gateways are key switching nodes in the global public network. For more information on telephone technology, please refer to [Noll, 1986]. As a point of reference, we next review the manner in which telephone service is rendered to subscribers in North America.

MTS and WATS

The following discussion reviews the function and operation of the two most popular forms of basic telephone service. The universal form of telephone service available virtually anywhere in the world is commonly referred to as *message telephone service* (MTS). In the United States, MTS is offered by the local telephone companies and by the long-distance companies as well. The process of *direct distance dialing* (DDD) was introduced in the United States following World War II, allowing subscribers to place long-distance calls without the assistance of an operator. The domestic telephone numbering plan allows a given subscriber to be identified by the area code, followed by the exchange number and the final digits of the particular subscriber. The subscriber can place domestic calls automatically by dialing the appropriate digits. Routing over the network is accomplished automatically by the local and long-distance telephone exchanges, which are maintained by the domestic common carrier or *common carriers*. Charges for calls are determined by the switching equipment, typically based on call duration and distance traversed. Consequently, MTS is said to be a "metered" service.

Not very many years ago, all international calls were placed by operators. With that type of scheme, an international numbering plan would not be required because connections between countries could be identified by the names of the countries themselves. Gateway exchanges were nothing more than manual switchboards and subscribers needed to "book" calls with the operator before they could be placed. Conversion of gateways into automatic exchanges has advanced rapidly in the industrialized countries, allowing DDD types of service to be available to subscribers wishing to use international dialing. The international numbering plan (discussed in Chapter 3) provides a unique code number for each country in the world. The rest of the number would include area codes in North America or city codes for locations in the rest of the world.

Wide Area Telecommunication Service (WATS) is an innovation of AT&T designed to provide reduced costs to business customers who use a large amount of telephone service. Furthermore, WATS allows a potential customer to call a particular business on a toll-free basis (of course, the charge is automatically transferred to the called party, much like reversing the charges). Telephone numbers in WATS are identified with a universal area code, such as 800. A typical 800 number does not identify the specific local subscriber in the same way that the conventional telephone number does. Rather, the 800 number identifies this as a WATS call to a business customer who has contracted for this service. The telephone network and its associated automatic call routing systems convert the 800 number into the standard area code–exchange number–subscriber number so that the call reaches the correct destination. AT&T relatively easily created WATS before divestiture, when they controlled the long-distance network and most of the local exchange companies. Today, WATS is available from all major long-distance companies and the local exchange companies as well. For MCI and US Sprint to create their own WATS, they had to coordinate with every local exchange carrier in the country so that their unique set of 800 numbers could be correctly converted into standard dialing digits.

Wide Area Telecommunication Service is now available in parts of Western Europe and Japan. How useful it is to provide toll-free calling between countries that conduct a lot of business among each other can easily be visualized. International WATS is still rather new, and probably some time will pass before it is universal in its scope.

Software Defined Networks

First introduced in the United States in 1987, *software defined networks* (SDN) are automated telephone dialing systems offered to major customers, allowing simplified dialing. The offerings of the major long-distance carriers have varying names, including *virtual private networks* and *Vnet*. They rely on the common channel signaling system, which the carrier uses to control its long-distance exchanges around the country. A major customer could have many locations around a region or the country, and would wish to tie these locations together in a private telephone network. Historically, this would be done with dedicated private lines and PBXs. Calling between locations would use a simplified numbering plan, perhaps with five or seven digits. Over the years, AT&T introduced calling schemes that allowed major customers, such as General Motors, American Express, and the U.S. government to create a private network without relying upon dedicated private lines. Called Electronic Tandem Network (ETN), one of the more recent of these systems allowed the customer to connect PBXs to AT&T's long-distance exchanges with dedicated lines (trunks to specific destinations) and use a simplified numbering plan. The actual routing would be done automatically by AT&T without the customer knowing how it was accomplished (such a process is said to be "user blind"). SDN works in much the same way, but gives the customer greater flexibility in building and operating the private network.

Software defined networks are technically feasible on international circuits. U.S. long-distance carriers and major foreign international service providers including BTI, France Telecom International, and KDD have developed business offerings that approach the capabilities of SDN. Along with the Atlantic and Pacific fiber optic cables, the carriers are expanding the sophistication of the gateway switches that are used to provide international telephone services. Demands of users along with the highly competitive environment in existence in the 1990s are the driving forces that result in the introduction of capabilities like SDN.

1.3.1.2 Telex

For several decades, the teletypewriter was the common denominator of international record communication. The basic device is similar to a typewriter, but with several peculiar features and limitations. Principle among its benefits is the ability to send character data directly between machines that may be located in virtually any major city in the world. Telex, a contraction of *teletypewriter exchange,* employs switching equipment much like the telephone network. Worldwide, there are not even one million machines, but their utility for business and government communication greatly amplifies the importance of this somewhat limited connectivity.

One of the two main reasons that telex service has been so popular in the global environment is its relatively efficient communication across multiple time zones. The other reason is that telephone networks in many countries do not deliver adequate quality to allow reliable international calling, although this deficiency is fast disappearing. The venerable telex machine leaves much to be desired, with its limitations on print quality and speed. It is, however, capable of automatic operation. The sender can prepare the message, dial the destination in a foreign country, and, when the distant machine answers automatically, send the message. This can occur in the middle of the night at either end. Upon arrival in the office the next morning, the addressee will see the message that has been printed by the telex machine. This facility allows a form of efficiency which may have been hard to achieve domestically, whereby one office works all day, prepares and sends instructions to the distant office, which receives and acts upon the message the next day while the sender is asleep.

In the West, the heyday of telex has already passed. The speed of transmission on telex networks is approximately 60 characters per minute. Approximately one minute will be taken to transmit the text on this page. A standard fax machine will accomplish the same task in approximately 20 seconds, complete with figures. Using a 9600 b/s data circuit, the text in ASCII form will completely flow in under three seconds. Therefore, record communication within industrialized countries and between major centers has moved from telex to fax and E-mail. Still, telex is employed extensively within Eastern Europe and in developing countries. Companies wishing to do business in these regions still rely on telex. Fax machines, however, can be easily transported because devices small enough to be placed in an attaché case are available on the commercial market.

1.3.1.3 Videotex and Teletext

An outgrowth of time-sharing computer systems, the *videotex* and *teletext* services provide a new class of subscriber with access to information databases. The service is operated as a business, of course, so that subscribers pay either for "connect time" or by the information bundle. The distinction between videotex and teletext is a narrow one. Videotex is a service accessed via the public telephone network or a separate data network. The key point is that videotex is a two-way interactive service, allowing the subscriber to make requests, to search, and even to create unique reports based on information within the database. Teletext, conversely, is broadcast over some point-to-multipoint means. The user must select the desired information from the broadcast, either scanning everything as it is received or selecting it after the information has been collected by a local database. Examples of broadcast media include satellite transmission, the vertical interval of a normal television broadcast signal, a subcarrier on an FM radio station, and possibly the terrestrial telephone network. Teletext formats contained in CCITT Recommendations include predefined graphic display capabilities. As discussed in Chapter 3, the Group 4 facsimile standard includes provisions for teletext reception.

Perhaps the best known videotex service in the world is the French Minitel, an information system provided by France Telecom via the public telephone network. Subscribers in France use Minitel for such basic purposes as an on-line telephone directory and banking at home. Minitel is open to literally thousands of third-party information providers and other businesses. Subscribers can therefore access these third parties to place orders for products and services, which are subsequently delivered directly to the home or office.

Videotex and teletext are still developing as potentially viable services to businesses and individuals. The important point is that these services employ both public and private telecommunication networks to reach their customer base. Internationally, videotex could well facilitate trade between countries whose networks are interconnected.

1.3.1.4 Facsimile

The facsimile or telecopier machine (commonly called *fax*) is ubiquitous in business and government communication. Relying on the analog telephone network for carriage, fax messages can be transmitted literally anywhere in the world. Before long, fax machines will reside in most homes, allowing consumers to place orders for merchandise and even fast food. Facsimile has been used to transmit maps by the weather service and military. Also, a form of fax machine has been used for decades by the wire services to send photographs to newspapers. More recently, Japanese companies found fax to be very effective in communicating their complex written language, which is a combination of Chinese ideogrammatic characters (Kanji) and a phonetic script (Hiragana and Katigana). In fact, the Japanese really reduced fax to an affordable level, putting it within reach of the entire business world.

The analog telephone network is capable of passing CCITT Group 3 facsimile, which can send a page in approximately 20 seconds with reasonably good quality. High definition fax would employ the Group 4 standard, using a high-speed digital line operating at 56 kb/s or greater. Both the speed and quality of transmission are greatly improved with the Group 4 digital system. One of the greater promises of ISDN is the uniform availability of switched digital circuits able to support the full capabilities of Group 4 fax.

There are fax systems with the capability of reproducing graphic images of the highest quality. A very successful application of this is the transmission of newspaper pages for remote printing of national and international editions of the *Wall Street Journal,* The *New York Times,* and the *Financial Times,* among others. Color advertisements for reprinting in magazines and newspapers are also sent electronically by facsimile. With the greater resolution of this class of fax, the transmission line must be capable of 56 kb/s or greater. A common rate of transmission is T1 (i.e., 1.544 Mb/s). Quite obviously, high resolution fax has limited application, only where the expense of the equipment and transmission are justified.

1.3.1.5 E-Mail

The electronic mail or E-mail medium has gained much attention and interest during the late 1980s. A review of the entire range of E-mail systems and standards can be found in [Caswell, 1988]. Focusing on international telecommunication, E-mail is becoming the next-generation standard for record communication. The basis for this is the CCITT Recommendation X.400 protocol standard for the interconnection of E-mail systems in different countries. Already, in 1988 and 1989, leading providers of public data networks and E-mail systems such as SprintMail (formerly Telemail), MCI Mail, and Dialcom were offering X.400 services and interfaces. Some of the key message protocols are still being worked out at the time of this writing, so a global network is yet to be realized. Of particular need is a workable approach for directory services (designated as the X.500 protocol), wherein a sender can use the E-mail network to locate the "address" of a distant recipient. This is more complicated than it seems, because many E-mail systems are private and the actual identification of internal subscribers is kept confidential. Public E-mail systems would work like the telephone or telex networks, with the domestic network provider maintaining the common directory. The X.500 protocol is intended to allow one subscriber to reach any other by inquiring at each E-mail network directory database until a match is found. If not, the sender is informed that the message cannot be delivered.

Perhaps the most promising E-mail application for business is *electronic document interchange* (EDI). This uses the E-mail network to send standard message formats for commercial documents such as bills of lading, purchase orders, and invoices. The idea behind this is to transmit only the specific information that is contained in the standard forms. The forms themselves are printed, if ever, at the destination by software in the local computer. Perhaps the most successful international EDI system is that used for airline reservations over the network of SITA. In the area of the X.400 system, subcommittees are working to set standards for many other EDI formats. This will facilitate international trade and commerce because exchanges of information will be much more efficient than is possible by using telex or postal mail. Also, EDI can be independent of the local language as the forms themselves can be printed in whatever the local language happens to be.

1.3.1.6 Data Communication and Information Technology

We have touched on the role of data processing and data communication in international business. Fundamentally, applications of information technology are usually the key to business organization and automation. As stated by Michael Porter in his informative work, *Competitive Advantage* [Porter, 1985], "Information systems technology is particularly pervasive in the value chain, since every value activity creates and uses information." Porter uses the term "value chain" to refer to the set of processes and procedures that a company or organization uses to implement its strategy or business. You find information technology being used potentially everywhere in an organization to process orders, manufacture products, assist customers, arrange deliveries, interact with banks, or report to management. Extending a firm's value chain across international boundaries today means developing a global data communication network capability.

Data communication is different from voice, video, telex, and E-mail in that the information is being transmitted between computer devices. Human interaction is usually involved, but the key is that machines are primarily originating and using the information. It represents nearly total automation. As discussed in our previous work [Elbert, 1989], data communication is really an environment where machines interact with one another according to a set of predetermined rules. The most effective environments have been built and are offered by the leading data processing firms like IBM and Digital Equipment Corporation (DEC). Most major corporations and government agencies make a commitment to one or two of these firms and develop data processing systems based on their product lines. In fact, IBM has well over 50% of the data processing and data communication market worldwide. This leadership is a direct result of the ability of IBM to deliver products and software to do the automated functions that businesses want. Considerable information on international networking from an IBM perspective can be found in [Vignault, 1987].

There is an increasing possibility that data communication and networking can be approached without committing to the products and services of one vendor. Considerable progress is being made in developing an international set of data communication standards, under the sponsorship of the International Organization for Standardization (ISO) and the CCITT. The Open Systems Interconnection (OSI) protocol "suite" is evolving as the standard data communication environment of the future. Many vendors of hardware and software, including leaders like DEC, have developed interfaces with OSI. Although it will never replace the proprietary systems that now dominate the major applications in data communication, OSI will play the key interfacing role in the increasingly multivendor environment of the 1990s.

1.3.1.7 Business Video

Video communication has found its way into the business environment as a tool for bringing people together without them having to travel to a common location. (International television for broadcasting is introduced later in this chapter.) In the international network environment, business video has particular value because it can save travel time and improve the quality of communication between people of different cultures. Domestic uses of the medium in North America include video education, teleconferencing, and private broadcasting (see [Elbert, 1989] for more detail). Because of the wider transmission bandwidths involved and the cost of the terminal equipment, the business video market has not yet developed, even in the United States. The international version is available on a limited basis between North America and Europe, and between North America and Japan.

1.3.1.8 Private Telecommunication Networks

The elemental services and applications discussed above are the ingredients of an overall telecommunication environment. For major businesses and government agencies, we refer to such integration of capabilities as a *private telecommunication network*. Bringing the private network to the international or global level is a major task and forms the primary focus of this book. One's ideal concept of a private telecom network is an all-pervasive cloud such as that illustrated in Figure 1.2. According to this model, users can simply connect any type of device to the network, address a distant user or users, and enter into a dialog. This dialog would be in real time with the distant user, or could be carried as a complete message for delivery at some future time. There are integrated voice-data networks that provide a great deal of flexibility, as implied by the ideal case. One must, however, be realistic when setting goals that are to be met to conduct the business at hand. Flexibility is the key, both in terms of the capability of the network and one's expectations.

A private telecommunication network can be used for purely administrative purposes within the structure of an organization. In this mode, the private network does not give a business an advantage over its competitors. A modern telecommunication network is more likely needed just to stay competitive in a particular industry, allowing employees to conduct normal voice, data, and message communication among one another. Many businesses use telecommunication as the primary contact with customers, propelling the network into a more strategic role. The introduction of information technology can give a business a strategic advantage over competitors; hence, the concept of the *strategic network*. These concepts, along with supporting information on network architecture, can be found in earlier work [Elbert, 1989].

1.3.2 International Television

The medium of commercial television has grown tremendously, even by international proportions. In 1988 alone, over 100,000 hours of programming material were transmitted between the major countries of the world, primarily through the use of satellite links. Many of the major networks in North America, Europe, and Japan maintain full-time satellite connections for use as pipelines for news, sports, and program exchange. One of the most innovative links is maintained by the Japanese national network, NHK, to carry nearly 24 hours of continuous material to Japan from Europe and North America. The feed is routed to Japan by a U.S. domestic satellite operator, Hughes Communications, and relayed over the Pacific Ocean by an INTELSAT satellite, where the signal is received by KDD near Tokyo. With this pathway, NHK provides programming material for its DBS system, the first one in the world to offer services to dishes of less than one meter in diameter. We will briefly review these and other applications in the following paragraphs.

1.3.2.1 Television Networks

Television networks typically operate in their respective home domestic markets, obtaining and delivering program material for over-the-air broadcast in the VHF and UHF frequency bands. The signals are transmitted locally by over-the-air television stations for reception by the home. Local stations usually do not have the financial resources or ability to generate enough quality programming to fill their schedules. Instead, programming material is created and distributed by a much larger organization called a *network,* of which a station is an affiliate. Networks use their greater buying power to obtain the rights to distribute movies and sports events, and to create specialized programming in the form of news and entertainment. For example, in 1989, CBS paid approximately $2 billion for the right to broadcast major league baseball games for a period of two years. The other important aspect of a network, the one significant to this book, is its role as the distributor of the programming material, for which networks make extensive use of telecommunication facilities and services.

In commercial television, the programming is supported by advertising (i.e., commercials), which uses blocks of air time to promote commercial products and services. Consequently, commercial television is really an advertising medium similar to newspapers and magazines, competing for a sizable portion of the advertising budget of large and small companies around the world. At the top of the list of advertisers one finds automobile manufacturers, food processing companies, cosmetic and drug companies, and retailers. Because funding is provided in this manner, the viewing public does not pay directly for the programming. In many countries, however, citizens are levied a tax on each television set to support the network and stations. Interestingly, commercials have even appeared in the Soviet Union. Pepsico, Inc., Visa International, and Sony bought commercials that aired in the Soviet Union in 1988 during programs about the United States and during the Olympic Games. Revenues from commercials are not to be slighted, even in a purportedly socialist land.

There are forms of over-the-air broadcasting for which the public is expected to pay. In the United States, public television is an important noncommercial broadcasting medium. It shares the same frequencies and systems as are used by commercial television. The difference is that a significant proportion of funding is obtained through voluntary contributions. Local public television stations hold membership drives on their channels to collect money directly from viewers, offering the promise of "TV worth watching." This is augmented by contributions from corporate sponsors. Programming is generated by stations or acquired from outside sources such as the BBC. Distribution of programming is the responsibility of the Public Broadcasting Service (PBS), a nonprofit corporation that operates a satellite network for this purpose. A significant part of the funding for the distribution network is provided by the U.S. Congress, channeled through the Corporation for Public Broadcasting (CPB). Public television in the United States is perhaps unique because in other countries, such as Japan and the United Kingdom, the government networks obtain their funds through taxes levied directly on owners of television sets. In an interesting merging of Japanese and U.S. noncommercial activity, WGBH of Boston and several other public television stations in June 1989 began to carry a live half-hour English broadcast from NHK Tokyo, solely covering Japanese news. This daily transmission from Tokyo follows the reverse course from the DBS programming channel discussed at the end of this chapter.

Radio networks, which existed long before television and still thrive in the United States, also extensively use international telecommunication. ABC Radio Network was present at the 1988 Seoul Olympic Games, employing a digital satellite link to provide live coverage of events and news summaries. The radio counterpart of PBS, called National Public Radio (NPR), carries a daily news broadcast from the BBC, which is distributed by a domestic satellite network to public radio stations, including KUSC in Los Angeles.

Commercial and noncommercial networks have become increasingly involved with international activities. The challenge is the carriage of real-time programming from major events around the world. Because the viewing public enjoys watching events in foreign locations, the networks must obtain and carry programming over great distances. Tremendous logistical problems were encountered when covering the Olympic Games in Seoul, for example. Some 30 different venues needed to be covered during the games and INTELSAT had orders for as many simultaneous satellite video channels. Television is a quite demanding type of signal, requiring very wide bandwidths and high quality of transmission. Camera and studio equipment must be transported great distances and is often employed under hostile conditions. Cable, microwave, and satellite connections are needed just to coordinate the programming material at the distant end. Then, a stable international link must be arranged so that the valuable programming product can be transported to the home country.

1.3.2.2 Cable Television

Originally devised as a means to bring distant television signals into a community with poor reception, cable television has evolved into a powerful medium in its own right. Countries with well developed CATV infrastructures include the United States, Canada, Belgium, and the Netherlands. A typical CATV system serves a city or community with from 25 to 100 television channels, employing a cable distribution network of either coaxial or fiber optic cables. What has made CATV exciting in North America is that, in addition to the normal over-the-air signals from broadcast television stations, the cable system offers specialized programming. The first such service appeared on Home Box Office (HBO) in 1975, in the form of the "Thrilla in Manila" broadcast of a live boxing match featuring Muhammed Ali and Joe Frazier. Now, HBO is the largest and most successful pay-television service in the world, emphasizing movies, dramatic productions, concerts, comedy, and special events such as tennis from Wimbledon.

The key to the success of CATV, which is now an industry with total revenues well in excess of $10 billion, is the use of satellite transmission. The basic application of satellite links is to distribute programming to cable systems around the country. In the United States, more than 100 transponders on five C-band satellites are dedicated to this purpose. Receiving systems are relatively inexpensive with antennas in the range of three to five meters being common. There is a well established direct-to-home business in the United States because antennas of this size are being used in two million backyards. The television programmers who serve the cable systems also sell their services to this backyard market.

Satellite communication is also used by sports and news services to gather programming material. This technique is called *backhaul.* One of the most dramatic examples of backhaul was the coverage of the Goodwill Games in Moscow by the Turner Broadcasting System (TBS). Employing INTELSAT and INTERSPUT-NIK, TBS almost single-handedly provided two weeks of Olympic-style programming for its cable channel, Super Station WTBS. The telecommunication manager for TBS, James Kitchell, worked long hours, traveling back and forth between the United States and Moscow, to organize the temporary video network and to obtain the needed approvals. As an interesting sidelight, the basketball finals were conducted in Spain, further complicating this already monumental effort for an organization of TBS's modest size.

Cable television has gone international. One of the leading CATV services, Cable News Network (CNN), advertises itself as the "world's most important network." This is because TBS, CNN's parent, has undertaken to make the channel available in every major city of the world. Already, one can watch CNN from a hotel room in London, Paris, Singapore, Tokyo, Hong Kong, or Sydney. As a North American outside of his or her country, being able to stay in contact with home and the rest of the world at the same time can be gratifying. Other cable services appearing around the world include ESPN, Music Television (MTV), and the Disney Channel. Counterparts to these services have appeared in Western Europe, with Sky and Super being the most recognized as of this writing.

Although well developed in North America, CATV is only beginning to gain acceptance in Europe and Japan. Where CATV has developed, the systems have been immensely successful. For example, the cable "penetration" of U.S. homes is now 55%, whereas for Canada and Belgium, the rate is over 80%. As more of these systems are installed and homes are connected, the demand for a greater variety of programming increases. The U.S. market is the largest and most diverse in the world. Despite the fact that the United States is known for such entertainment series as "Dallas" and "Star Trek," there are other programming formats that appeal to much narrower audience segments. Cable television has been found to offer the opportunity to "narrowcast" to smaller segments of the public than can be economically served through over-the-air television stations. Many of these services use material gathered from other countries. This leads to the need for international television transmission for backhaul of real-time events. News programs from world centers like London, Hong Kong, and Tokyo are regularly transmitted to other countries for inclusion in local channels. Other events, such as the Live Aide concert, provide the opportunity to arrange worldwide interconnection.

1.3.2.3 Satellite Broadcasting

Direct broadcast satellite systems were devised as a means to transmit programs from space to the home without passing through studios, television stations, and cable systems. From the standpoint of reaching the largest possible audience without intermediaries, DBS represents the ideal medium. The high power of the satellite transmitter allows reception by an antenna of nearly one foot (30 cm) across. The technology is very appealing because antennas of this size are no bigger than a normal television aerial or a small window. In addition to television channels, DBS systems can deliver stereo sound programming and broadcast data for teletext.

The International Telecommunication Union allocated separate frequency bands for this sole purpose. In the United States, an effective direct-to-home market has evolved for the reception of CATV programming from existing C-band satellites, but true DBS systems have been slow to gain popularity in North America, where television services are of high quality and plentiful.

The true DBS system can employ frequency channels and orbital positions that have been assigned by the ITU during previous World Administrative Radio Conferences (WARCs). International regulation is covered in detail in Chapter 9. Japan, France, and the United Kingdom have launched direct broadcasting satellites that fit these new assignments. The associated businesses, however, are still in the nurturing phase.

The approach taken by NHK with the Japanese broadcasting satellite is particularly germane to international television. NHK programs the DBS with nearly 24 hours of live news, sports, and special events gathered from around the world. The New York office of NHK selects from a variety of U.S. and European programming, and routes it directly to Japan via a Pacific INTELSAT link. Japanese citizens purchase a small dish or flat-plate antenna and a satellite video receiver unit. Then, viewers tune to world news from CNN, economic news from Financial News Network, sports from CBS, and entertainment from the BBC. Prior to the official start of service in the summer of 1989, Japanese viewers could have the programming at no cost. This was due to the experimental nature of the programming and DBS itself. Now, citizens must pay an additional DBS usage fee to NHK to help offset the cost of the service. That people are willing to pay, clearly demonstrates a real demand (in an economic sense), instead of just a curiosity.

This chapter is an overview of international telecommunication management. We have considered the structure of the media used for international services. Clearly, there is considerable variety, when we consider fiber optic cables, satellites, and transborder links. The domestic and multinational organizations that provide these facilities employ the profit motive; many of the providers, however, are monopolies in their home countries. Much of what follows considers how the telecommunication manager can deal with these entities around the world. Employing these media, the telecommunication manager can supply users with a variety of voice, data, facsimile, and video service options.

Chapter 2

THE INTERNATIONAL NETWORK ENVIRONMENT

The world telecommunication infrastructure as it exists today is an incredibly diverse set of domestic networks within countries and international links to access them. In Chapter 1 we identified the basic elements of international telecommunication, setting the stage for the more detailed discussion in the remaining chapters. Telecommunication service is, after all, an international business. Through investment in technical facilities, telecommunication service providers create the means to carry voice, data, message, and video traffic — transiting borders, traversing land masses, and reaching across bodies of water.

In this chapter we review the two archetypical public telecommunication service providers: the domestic PTT and the international carrier. A summary of the leading international carrier organizations in the industrialized world is presented in Table 2.1. The PTT is the domestic telecommunication authority responsible for the operation of the national infrastructure and the delivery of most of the services therefrom. (The terms PTT and TA are used interchangeably, although there is a slight distinction in that the PTT is always part of the government structure and the TA may be a corporation.) Whether a government agency or a private company operating in a monopoly or competitive environment, the TA is a vital source of networking capabilities that the telecommunication manager must employ. Understanding the local PTT environment thus is a vital requirement. Hence, we profile a few of these organizations to provide the reader with an appreciation for the range of possibilities.

Connections between these domestic networks are with international long-haul links offered by international carriers; they, in turn, obtain cable or satellite capacity from a variety of consortia and cooperative structures.

Table 2.1 Major International Telecommunication Carriers

Headquarters (Country)	Carrier Name	Service Emphasis
United States	AT&T	Satellite and cable, full service
	MCI	Satellite and cable, full service
	US Sprint	Satellite and cable, full service
United Kingdom	British Telecom International	Satellite and cable, full service
	Cable & Wireless	Satellite and cable, PSTN and private line
France	France Telecom	Satellite and cable, PSTN and private line
	France Cable & Radio	Satellite and cable, PSTN and private line
Federal Republic of Germany	DB Telekom	Satellite and cable, PSTN and private line
Netherlands	PTT Telecom	Satellite and cable, PSTN and private line
Japan	KDD	Satellite and cable, full service
	International Telecom Japan	Satellite communication, international PSTN and private line
	International Digital Communications	Satellite and cable, international PSTN and private line

2.1 PUBLIC TELECOMMUNICATION AUTHORITY

Domestic telecommunication authorities usually do a reasonably good job of meeting the needs of home populations and businesses. There are exceptions, primarily among the lesser developed countries, but the domestic TA represents a valuable resource due to the importance of telecommunication to the national economy and the revenues which are easily collected from users.

The domestic TA can make you or break you, so to speak. Consequently, the telecommunication manager contemplating an international link or network must take steps to learn about the environment within each host country. To generalize when considering these organizations as a group is difficult, if not impossible. The correct way to approach the problem is to investigate the particular countries and service providers involved. Consider also the fact that the character of the environment in a given country can change over the years or even months, rendering our working assumptions invalid. In particular, numerous countries experience changes of government at irregular intervals, resulting in a change of people and policy within the telecommunication organization. A detailed discussion of approaches for dealing with the PTTs is presented in Chapter 8.

We profile a few of the TA organizations, some of which are government-controlled monopolies and others are private companies under limited regulation.

Much of the discussion of telecommunication business in the 1980s has to do with the "privatization" of the government-controlled authorities, such has been done in the United Kingdom and Japan. This is a trend insofar as the industrialized world is concerned. We examine the situations in the United Kingdom and Japan, two countries where privatization has been accomplished and some competitive entry has been allowed. Discussion of telecommunication in various regions of the world was provided in Chapter 1.

2.1.1 Telephone Service in the United States

Deregulation is a trend in the United States primarily instituted under the administration of President Jimmy Carter during the 1970s. Deregulation has had a profound effect on the U.S. economy and is still redefining the role of government. This trend of reduced governmental involvement in the economy has carried over to the telecommunication service industry. Competing long-distance carriers and the breakup of the Bell System by the U.S. Federal District Court are key examples. Prior to breakup, the Bell System of AT&T could be viewed as the TA of the United States because both local and long-distance services were largely under the control of one entity. The management of the telephone network was by a commercial company, the objective of which was to make money (which it did, but under the watchful eye of the FCC).

The provision of local service at reasonable cost to consumers was a mandate of AT&T; consequently, the overall rate of return of the system was maintained through subsidization of local service by long-distance service. This is known as "cross subsidization," which is often practiced by PTTs in other countries. More affluent individuals and business customers are the dominant users of long-distance, and therefore they pay more so that the average citizen pays less for local service. Competing long-distance carriers did not — and do not — bear the burden of local service, hence they could sell long-distance services at a lower price. Not long after divestiture, AT&T cut long-distance rates to the point where little difference in price existed, particularly for business customers. Rates for long-distance service have dropped while those for local service have generally risen (the effect of the continued monopoly on the local loop).

At the beginning of the 1990s, the domestic telecommunication environment in the United States is the most liberal and competitive in the world. Policies of the U.S. government, notably the Federal Communications Commission, promote competition among rival long-distance carriers and providers of value-added services (particularly public data networks). Local telephone service remains a regulated monopoly, primarily under the jurisdiction of the public utility commissions of the 50 states. In addition, the Federal District Court in Washington, D.C., still is a controlling force over the seven Regional Bell Holding Companies, which resulted from the divestiture of local telephone service by AT&T.

Telephone service in the United States is therefore broken up into local telephone services, provided on a regulated monopoly basis by companies called "local telcos," and long-distance service, provided by largely deregulated long-haul carriers. The one exception to the latter category is AT&T, which continues to operate under FCC regulation because of its status as the "dominant" carrier in the United States. In 1989, the FCC simplified regulation of AT&T by changing from control based on rate of return on invested capital to "price caps." This change is intended to allow AT&T to compete more easily with its rivals, particularly MCI and US Sprint. AT&T still must file tariffs before service can be initiated or prices altered, but the FCC need not review the costs to be incurred by AT&T in the institution of the proposed service. Although designated as the dominant long-distance carrier, AT&T's real position of domination has eroded along with its market share, which stood at slightly less than 75% in 1989. The estimated total investment of AT&T in telecommunication operating facilities is now exceeded by the sum of the investment of MCI and US Sprint.

Residential service is provided by the local telco, which also permits access to the long-distance networks as selected by the customer. Under equal access provisions, a subscriber preselects the preferred long-distance carrier, to be reached when the numeral 1 is dialed before the area code. The area codes themselves are used by the long-distance networks to route calls between the regions of local telephone services. The RBHCs have local service monopolies, but they are prohibited from providing long-distance service, thus ensuring a market for the long-distance carriers. Local telcos provide a variety of other services for businesses. One of the most important for businesses is private switching under what is generally known as CENTREX (a contraction of *central exchange*). Private leased lines for analog and digital transmission are available on normal telephone facilities. In fact, the local telco continues to have a near monopoly on the termination of private lines between regions due to having the lion's share of local loop circuitry. Bypass of the local telco is available to some business addresses in large cities like New York and Boston by teleport companies with their own fiber optic cable.

Regional Bell Holding Companies and independent local telcos are entering new competitive fields where those companies do not hold monopolies. Some RBHCs allow subscribers to access information services like CompuServe and Dow Jones News/Retrieval® through packet-switching facilities of the local telco. Another popular activity comprises business equipment and computer sales and service. These activities are encouraging some PUCs to allow competitive entry into the more basic services of telcos. The types of organization capable of competing with the telcos are cable television companies and even the long-distance carriers. Experience with this kind of opportunity is being gained in the United Kingdom, where the government has already allowed competition in the local service markets.

2.1.2 TA Profiles

The general characteristics of domestic public telecommunication service capabilities in various regions of the world were considered in Chapter 1. The situation in the United States (reviewed above), the United Kingdom, and Japan (the latter two discussed in the next section) indicates the structure of future deregulation. The telecommunication manager involved in the general problem of international communication, however, must be prepared to deal with the kind of government monopoly profiled in this section. Because of their outgrowth from the postal service (uniformly a government agency throughout the world), the generic name used is the Post, Telegraph, and Telephone administration. We have mentioned that there is a danger in generalizing about the manner in which PTTs conduct business. Being a government-managed monopoly, there are characteristics that the PTTs have in common. There are enough differences, however, to recommend that every country be approached as a new case.

In [NTIA, 1983], the telecommunication policies of 17 countries are characterized. This document provides detailed profiles of PTTs in Europe, Asia, South America, and the Pacific, and is perhaps the best work of its kind. The specifics for each of these countries have changed since its publication, however, and will continue to evolve. Six market structure models are presented, where the two primary differentiators are whether the government is providing services and the degree to which entry is permitted into the domestic telecommunication market by private companies.

1. Government Monopoly — The classic case of a monopoly vested in a government ministry or department. This monopoly may be based in law or on administrative fiat or convention.
2. Public Corporation Government Monopoly — A government monopoly, but one instituted through a public corporation rather than a ministry or department.
3. Government Competition — A structure wherein both the government and private-sector entities compete in the market.
4. Regulated Monopoly — The classic case of private ownership of facilities (one entity) with regulation by a government department.
5. Regulated Competition — The extent of competition, number of competitors, and products or services that can be offered by each supplier are subject to authorization, licensing, type approval, *et cetera,* by a government department or agent of the government.
6. Liberalized Entry — A market situation characterized by the absence of all official government rules and regulations that serve to regulate market entry, structure, or conduct, coupled with complete reliance on private-sector entities for the provision of goods and services. This situation can include

minimal government requirements for, let us say, notification or meeting independent (non-governmental) standards. If the requirements are more than minimal, however, or used to influence or control entry, structure, or conduct, the case is one of regulated competition and not liberalized entry.

The first category, government monopoly, is the most common form of the TA. A government agency makes the investment, operates the facilities, and provides users with access to domestic and international networks. Telecommunication is viewed as part of the national infrastructure, much like roads, sewer systems, and the postal service. Unlike these physical examples, however, telecommunication is a dynamic activity, requiring a great deal of attention to the details of service. Generally recognized is that a private company with a profit motive will tend to be more effective in the TA role, thus leading to the introduction of the successive market structure models. In moving from monopoly to competitive entry, the domestic telecommunication environment opens itself to faster innovation and price cutting. The disadvantage of this model is that competition can render facilities obsolete although they are in working order, but cannot match what may be used by a better positioned competitor. For example, the market entry of US Sprint and their digital fiber network eventually forced AT&T to write off several billion dollars worth of analog switching and transmission facilities. The case of liberalized entry is what Adam Smith would view as ideal, where TAs are protected by the government from neither new entrants nor the competitive actions of one another (except possibly for antitrust laws, as in the United States). This model exists nowhere on the planet and represents a theoretically extreme case.

We will discuss the policies of three of the countries in the NTIA work. They fall within the range of cases 1 through 4. The countries are the Federal Republic of Germany, Spain, and Brazil. All are monopolies and can be considered to provide modern telecommunication service capabilities. We have updated the NTIA information as of 1989.

2.1.2.1 Deutsche Bundespost (West Germany)

The Deutsche Bundespost (DB) is the PTT of West Germany, directly under the Minister of Post and Telecommunication. The DB has a government mandate to provide all domestic telecommunication services, including telephone, data, facsimile, and video. Under the federal constitution, the government is responsible for the provision of all post and telecommunication services that are provided to the general public. In 1989, DB was restructured by federal law, forcing the separation of the TA from the postal service. This formed DB Telekom as the sole telecommunication provider in the domestic and international markets.

DB Telekom takes its responsibility very seriously, and is dedicated to the concept of universal service at pricing that is fair and equal. With the total backing of the government and German federal law, however, there can be no competition. We can conclude that DB Telekom is very much in the first category of a true government monopoly. Having rebuilt the infrastructure following World War II, the Deutsche Bundespost has reached the goal of having a telephone in essentially every German household.

The regulation of telephone and data networks in West Germany is very strict, with interface specifications and protocols rigidly set and enforced. From a business standpoint, there are some important advantages to this kind of policy. DB Telekom does not manufacture equipment, but all hardware connected to the network must operate in the appropriate manner. West Germany is very active in international bodies such as the CCITT and INTELSAT, and the recommendations and standards of these organizations are typically strictly followed.

DB Telekom does an excellent job of operating the national networks. DB Telekom's facilities are high quality and properly maintained. In touring the country in 1988, this author was particularly impressed by the elaborate microwave towers that appear to rise above every hilltop. Because of this investment, the full range of voice and data services is no doubt available everywhere in the country. The mesh network formed by these microwave links and fiber optic cables connects to the diversity of large cities, towns, and villages throughout the Federal Republic of Germany.

West Germany implemented the DATEX-L circuit-switched data network in 1975, and therefore was one of the first countries to "go digital." Normally, telex operates at less than 100 b/s, but DATEX-L provides data communication at speeds of up to 9600 b/s for synchronous and asynchronous transmission over the same network. DB Telekom is now the operator of the largest switched telex-data network in the world. Extension of DATEX-L to switched 64 kb/s service will provide an early demonstration of ISDN-like capabilities on the European continent.

The public data network, using packet switching, in the Federal Republic of Germany is called DATEX-P. Employing the same types of packet-switched nodes and protocols available in the United States through SprintNet, BT Tymnet, and others, DATEX-P is a service for which charges are rendered on the basis of actual quantity of transmitted data. An important capability of DATEX-P is that of protocol and speed conversion, a standard of public data networks in the United States. In West Germany, however, there is only one provider of a value-added network capability, and, of course, that is DB Telekom. An important feature of DATEX-P is that it directly interconnects with public data networks in North America, Japan, and other Western European countries.

There is an important constraint with regard to transmitting data between West Germany and other countries (i.e., international data flows, considered further in Chapter 9). In 1989, the policy of DB Telekom is to require that any data to be passed over an international leased line from West Germany must emanate from some type of computer device. Thus, a user cannot make a direct connection from the switched telephone or DATEX-L networks to such an international leased line. Data transmitted within the borders of the Federal Republic are not restricted as to content or connection. If any of the data are to be routed via an international leased line, however, a computer "processing" device must be the interface. One of the consequences of this policy is that it encourages the use of DATEX-P and the PDNs of foreign countries to which the former is connected. This result is consistent with the local policy of investing in facilities to provide the full range of services — a policy which places considerable demand on DB Telekom to develop as much paid usage as possible.

As with domestic service, DB Telekom is the sole provider of international telecommunication services for the Federal Republic of Germany. DB Telekom operates high-quality earth station facilities for access to the INTELSAT system. Transborder links within Europe are rather extensive, tying into undersea coaxial and fiber optic cables, which land in France and the United Kingdom. As will be discussed in Chapter 3, DB Telekom has made convenient the use of INTELSAT Business Service (IBS) through its own dedicated facilities. Customer premises earth stations may be justified under certain circumstances, but ownership will always reside with DB Telekom. As this discussion intimates, DB Telekom is the "first and the last word" in telecommunication within West Germany and between it and other nations.

Liberalization of telecommunication in West Germany first took the form of allowing users to connect their own devices to the national networks. DB Telekom, however, dictates interface specifications and new equipment requires type approval. Private networks can be established by using leased analog or digital data lines. Regarding digital, users can obtain full-time and part-time terrestrial lines at 64 kb/s and 2.048 Mb/s, which are standard in Europe. DB Telekom has already demonstrated ISDN services, and has committed to international connectivity within Europe by 1992. Otherwise, that substantial competition in voice or data networking services will be permitted in West Germany is highly unlikely for some time.

2.1.2.2 Telefónica (Spain)

The telecommunication infrastructure of Spain is owned and operated by Telefónica, a private company whose stock is nationally traded. Formerly known as Compañía Telefónica Nacional de España (CTNE), Telefónica provides all domestic and international telephone and private line service for Spain. A second organization, Dirección General de Telecomunicaciónes (DGT), is a government agency with responsibility for certain messaging services like telex and facsimile. The actual data transmission backbone is provided by Telefónica.

The Spanish government owns only 25% of Telefónica's stock, yet exercises considerable control over the enterprise. For example, the chairman of the board is selected by the government, although the majority of directors are elected by the stockholders. Along the continuum of PTT profiles, Telefónica is categorized as case 2 because of its government control and status as a corporation. The distinction, however, is that three-quarters of the stock is widely distributed among the population. This provides considerable motivation for Telefónica to seek profits, and also gives it access to capital markets inside and outside of Spain. Telefónica has moved aggressively to diversify into manufacturing of telecommunication equipment by participating in joint ventures with foreign companies. To aid in the development of the Spanish economy, the government has adopted a policy of encouraging foreign companies to develop new products and to manufacture them in Spain.

Telefónica has monopoly privilege for the provision of all basic telecommunication services and is rapidly modernizing the network with digital switching and transmission technology. Their monopoly extends to common terminal devices, such as telephone sets and modems, but other types of equipment for private network applications can be commercially obtained and connected to the network. Approval of any customer-owned equipment must be obtained from Telefónica before being used. As a basic requirement, equipment must meet applicable CCITT Recommendations, which are supplemented by domestic specifications. Spain's electronic industry is developing, so it will become easier to obtain more sophisticated devices in time. This may even result from a joint agreement with the user wishing to manufacture or assemble the necessary equipment in Spain. Importing equipment into Spain is possible, provided that it is not available on the domestic market. A good example is satellite communication earth stations and their associated specialized electronics.

Telefónica purchases the bulk of all telecommunication equipment sold in Spain, which represented a market of $1,372,000 in 1987. The largest manufacturer of telephone equipment in Spain is the former ITT subsidiary, which now operates as part of the pan-European giant, Alcatel. Note that Telefónica has a minority interest in this company.

Both circuit-switched and packet-switched data services are available in Spain from Telefónica or DGT, as appropriate. The Spanish are active in European and international regulatory affairs as well as in the development of fiber optic cable facilities on land and undersea (see the discussion on TAT-9 to follow).

Investment in domestic facilities has grown substantially in recent years as Telefónica modernizes its network. An area of urgent need is to increase the percentage of residences with basic telephone service. Approximately half of Spain's 40 million citizens have telephones in their homes. Essentially all service is rendered through automatic exchanges, a major accomplishment after only one decade of rapid development. In some ways, for businesses to obtain modern services is easier because the revenue exists to support the investment.

As mentioned at the beginning of this section, telex and message services are provided by DGT, a government agency. The transmission paths for such traffic are provided over the domestic network of Telefónica. In particular, Telefónica operates the Spanish data transmission network, called IBERPAC, which is used by DGT for telex transmission and switching.

Telefónica began providing INTELSAT Business Service to companies through community earth stations in 1988. Similarly to most other European PTTs, earth stations are owned and operated by Telefónica. There is a preference for using the existing four main INTELSAT earth stations.

With its temperate climate, stable political situation, and low labor costs, Spain is developing into what has been called the "California of Europe." Spain is undergoing remarkable change, benefiting from the introduction of technology and a keen governmental interest in expanding the domestic economy. Spanish companies can provide a springboard for outsiders wishing to gain access to Spain's markets and manpower. The telecommunication infrastructure will remain the domain of the monopoly service provider, so competitive entry is not expected during the 1990s. Having a modern digital network and close cooperation with other nations of the European Community, however, Spain should not present major obstacles to the application of information technology within its borders and along international pathways.

2.1.2.3 Embratel (Brazil)

Telecommunication services in Brazil are provided by government-owned enterprises. The local telephone companies, which were privately owned at one time, have essentially been nationalized and are under the control of the government holding company, Telebras. Today, the Telebras system consists of 25 companies, which serve the telecommunication needs of the Brazilian states. All interstate long-distance and international services are provided by Empresa Brasiliera de Telecomunicaçoes (Embratel). The headquarters of Telebras is in the modern capital of Brasilia and that of Embratel is in the attractive seaside city of Rio de Janero.

The combined government-controlled network has grown rapidly since the 1970s. Long-distance trunks and telephone service in major cities equal those of most European nations. As in other industrial sectors, Brazil maintains the most advanced infrastructure in South America. Also, this author's experience is that the technical staff of Embratel is highly competent and familiar with virtually any technology that can reasonably be applied for commercial telecommunication. For example, Embratel operates the only domestic satellite system on the continent and has already embarked on a second-generation replacement project.

We can conclude that services are available in Brazil by monopoly service providers, placing it in the category of case 2 among PTTs. There is little chance that competition will be introduced because of the perceived need for fully utilizing the invested capital. Embratel is a successful venture for Brazil, and is likely to be in a position to continue to serve the growing needs of domestic and international business.

All equipment purchased by users must meet technical standards set by Telebras. Typically, this equipment is supplied by the local telephone company or by Embratel. Obtaining permission to use customer-owned terminal devices is difficult. One reason for this is the government's policy to encourage domestic industry, which is generally capable of manufacturing even the most complex electronic products. For a Brazilian manufacturer to introduce a technology or specific device that is not available on local markets, however, can be difficult. On the national level, there are several government-supported research facilities that develop technologies for applications within Brazil or for potential export in the form of manufactured products.

Embratel offers a variety of data communication services to domestic businesses. The equivalent of leased lines for data (DDS in the United States) are available via the network. A store-and-forward messaging network operates in conjunction with the telex network. Also, implemented in 1989 is a public packet-switched network. The government is encouraging the development of value-added and information services through the public network. The entry of foreign companies in this market, however, is restricted.

In addition to the limits on entry into Brazilian markets, the government has usually been very supportive of foreign companies that can transfer technology to the local economy. The satellite project previously mentioned was viewed as a national endeavor, akin to the space program in the United States and Western Europe. Joint ventures, usually controlled on the Brazilian side, are developing and manufacturing Brazilian versions of popular computers, communications devices, and control technologies. This bodes well for the future because advanced information technologies can become available in this South American country well before other parts of Latin America or even other Western countries.

2.1.3 PTT Privatization and Competition

The spectrum of competitive entry into telecommunication markets as represented by the six model cases includes forms of competition that are only beginning to appear. In fact, the last category of liberalized entry without government regulation is many years away. Competitive entry is allowed under government regulation and oversight in case 5, the model first employed in the United States. This form of competition is being adopted in some Western democracies.

In the paragraphs that follow, we review the situations in the United Kingdom and Japan, two countries which have moved rapidly to privatize their PTT operations and allow some competition in domestic and international telecommunication service markets. Even the most capitalistic of governments is concerned that open entry will undermine the basic structure as competitors gain control of the more lucrative sectors, leaving the less attractive subscribers either without service or in the hands of a badly weakened national networks provider. The first step in each case was to privatize the government controlled PTT monopoly to introduce a strong profit motive into the rendering and development of services. A privatized TA is also in a better position to invest funds by exploiting the private capital markets. For example, BT can offer more of its stock to individuals and can sell bonds to banking institutions in Europe, the United States, and Japan. After privatization, the market is opened to a limited extent so that one or more "qualified" competitors may engage in the public telecommunication service business.

2.1.3.1 The United Kingdom

The Conservative government of the United Kingdom under the leadership of Prime Minister Margaret Thatcher has dramatically altered the telecommunication business environment in that nation. As with all other European nations, the United Kingdom operated under the PTT system with all telecommunication services (domestic and international) being provided by the British Post Office (BPO). First, British Telecom (BT) was separated from the BPO and subsequently privatized into a public limited corporation (PLC) with publicly owned stock. Overnight, BT was a success with British investors who saw it as an opportunity to take part in a rapidly growing and profitable business. BT has expanded from its solid base as the principal TA of the United Kingdom into a world leader in telecommunication equipment and international services. A subsidiary called British Telecom International (BTI) was created as the vehicle to diversify and capitalize on BT's strengths in telecommunication technology. BTI owns international information networks (Dialcom and Tymnet) and undersea fiber optic cables in partnership with other international carriers.

A second government-owned carrier, Cable and Wireless (C&W), has been privatized. Cable and Wireless is a highly respected organization with telecommunication activities throughout the world. One of the more familiar operations of C&W is the Hong Kong telephone company. Recently, the assets of the Hong Kong telephone operation have been placed under a separate company there. Many of the former British Colonies are served through INTELSAT earth stations owned and operated by C&W, which participates in several undersea cable systems as well. Recently, C&W entered the U.S. domestic telecommunication market by purchasing a fiber optic carrier called TDX. As a worldwide telecommunication provider, C&W has gained a substantial customer base and has effectively supplanted ITT from its formerly unique position.

The Thatcher government decided to allow competitive entry in the domestic market on a limited basis. (For a detailed discussion of these actions, see [UN, 1987].) Two acts of Parliament in 1984 had a direct bearing on the structure of the domestic telecommunication industry. The Telecommunication Act of 1984 established a policy-making ministry called the Office of Telecommunications (OFTEL). Acting much like the FCC, OFTEL is responsible for licensing service providers and setting the rules of competition within the United Kingdom. The objective is to be fair to consumers, purchasers, and other users of domestic public telecommunication services. Another body, the British Approvals Board for Telecommunications, was created to evaluate subscriber devices before they can be connected to the network.

A key measure of this legislation was the conversion of BT from a government-owned monopoly into a privately held company with stockholders. BT is organized into five divisions: local communication services; national networks; British Telecom International (BTI); development and procurement; and BT Enterprises. As a truly commercial enterprise, BT is highly competitive at home and overseas, and has proved to be very profitable as well. British Telecom completed a program in 1989 to convert the U.K. long-distance network from analog to digital technology. Local exchanges are being converted more slowly, but BT has the objective of being one of the first in the world to provide ISDN services on a nationwide basis.

British Telecom's protected monopoly, which existed prior to privatization, was diminished through controlled entry of a limited number of competitors. Principal among these is Mercury Communications, Ltd, a wholly owned subsidiary of Cable and Wireless. (Another local telephone service provider, Hull Communications, has had a government franchise for some time, but is not a major player in the domestic or international markets.) Mercury moved aggressively into the field of business communications by installing a modern digital fiber optic network around London. This was expanded to include an islandwide loop of fiber optic cable. Business customers can be connected to this backbone, having available voice and data services. One of the more popular applications on Mercury's network is that of private lines for data communication.

Having a solid base in international communications, both BTI and Mercury-C&W are positioned for future growth. The former British Post Office earth stations in the INTELSAT system are operated by BTI, providing major gateways to the British Isles. Furthermore, BTI holds a significant interest in copper and fiber optic cables to Europe, the Americas, and across the Pacific. BTI is a leading international carrier, even in markets outside of the United Kingdom. Cable and Wireless also has a sizable presence in these markets by virtue of their involvement with public telephone services in the former British Colonies. For the business customer, BTI and C&W operate IBS sites in the United Kingdom, offering the full range of voice, data, and video services permitted under British law.

International television is a market area which BTI has targeted, with sales offices in key cities throughout the world.

The developments in the United Kingdom are exciting to watch, as the government moves aggressively to deregulate industries such as telecommunication services. Because of its financial and technological base, British Telecom will easily expand its domain as an international leader. BTI and C&W are impressive giants with accomplishments in performance and profits. Other companies are entering specialized markets such as value-added networks and electronic mail. In 1988, one-way video and data broadcasting via satellite to businesses was opened to competition within the United Kingdom. Among others, the Electronic Data Systems (EDS) subsidiary of General Motors obtained a license to enter this market, being perhaps the first time that a foreign company has been allowed to enter the domestic U.K. market.

2.1.3.2 Japan

The Japanese government has fostered the development and orderly growth of the domestic telecommunication industry. Prior to privatization in 1985, Nippon Telegraph and Telephone was the government-owned corporation with a complete domestic monopoly. As the PTT, NTT had a free hand in deciding on technology, investment, and service policy. The Ministry of Post and Telecommunications (MPT) had little influence on the operation of NTT. All international services were provided by another government corporation, Kokusai Denshin Denwa. KDD plays an active role in INTELSAT, with a significant share of the investment. KDD also invests in undersea cables, which extend from Japan to other points in the Pacific and Indian Oceans. Both KDD and NTT conducted extensive research and development programs, with elaborate laboratory facilities populated with the top graduates of Japanese universities (notably Tokyo University). As Japan developed into a world economic power, the two organizations created a first-class telecommunication infrastructure.

Interestingly, when NTT was privatized in April 1985, the value of its stock soared to the point where the total investment far exceeded that of any other corporation in the world. This was due mainly to the value of the real-estate holdings of NTT. At the present time, it is permitted to enter virtually any market in telecommunication or information technology. NTT is a primary provider of electronic funds transfer for Japanese banks, a business that RBHCs and AT&T are prohibited from entering in the United States. NTT has also moved aggressively into public data networks using packet switching, and is implementing a videotex service similar to Minitel. Another area of development and expansion is ISDN, where NTT was one of the first PTTs to run trials with live customer traffic. Among the beneficiaries of this activity were telecommunication manufacturers like NEC and Fujitsu, the equipment of which was employed for the trials.

The domestic networks of Japan were primarily based on line-of-sight microwave radio systems prior to the introduction of digital fiber. Even in the 1960s, NTT and its Japanese support industry pursued digital technology, making Japan the world leader in digital multiplexing and transmission. Graphic communication techniques were well developed, and the success of facsimile communication likewise is the result of Japanese activity.

Along with privatization, the Japanese government opened the domestic and international markets for limited competition. Three areas were authorized and established by MPT. The first was for value-added networks, which employ the facilities of NTT to provide specialized services. These include reselling of private lines and packet-switched data services. Restrictions still exist on the application of private lines for businesses in Japan. For example, a private telephone trunk cannot access the public network through a PBX. This protects NTT from customers using private trunks to bypass the long-distance network for calling third parties.

The second domestic overture by MPT was to authorize public telecommunication services by competing carriers. Terrestrial fiber, digital microwave, and satellites are owned and operated by new entrants from the Japanese industrial community. These firms are designated as Type 1 carriers, and may own and operate the facilities for the purpose of making a profit from switched and private line services. Japan Telecom, which was formed by the largest railway companies, built a network by laying fiber optic cables along the railroad rights-of-way. Private highway companies have provided their rights-of-way to Japan Telway, the fiber optic carrier that they in turn own. Both optical fiber carriers offer digital trunk services, including long-distance switched services in direct competition with NTT. These national entities have been joined by regional and local carriers, which have the ability to connect directly to the business customer. The Tokyo Electric Company is able to access business customers by entering buildings in much the same way that electrical service is provided. Tokyo Electric's fiber optic telecommunication carrier is thus able to service the Tokyo business subscriber and interconnect with competing long-distance carriers. NTT can now be bypassed on both the long-haul segment and the local loop.

There are two competing satellite ventures authorized as Type 1 carriers that are now both in operation. First to launch a Ku-band satellite was a joint venture company backed by Hughes Communications of the United States, C. Itoh, and Mitsui and Company. The other venture is led by Mitsubishi. Both systems employ satellites built in the United States and designed to provide commercial services similar to those offered by U.S. satellite carriers. The two companies, JC Sat and Space Communications Corporation, respectively, are providing their transponders to video enterprises to bypass the NTT microwave network. Even with all of the activity in terrestrial fiber and microwave, VSAT networks for data communication and business video are being pursued with vigor by several new ventures. One of these is JSnet, a joint venture of NEC, C. Itoh, Mitsui, Hughes Communications, and several other Japanese companies. Interestingly, NTT is taking a significant block of each system's resources to provide alternate capacity and to enter new markets using VSATs.

The third area of competitive entry relates to the focus of this book, international telecommunication access to Japan. Initially, there was a concern about allowing foreign companies to participate in this market in competition with KDD. An attempt at so restricting foreign entry was defeated, and at least one venture has participation from abroad. The two authorized new entrants are International Telecom Japan (ITJ) and International Digital Communications (IDC). Both are partnerships involving dozens of Japanese industrial companies, each hoping either to make or to save money (i.e., by getting international service at a discounted price). Both ventures are currently using INTELSAT capacity through their own earth stations near Tokyo, offering international direct distance dialing to the United States and other countries at a discount of 20% relative to KDD. In the near future, IDC will transfer its services to the North Pacific Cable, as this particular company is partially owned by Cable and Wireless.

Under the new regime, MPT has taken on much greater responsibility for setting and implementing telecommunication policy. The MPT's view is that the market should be orderly, with reasonable amounts of competition. Any price or service must be reviewed and approved by MPT. As an example of how it views its role, MPT refused to allow a joint venture of Sony and GE to become a third domestic satellite operator. Having a third competitor would have created a capacity glut, making it difficult for any operator to become profitable. MPT is also very concerned that prices be consistent and service quality not be sacrificed. Still, the number of competitors as well as the degree of competition are remarkably high.

2.1.3.3 Western Europe

As the traditional domain of the monopoly PTT, Western Europe has been slow to move toward deregulation of telecommunication services. As is well recognized, trade barriers will be greatly reduced by the end of 1992 between countries that are members of the European Community. The PTTs recognize that major users of international telecommunication facilities are watching closely to see the form that liberalization will take. As of 1990, there had been many meetings and press releases, but we were yet to see concrete actions to open markets and to allow users to implement information technology capabilities such as in the United States. Perhaps this was due to the different role that the PTT takes on in a typical Western European country, as demonstrated by the previous discussion on DB Telekom. In any case, liberalization will come because domestic forces have the political power to cause the government to change its policies (whether the PTT concurs or not).

A 1988 resolution of the telecommunication ministers within the European Community has as its purpose the deregulation of data communication services. Members of an industry organization called *Unice,* for the EC federation of employer's organizations, have welcomed this initiative. The industry wants cheaper and more flexible telecommunication services offered by the plan, which seeks to break public authorities' monopolies over basic data transmission. The supply of value-added services, such as electronic mail and videotex, would also be affected by the plan. If these steps were taken, pan-European data services should be as diverse and attractive as in the United States and Japan. The concern is that some of the PTTs are less inclined to implement their portion of the plan due to concern that domestic policies will not be maintained.

Cooperation among PTTs is natural on the continent, and all participate in areas of mutual interest. Primary among these are efforts toward standardization of interfaces between networks. Likewise, agreement is obtained on the setting and sharing of tariffs for transborder services. The organization that traditionally dealt in these areas is the European Conference of Postal and Telecommunications Administrations (CEPT). An administration is the local government body with the responsibility to regulate and provide telecommunication services. In Europe, the administration and PTT are usually the same entity, such being quite convenient for an organization such as CEPT to function. CEPT is not part of the International Telecommunication Union, but is a powerful block of administrations that strongly influences the directions taken by the ITU. As an association of PTTs, however, CEPT is not a force for change and deregulation because it is mainly concerned with protecting the PTTs' interests. The move toward deregulation is reducing the influence of CEPT, and new organizations like the European Telecommunication Standards Institute (ETSI) are gaining in importance.

The PTTs of Europe are seizing ISDN as the strategy to give international users the advanced digital services that they say they want for private networks. In 1989, a decision was made to have a measure of ISDN capabilities in operation between European countries. This would provide circuit-switched 64 kb/s data services in addition to some advanced features for telephone calling. Leased line facilities at a variety of data rates would also generally be available. The three-year time frame represents a considerable challenge to the PTTs because not all of them have digital network facilities in place at the start of the period. An ISDN plan, adopted by a working group of CEPT, has two purposes: to ensure that the network becomes sufficiently widespread to meet the needs of international users; and to create ISDN applications and equipment markets in time to maximize the benefit to Europe's manufacturing industry. Another objective for the participating PTTs is to offer full international interconnection and unified access for terminal equipment by the end of 1992. Agreement is being reached on the technical details of the interface with terminal devices and on the structure of *signaling system number 7* (SS-7).

At an exhibit in West Germany in March 1989, DB Telekom demonstrated the transmission of voice, Group 3 Fax, video, videotex, teletext, packet-switched data, and high resolution still images over ISDN links. Services have also been launched in France and the two countries are expected to make the transborder connection in 1990. Note that SS-7 is already in use on several domestic networks in North America and Europe, but some necessary details need to be resolved before they can be interconnected. Also, meeting the goal of having ISDN capabilities across national borders in Europe by 1992 is a monumental task. Eventually, however, the goal will be met.

An area where liberalization has begun is that of customer-owned terminal devices. PTTs, for the most part, allow users to acquire telephone and computer terminal devices and attach them to the public networks. This would be subject to type acceptance and specific approval of the PTT, which has bureaus specifically set up for this purpose. The flow of information between these devices and other countries may be restricted, as discussed for DB Telekom. More information on transborder data flows is presented in Chapter 9.

Major private network users in Europe share another concern about the availability and cost of leased lines for voice and data communication. In some countries, these lines are offered at reasonable prices, using their cost to the PTT as a base. The United Kingdom is a notable example, where the competition between BT and Mercury is having the desired result. Other PTTs adopt the policy of charging whatever the user will bear. In many cases, this is more than four times the price that users expect to pay, based on the cost of implementing the service. The discrepancy is even greater on international private lines within Europe, with rates multiplying 10 to 20 times, depending on the countries involved. The PTTs concerned must face this disparity and its effect on trade, particularly in the post-1992 era. Many business people believe that telecommunication services ought not be excluded from steps being taken to remove trade barriers. This position is an important element in the policy of the European Commission. Both CEPT and EC agree that realistic pricing will enhance consumerism and competition within a more unified Europe.

2.2 FIBER OPTIC NETWORKS

The timetable within which fiber optic cable systems have been implemented in the developed world is literally a revolution. That this technology is superior in terms of quality and economics to other approaches for point-to-point communication services is undeniable. Any international private telecommunication network will employ fiber optic cable systems, either on land or under the seas. The reader should bear in mind, however, that fiber optic cables would not be a replacement for all other modes because of two factors. First, fiber is a point-to-point medium and thus is subject to breakage due to cutting and equipment failure. The now-recognized phenomenon of the "backhoe fade" and its undersea cousins, the "shark attack" and the "fishing trawler," are ever-present threats. Therefore, alternate paths or media need to be available. The second consideration is that fiber is very expensive to construct and maintain. This is countered by the extremely high capacities that can be carried, resulting in the lowest possible cost per channel. Long-haul fiber optic cable systems are primarily the domain of the wealthy international carrier or PTT. Applications of fiber on the local loop and within buildings or campus-style settings are also attractive, but are not considered to be an international telecommunication application.

We classify fiber optic systems as *land-based* and *transoceanic*. In the following paragraphs, we review some of the systems in operation as of 1989 or planned to be ready for service early in the 1990s. The land-based systems are those used for long-distance and interexchange transmission. Several of these networks are in operation in the United States; NTT has pioneered their use in Japan; and several European PTTs have significant fiber capabilities. Specific transoceanic cables are described. The area is developing particularly fast, as the major carriers implement these systems to enhance their revenues and market share. As recently as 1984, none of these systems was in operation and only a few were actually under construction. The total investment in these long-haul systems as of 1989 is in excess of $2 billion, and is expected to double by 1992.

2.2.1 Land-Based Fiber Networks

Fiber optic systems are an ideal medium for the terrestrial long-distance network. Employing this technology as a marketing weapon, United Telecommunications, a Kansas City–based independent telco, entered the U.S. long-distance market. The firm's lead over MCI and AT&T in the application of digital communication via fiber allowed it to gain an adequate market share in the highly competitive long-distance market. In 1986, the long-haul networks of United Telecommunications and GTE Sprint were merged, resulting in US Sprint, the long-distance network of which (illustrated in Figure 2.1) was essentially completed in 1989. Industry leader AT&T along with MCI played "catch-up," installing digital fiber and switching systems to replace less marketable analog systems that were in place and producing revenue.

This type of behavior and performance could have been expected in an environment of competitive entry in long-distance service. At present, only AT&T is subject to rate-of-return regulation by the FCC; MCI, US Sprint, and other smaller regional carriers are free to set prices and define service offerings at will. Under the direction of former FCC Chairman Dennis Patrick, regulation of the economic side of AT&T is being changed from rate-of-return to price caps. The

Figure 2.1 US Sprint's nationwide digital fiber optic network. (Courtesy of US Sprint.)

In-Service

Network — through 1988

Network — (1/89 +) Currently served by leased facilities.

purpose of the price cap approach is to allow AT&T to meet the competition from the other unregulated carriers. AT&T's dominant position (which initially justified regulation) has been eroded, with its market share in long-distance decreasing from 85% to less than 75% over the six-year period since divestiture.

The result of this activity in fiber cable is that most long-distance calls within the United States travel over digital fiber. Similarly, fiber optic networks have appeared in Western Europe. The United Kingdom, France, and most of Scandinavia employ nearly all fiber for long-haul with the other major countries soon to follow. By the end of 1992 or shortly thereafter, these fiber networks will be interconnected at the borders to produce a nearly uniform infrastructure based on the standards for ISDN and SS-7. The quality of voice service on fiber is decidedly better than that of any previous analog techniques, with speech sounding as if the distant party were in the same city. Subscribers have already become accustomed to this, and no one cares to regress to the analog world. Digital satellite links have long been available, providing speech quality superior to that of analog long-distance circuits. Satellite voice (even with appropriate echo cancellation) still suffers, however, from the quarter-second delay produced by 45,000 miles of relay path. The majority of callers may not mind, but a significant percentage are disturbed by this delay and prefer the shorter path length afforded by terrestrial fiber. The reader should bear in mind that a fiber circuit of greater than approximately 500 miles in length must be fitted with a digital echo canceller. For data communication, the effect of the delay can be controlled, as will be discussed in a subsequent section.

We have already mentioned the expansion of terrestrial fiber optic networks in Japan and the United Kingdom. In the case of Japan, the motivation has more to do with technology than domestic competition. By gaining an advantage with fiber optic systems, Japanese technological giants like NEC, Fujitsu, and Hitachi are now in a position to dominate the worldwide industry. This is one of the reasons why NTT and the Japanese government aggressively pursued fiber. Another reason has to do with the environment in Japan, where microwave links have been heavily employed. Eventually, the microwave frequency spectrum will have become saturated. As stated previously, fiber optic cable systems have very attractive economics because of the large amounts of traffic that can be accommodated. The rapidly growing Japanese economy is positioned to employ this capacity.

A well interconnected fiber optic network offers a homogeneous environment for the development and provision of digital services. The basic capacity of one fiber pair is greater than one billion bits per second (one gigabit), which can support 30,000 voice circuits, 100,000 9.6 kb/s data circuits, or 20 color television channels. Through multiplexing technology, any combination of these services can be provided. Even more importantly, the network can be made to adapt to traffic loading for a wide variety of bandwidths. Both switched services and private line services are easily accommodated. At each node in the network, capacity can be reallocated for traffic on demand because a common digital hierarchy is used.

Another important dimension of land-based fiber networks is the signaling and control of services. As mentioned above, SS-7 is a standard system for conveying call-processing information between switching centers on the network. SS-7 employs the OSI model for the protocol structure of a data communication network. The signaling system is actually a packet-switched network that does not carry user traffic. Rather, packet switching provides a highly reliable means of setting up and taking down calls. Some of the other features include calling party *automatic number identification* (ANI), credit card verification, 800-number conversion, automatic call diversion, and collect calling.

In summary, a full-mesh domestic fiber network is the backbone for most telecommunication services that are interactive in nature. The facilities for wide bandwidths, complex switching of traffic, and SS-7 make fiber networks ideal for widespread introduction of ISDN in the 1990s.

2.2.2 Transatlantic (TAT-8 and TAT-9)

On June 30, 1988, the first transoceanic fiber optic cable was put into service. The Transatlantic Telecommunication Cable (TAT-8), approximately 4000 miles long, was a joint undertaking of 29 telecommunication administrations from Europe and North America, costing approximately $335 million. AT&T, as the lead carrier and driving force, constructed the majority of the link with a smaller section constructed by Standard Telephones and Cables of the United Kingdom and Alcatel of France. Simplified routing of TAT-8 and other transatlantic fiber projects are shown in Figure 2.2.

The cable connects from New Jersey across the Atlantic to a point south of Ireland, where it branches to France and the United Kingdom. The underwater branching repeater can be powered from either the French or UK terminus, thus enhancing redundancy. The reinforced cable contains three fiber pairs: two working pairs with total capacity of 560 Mb/s, and one spare pair for backup. The actual cross section of this type of cable is shown in Figure 2.3. The total bandwidth is subdivided into *minimum assignable units of ownership* (MAUOs), each of which is equivalent to 64 kb/s. Investment and use of the installed cable is allocated with the MAUO standard of measure. When used in telephone service for voice traffic, the link can support a total of 40,000 voice conversations by using *digital circuit multiplying equipment* (DCME). Technically, DCME takes a multiplexed stream of standard digital voice channels (i.e., 64 kb/s PCM) and compresses it by a factor of 4 or 5 [Seymour, 1988]. The estimated capacity for the cable of 40,000 channels assumes a factor of 5 in this compression process. The particulars are beyond the scope of this book, but techniques such as compression algorithms (e.g., ADPCM or RELP) and *digital speech interpolation* (DSI) are usable [Elbert, 1989]. In the absence of DCME, the capacity of the cable in digital terms can be subdivided to support a variety of data and digital video applications.

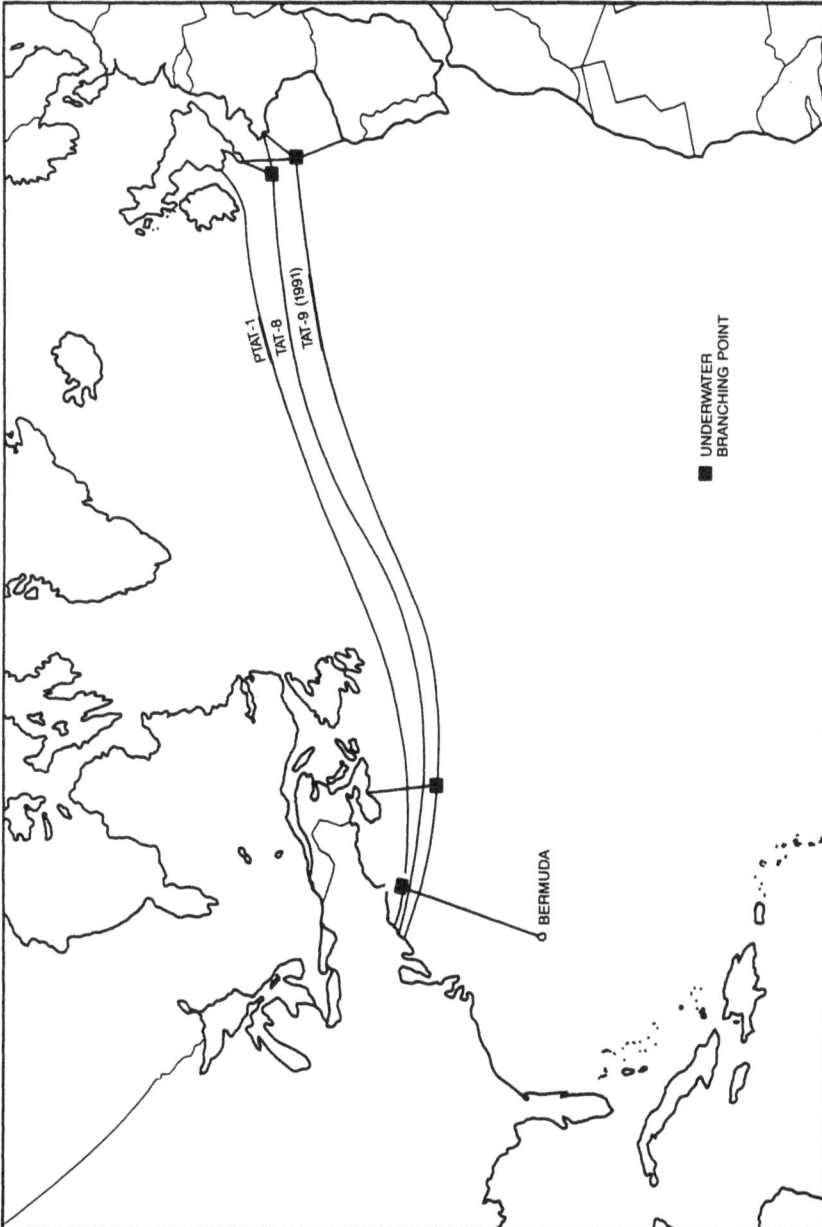

Figure 2.2 Atlantic Ocean region fiber optic cables in transatlantic service.

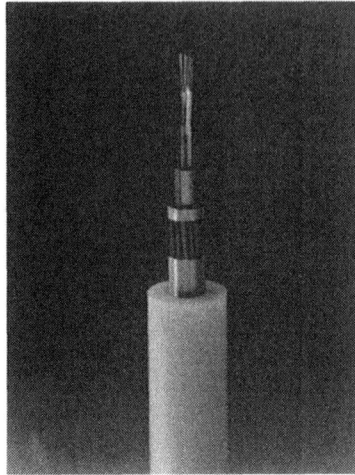

Figure 2.3 The constituents of a typical transoceanic fiber optic cable (photograph courtesy of KDD).

Fiber optic technology for TAT-8 was proved in a variety of field trials before it was committed to an operational system. Cable and repeater technology must be fully tested and its reliability established prior to installation under the ocean. This is particularly advisable because undersea cables are extremely expensive to repair and the duration of outages due to failure are often quite long. Great depths and rough seas can hamper repair operations aboard ship. For a cable to be out for a week or more is not unusual, necessitating the use of alternate routing. If other cable capacity is available, services can be maintained with minimum time delay. More likely, satellite communication will be employed to restore a failed high capacity cable. INTELSAT maintains spare capacity on satellites and through gateway earth stations for the eventuality of cable restoration.

To demonstrate the viability of the technology, AT&T and the other co-owners installed a prototype system between two Spanish islands. The prototype proved the reliability of the cable, and also was the vehicle for checking repair and maintenance techniques. The operating life of TAT-8 itself is projected to be 25 years. By that time, this cable of modest capacity (in fiber terms) will be obsolete as newer links with as much as ten times the capacity may be readily available. The average distance between regeneration repeaters will increase, which currently is approximately 60 km. Regeneration is a technique where the digital pulses are recreated within each repeater, which essentially eliminates noise and distortion from the intervening fiber path.

Considerable detail on services offered on TAT-8 by the various PTTs and international carriers can be found in [KJH, 1988]. As lead partner, AT&T was the driving force behind TAT-8 (and several other cable projects) and holds approximately 35% of the ownership. The heavy association of PTTs and U.S. service providers has resulted in the bulk of the capacity being committed to public switched telephone traffic. A small but significant portion is available for analog and digital private line services. Among the new offerings, BTI and AT&T are providing *switched 56 kb/s services*. In 1989, MCI International announced that Mobil Corporation would be their first private line customer on TAT-8.

After one year of service, TAT-8 had experienced three major disruptions. The first took out the entire link. The second took out the branch to the United Kingdom. Cleverly, traffic was routed along the branch to France and back to the United Kingdom by using terrestrial links without the need for satellites during restoration. What was most disturbing to users, however, was the subsequent failure of the French branch before the U.K. branch was repaired. This again took out the entire link. These failures are probably due to problems with the original installation, and reliable operation is no doubt forthcoming. The particular problem could involve a regenerative repeater, a fiber splice, or the cable itself, which could have been broken by natural causes or something being dragged from a ship.

As discussed in the next section, the owners of PTAT-1 intend to allocate capacity on their fiber optic cable for restoration purposes (and *vice versa*). Until such time as diverse routing is working satisfactorily, international telecommunication will rely on satellite restoration by the INTELSAT system. The management of INTELSAT enjoys the use of satellite communication for cable restoration for two reasons. First, the revenue is attractive because of the amount of capacity required and the duration of service. Second, cable failures demonstrate that a single communication medium is neither perfect nor impervious. Satellite systems themselves are also subject to difficulties, and a fleet of spacecraft is almost always preferable to a system dependent on one satellite.

The PTTs and U.S. carriers are proceeding with another fiber optic cable called TAT-9. This cable will have twice the capacity of TAT-8 and will branch to three European locations: Spain, France, and the United Kingdom. AT&T will own 28% of the link and be responsible for approximately half of its construction. In addition to providing expansion capacity, TAT-9 is earmarked as a diverse route across the Atlantic. The owners hope that with TAT-9 satellite restoration of the inevitable cable failures will not be required. A network consisting of TAT-8, TAT-9, and PTAT-1 will indeed be welcome to major users of international telecommunication who rely on the characteristics of terrestrial transmission for interactive applications in voice, data, and video communication.

2.2.3 Private Transatlantic Cable (PTAT)

A private business venture was undertaken to exploit the capabilities and market potential of a fiber optic cable between the United States and the United Kingdom.

The partnership was formed on the U.K. side by Cable and Wireless and on the U.S. side by a group of investors incorporated under the name Private Trans-Atlantic Telecommunications System, Inc. (PSI). The first end-to-end connection was completed in May 1989, and the cable was ready for service in August of the same year. The cable is routed from the New Jersey shore to Dublin, Ireland, and London, England. A spur to the island of Bermuda is provided to serve C&W customers. On May 9, 1989, US Sprint announced that it had purchased the U.S. half of the venture from PSI, putting it under a new corporate umbrella called Sprint International. Interestingly, NYNEX, one of the seven RBHCs, had an option to purchase PSI's interest, but was forced to withdraw by the Federal District Court that administers the Bell System breakup.

The FCC granted the original license to PSI to be a private international carrier as opposed to being a common carrier. Under this status, PSI would transact private deals with major users of bulk capacity. PSI could also offer capacity to common carriers, allowing them to provide alternative routing relative to TAT-8 and TAT-9. With present technology, PTAT-1 is capable of carrying 27 DS-3 circuits, amounting to a total of 1.2 Gb/s (nearly 100,000 telephone circuits using DCME).

US Sprint and C&W have already begun to develop the synergy of their respective domestic networks. Both carriers employ digital gateway switches made by Northern Telecom and the two can interconnect very efficiently. With a common architecture and software, the companies quickly will be able to deploy SS-7, bringing with it specialized services such as 800-number routing, credit verification, ANI, and eventually ISDN facilities. C&W will serve the British Isles through its Mercury Communications subsidiary. Mercury, as discussed previously, has developed a fiber optic loop in the United Kingdom, offering digital switched and private line services. Customers in the United Kingdom will use PTAT-1 for international digital private lines with Mercury as a single supplier.

Through the use of fiber links with Europe, C&W can extend networks across three continents. This capability takes advantage of U.S. fiber networks and the North Pacific Cable (discussed later in this chapter) in which C&W participates. In the United States, US Sprint will provide intelligent network services from there to the United Kingdom. The virtual private network capability now being used by a number of corporate customers is to be extended across the Atlantic. VPNs will allow simplified calling with seven digits or less between business locations on opposite sides of the Atlantic without the need for dedicated private lines. In addition, US Sprint will use the capacity on its long-distance network to capture its share or more of the expected growth in international calling. According to William Esrey, chief executive of US Sprint, it expects international traffic to continue to grow at the rate of 20% per year.

An important attribute of PTAT-1 is its positioning as an alternative route to TAT-8. US Sprint is expected to allocate some DS-3 capacity for common carrier services for use in public switched telephone networks. In this manner, both cable systems will be available to back up each other. The fact that TAT-8 has experienced more than its share of initial failures has caused key users to be concerned about the viability of the medium. Although satellites provide excellent circuit restoral, there are applications where satellite delay is not tolerable and alternate digital fiber will be extremely valuable.

2.2.4 Transpacific (TPC-3 and TPC-4)

Leading international carriers and PTTs, including AT&T, BTI, KDD, France Telecom International, and DB Telekom, participated in the construction of the first transpacific fiber optic cable, TPC-3. The cable was installed by KDD and AT&T and is similar in design to TAT-8. Figure 2.4 provides a photograph of KDD's cable ship, which was used in this installation. The connectivity of TPC-3, TPC-4, and other Pacific cables is shown in Figure 2.5. The section between the U.S. mainland (near San Francisco) and Hawaii is called HAW-4 with TPC-3 continuing on across the Pacific Ocean. The cable splits in the middle of the ocean, with branches going to Guam and Japan. This system entered service in April 1989, opening the way for complete terrestrial fiber circuits from Europe to Japan. The link cost approximately $700 million, is 8271 miles in length, and has a digital capacity of 560 Mb/s. In terms of simultaneous voice conversations, the cable can support a quantity of 40,000 voice conversations by using DCME. Time division multiplexing is used to split this capacity into a variety of other bandwidths, such as 64 kb/s (and multiples thereof), and T1. Only satellites are currently able to transfer information at comparable rates.

Figure 2.4 One of the cable ships used to install TPC-3 (photograph courtesy of KDD).

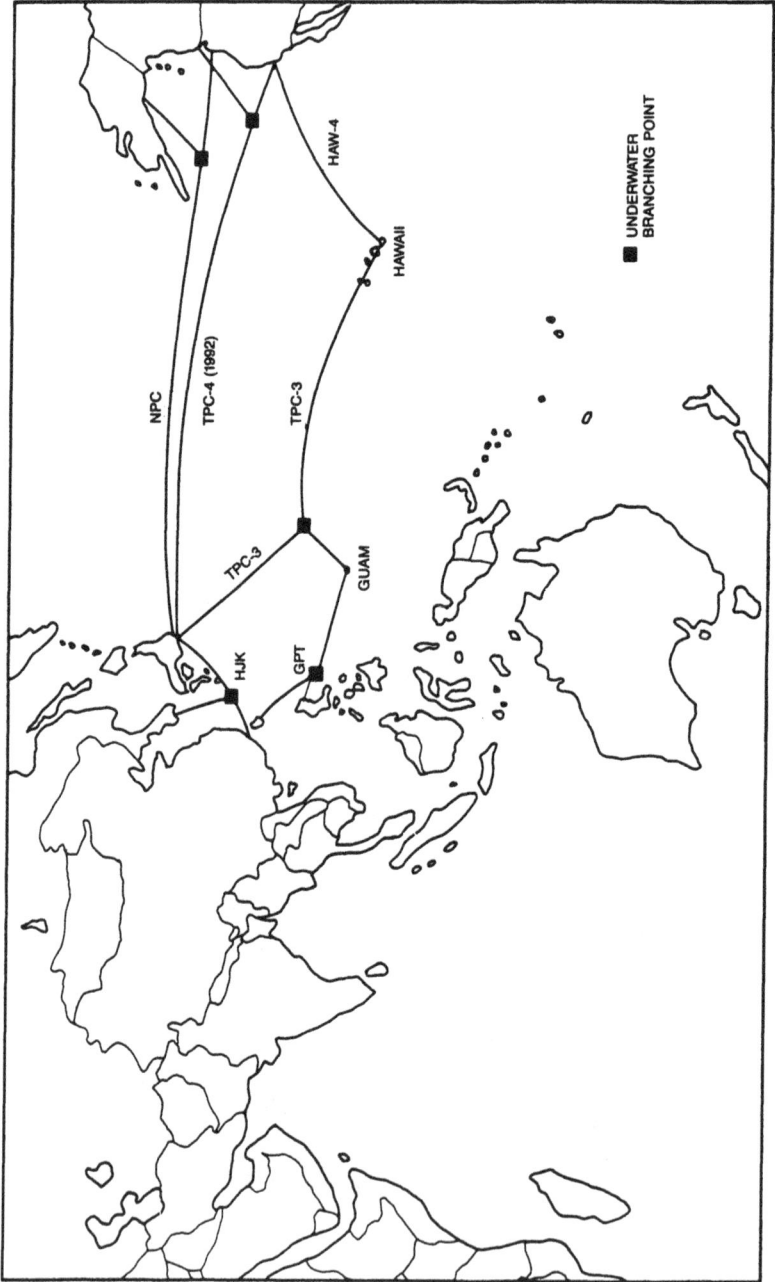

Figure 2.5 Pacific Ocean region fiber optic cables in transpacific and western-Pacific service.

AT&T, which has the largest interest in TPC-3 at 35%, and KDD are marketing AT&T's Accunet digital services between the United States and Japan. Accunet services include digital private lines at rates up to T1 and circuit-switched services at 56 and 64 kb/s and T1. The new cable allows KDD to offer terrestrially based switched 56 and 64 kb/s services to the United States and Europe. Once integrated into the digital land-based networks of the domestic carriers, switched 56 and 64 kb/s will usher in attractive applications like compressed video teleconferencing on demand and Group 4 facsimile. British Telecom International will also participate in the provision of international switched digital services.

Several other cable projects are in the pipeline for the Pacific Region, as shown in Figure 2.5. One of the first extends from Japan to Korea and Hong Kong (H-K-J) and is projected to be completed in May 1990. Planning for the fourth Transpacific Cable project (TPC-4) began in 1989, with AT&T and KDD again taking leadership roles. The consortium includes 31 common carriers from 18 nations. Other U.S. carriers participating include MCI International, Sprint International, and several other specialized communication service providers. The TPC-4 link is scheduled for service in 1993 and will connect Japan with Canada and the United States. TPC-4 will bypass Hawaii, taking the North Pacific route from Tokyo to Vancouver, British Columbia, and on to San Francisco, California. AT&T is constructing half of the cable, which will support twice the capacity of TPC-3. At the time that the project was formulated in 1989, more than 95% of the transpacific design capacity had been assigned to existing owners. All circuits on the HAW-4–TPC-3 cable system have already been fully subscribed.

With TPC-3 accomplished, the Pacific basin carriers are completing other important fiber links. Development of fiber optic facilities is proceeding within the Western Pacific and Southeast Asia region, which has long been dependent on international and regional satellite communication to provide backbone and thin-route links. Economic growth and prosperity among a significant number of these countries, however, is now able to support the kind of investment needed for undersea fiber. The benefits in terms of service availability and quality will surely contribute to more growth in the region. Six nations are already cooperating in such a regional cable project. The role of Singapore as a vital port and manufacturing center in the region is also key to telecommunication connectivity with fiber optic. Singapore, one of the investors in TPC-4, is an active and avid participant in several cable projects that employ the island nation as a junction point.

2.2.5 North Pacific Cable (NPC)

The second transpacific fiber optic cable is expected to be ready for service in 1991. Northern Pacific Cable (NPC) is being constructed directly between North America and Japan, bypassing Hawaii. Historically, this has been a difficult route for laying cable, but offers significant cost savings due to the short distance involved. The backers for the venture include Pacific Telecom, Cable and Wireless, Pacific Telesis, and one of the two new Japanese international carriers, International Digital Communications (IDC). A principal investor in IDC is C. Itoh and Company, the Japanese trading company that holds a major stake in JC Sat. With this type of support, the NPC cable project will have the financial and management talent needed to be a success. The business side looks attractive, with traffic growth in the Pacific projected to be at a rate of between 17 and 20% per year, according to Dennis Elliott, Vice President and General Manager of Pacific Telecom Cable [Elliott, 1989].

A study conducted by Bruno Miglio of Hughes Aircraft Company [Miglio, 1987] demonstrates that the cost of providing digital transmission capacity by modern satellite or transoceanic cable is approximately the same. In the case of the Pacific region, the balance could shift in favor of satellite because of the great distance involved. The NPC routing and cable technology offer lower investment and operating cost per circuit, however, possibly giving this venture an economic advantage over all other transpacific media. Some concern has been expressed that a price war may break out, possibly damaging the business base of the existing international carriers. Any harm would be short-lived because traffic growth in the Pacific should eventually eliminate a glut.

The cable system itself is being manufactured by NEC of Japan and Standard Telephones and Cables of the United Kingdom. With a basic capacity of 1.2 Gb/s, the links will be capable of supporting up to 100,000 voice circuits using DCME. There are three fiber pairs, each capable of carrying 420 Mb/s. Reliability is provided in the form of repeaters which can switch around a damaged section. Underwater repair can be effected by using submersible craft attached to a cable-laying ship. On the U.S. side, the cable has an underwater branching point and will land in Alaska and near Portland, Oregon. This will support the long-distance business of Pacific Telecom's principal operating subsidiary, Alascom, Inc. Much of Alascom's long-distance traffic in Alaska is actually between Anchorage and the contiguous 48 states, which is motivation to employ NPC when it goes into operation.

Cable and Wireless stands to make substantial gains through applications of NPC to the lucrative international market for digital services. Telecommunication users want fiber optic circuits available between North America and Asia. With NPC, Cable and Wireless can offer private lines and switched digital services on fiber facilities from the United Kingdom through the United States (via their U.S. domestic fiber carrier) and across the Pacific. Note that this was one of the original objectives of the founders of ITT Corporation, who saw the potential of an international system using cables to link the continents [Sobel, 1985]. ITT never accomplished this goal, but C&W most probably will.

On the Japanese side, IDC is a new company that was authorized by MPT as one of two competitors with KDD. As a participant in NPC, IDC will have access to excellent fiber optic facilities beginning in 1991. The issue of cable breakage, which is inevitable, can be handled through some type of sharing with TPC-3. After all, both cable links stand to gain by being able to back up each other, thus diminishing the need for satellite restoration. Customers will also prefer the situation where their time-delay-sensitive traffic can be guaranteed service by this type of diverse routing.

2.3 COMPETING SATELLITE SYSTEMS

The title of this section implies that users can select satellite services from a variety of sources, obtaining the best service at an opportune price. Whether competition in fact exists depends on many factors, such as overlap of satellite footprint, adequacy of downlink power level, and legality of direct access to the service provider. In the 1970s and early 1980s, the only viable international satellite system was operated by INTELSAT. In the 1990s, however, many regions of the world enjoy coverage from more than one satellite network. Whenever this is the case, users either benefit from, or can otherwise force, competition. The primary reason is that a satellite, once launched, is a fixed asset in space with a known lifetime (typically 10 to 12 years). Satellites are usually not full of traffic from the day that they are launched into orbit. Consequently, the operator is constantly looking for users and additional revenues. The money gained effectively comes at no additional cost to the operator, except for the marketing expense associated with finding the customer. When users have their own earth stations, they are in an even stronger position to bargain.

This ideal situation just happens to exist in practice in the United States, Europe, and parts of Latin America. In the following paragraphs, we examine the various satellite networks which offer satellite capacity and services to the public and private network users. This will give the telecommunication manager, news director, or network manager a feel for the alternatives.

2.3.1 INTELSAT

The International Telecommunication Satellite Organization is a commercial cooperative of 114 member nations. INTELSAT owns and operates the largest fleet of commercial geostationary satellites in the world, providing satellite transponder capacity to 177 countries, territories, and dependencies. Authorized users may employ capacity for international telecommunication services involving voice, data, and video transmission; in addition, transponders are used by some 30 nations for domestic telecommunication purposes. This latter usage is attractive for nations that lack sufficient demand to justify the launching and operation of their own satellite systems.

Membership in INTELSAT is open to nations that wish to invest in the system and use its satellite resources for commercial telecommunication services. A given country's amount of investment is basically determined by its relative utilization of INTELSAT capacity for international communication. Being the largest user of INTELSAT, the United States has the greatest investment share, amounting to approximately 26% in 1989. This share has declined from a majority stake over the years, and consequently the United States, represented mainly by the Communications Satellite Corporation (COMSAT), has seen its influence decline significantly from the late 1960s, when it maintained virtual control. Today, INTELSAT is very much independent, conducting its business through the administrative, financial, and operational activities of its four organizational bodies. The structure is discussed in subsequent paragraphs of this section.

2.3.1.1 INTELSAT System Overview

The following characteristics of the INTELSAT system are current as of 1989 [INTELSAT, 1989]. More than half of all international telephone calls and virtually all transoceanic television are carried by the system. In 1987, the system supported more than 100,000 equivalent voice channels, where one end-to-end circuit requires two channels. This capacity (double that of 1981) is distributed around the globe among 767 ground antennas and 1,738 earth-station-to-earth-station links. Very clearly, the strength of INTELSAT (and of satellite communication in general) is the high degree of connectivity that can be obtained without the use of physically separate pathways for each link. Fiber optic cables potentially have much greater capacity than INTELSAT can provide, but the connectivity of fiber is more constrained.

Owing to the wide bandwidths of the satellites and the variety of electronic equipment available at the earth stations, INTELSAT provides virtually any kind of analog or digital transmission service that one can imagine. Public switched telephone and private line network services have traditionally been the mainstay of the system. As mentioned previously, international television is totally dependent on the INTELSAT satellites to provide both full-time and occasional video feeds for use by broadcasters and closed-circuit business video applications. INTELSAT provides capacity for international computer and value-added networks for electronic funds transfer and worldwide airline reservations. The ability of rapidly deploying compact earth stations is exploited by the United Nations peacekeeping operations. INTELSAT's developmental programs, Project SHARE and Project Access, have enabled health, education, and social services to be provided to the most remote parts of the world.

2.3.1.2 *Structure of INTELSAT*

INTELSAT is unique among international carriers in that it is a cooperative, created by a treaty called the INTELSAT Agreement. The member nations are therefore bound to abide by its provisions. In 1989, the total membership was 114, with 3 being within Eastern Europe. The USSR was not a member at the time, but had direct access to the INTELSAT system through earth stations in Moscow. The organization has a four-tier structure consisting of the Assembly of Parties; the Meeting of Signatories; the Board of Governors; and the Executive Organ. This structure is reviewed in the following paragraphs to provide an appreciation of how INTELSAT accomplishes its objectives and deals with the changing international telecommunication environment.

At the highest level, the Assembly of Parties represents the member nations as a treaty-implementing group. This level deals with general policy and long-term objectives. One of the important tasks is to consider requests from members for review of plans to implement separate satellite systems (i.e., separate from INTELSAT). This important aspect, treated in Article XIV of the INTELSAT Agreement, is discussed in Chapter 9. By way of definition, a *signatory* is a government PTT, other TA, or designated international carrier that represents the country in the management of the system. Often, the signatory is also the ITU administration for the particular country. Exceptions include countries where telecommunication is privatized, such as the United States, the United Kingdom, Canada, and Japan. The Meeting of Signatories happens once a year to consider issues related to the financial, technical, and operational aspects of the system. Voting within the Assembly of Parties and the Meeting of Signatories is on the basis of one country, one vote.

The Board of Governors (BG) represents the interests of the signatories in supervising the running of the INTELSAT system. Important policies, approvals, and recommendations to higher levels are routinely made by the BG; voting by BG members is in proportion to their investment share. To be allowed to have a governor, a signatory must meet a minimum investment threshold set by the BG. Those with insufficient share can aggregate their votes under a single governor to meet the minimum requirement. As of March 1988, there were 28 governors representing 105 of INTELSAT's 114 signatories. An important advisory group composed of board members is the Advisory Committee on Technical Matters (BG/T). Among the duties of the BG/T are the setting of INTELSAT technical standards for transmissions and earth stations, and the creation of technical criteria for compatibility of separate satellite systems with the operation of INTELSAT (present and future).

In addition to the formal meeting structure, INTELSAT convenes an annual Global Traffic Meeting to forecast traffic, based on service needs anticipated by signatories and other system users and carriers. A typical meeting is conducted at the INTELSAT headquarters in Washington, D.C., and allows the traffic planning people from around the world to agree among one another in an atmosphere of cooperation. There are also annual meetings of operations representatives for the Atlantic, Pacific, and Indian Ocean regions. These meetings are attended by INTELSAT earth station operators to consider and agree on system operational plans and procedures.

A full-time staff, called the Executive Organ, manages the daily operation of INTELSAT. Its head is the Director General, the post assumed by Dean Burch of the United States on April 15, 1987. There are directorates for operations and engineering, finance, strategic planning, service development, and external relations. Importantly, the Deputy Director General of Operations and Engineering is responsible for system planning, research and development (INTELSAT has a sizable R&D budget, which it spends throughout the world), system implementation (buying spacecraft, launches, and other services), communication management, and satellite operations. Activities relating to satellite services to signatories and other users are supervised by the Deputy Director General for Service Development and External Relations. In many ways, the Executive Organ runs a mini-UN for the development and provision of international satellite communication services. The entire organization, including the other three tiers of the structure, performs remarkably well in its mission: "to continue the high quality and reliability upon which INTELSAT's members and users have come to rely."

2.3.1.3 INTELSAT Satellites

The satellites in the INTELSAT system are designed for long-life, reliable operation, each providing a capacity of approximately 50,000 telephone channels, 48 television channels, or combinations thereof. The particular value of capacity depends on the specifics of the spacecraft design, the size of the earth station antennas in the network, and the transmission techniques employed. In 1990, all operating satellites are constructed in the United States by two different aerospace electronic companies: Hughes Aircraft Company and Ford Aerospace and Communications Corporation. Interestingly, these electronic giants are themselves subsidiaries of leading automobile manufacturers, GM and Ford, respectively. The technology involved in INTELSAT satellites is probably the most advanced among commercial satellites. Launch services are provided by either the French company, Arianespace, or U.S. manufacturers such as General Dynamics and Martin Marietta. In-orbit lifetime of typically 10 years is primarily determined by the amount of station-keeping fuel left in the tanks after the satellite is positioned at its assigned orbital slot. A complete discussion of INTELSAT and other satellite systems and the technology can be found in our earlier work [Elbert, 1987].

The satellites of INTELSAT are grouped into three regions associated with the main ocean basins. As the largest region in terms of traffic and number of earth stations, the Atlantic Ocean Region (AOR) sets the pace for the rest of the system. Typically, there are three satellites in service in the AOR: two major path satellites to handle large cross sections of traffic between the developed countries of the region; and a primary satellite used to provide connectivity between all stations in the region. These satellites may be supplemented by one or more backup or auxiliary satellites, which serve the international television and IBS markets. INTELSAT has increased the capacity available in the AOR for occasional and full-time video services. These actions include the purchase of the unlaunched Satcom K4 satellite, which GE Americom has declined to use for domestic services.

The Pacific Ocean Region (POR) is similar, but smaller in its capacity requirements. In the case of the Indian Ocean Region (IOR), only one primary satellite is required. General views of these regions are provided in Figures 2.6, 2.7, and 2.8, indicating typical satellite footprints as well.

2.3.1.4 INTELSAT Earth Stations and Services

The earth stations in the INTELSAT system are owned and operated by the users. There are several categories of earth stations, which are tailored for particular network applications. This scheme is broken down as standards A through F,

INTELSAT VI EAST AND WEST HEMI BEAMS FROM 325.5°E

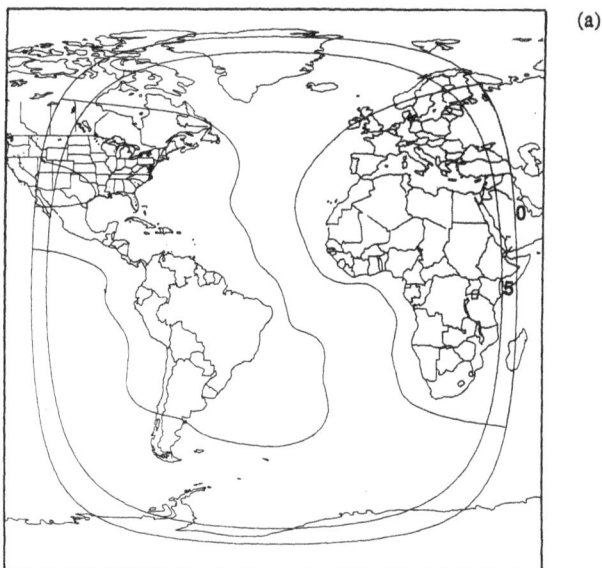

Figure 2.6 INTELSAT Atlantic Ocean region coverage patterns of the east and west hemisphere beams for the Intelsat VI satellite located at 325.5° east longitude (courtesy of COMSAT).

INTELSAT VI ZONE BEAMS FROM 325.5°E

(b)

INTELSAT VI EAST AND WEST SPOT BEAMS FROM 325.5°E

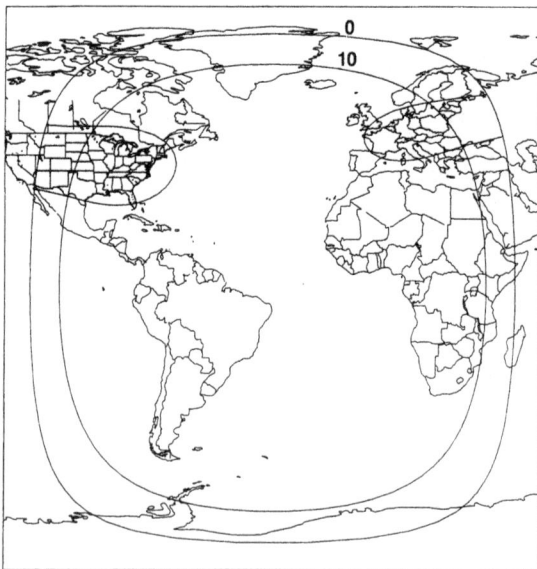

(c)

Figure 2.6 Continued

INTELSAT V/VA EAST AND WEST HEMI BEAMS FROM 180.0°E

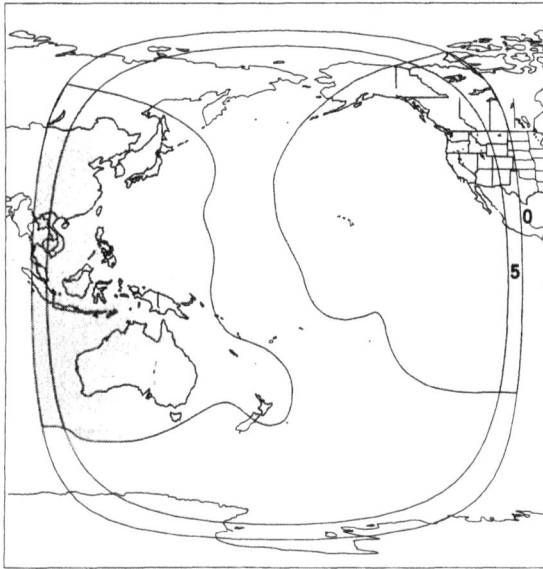

(a)

INTELSAT V/VA EAST AND WEST ZONE BEAMS FROM 180.0°E

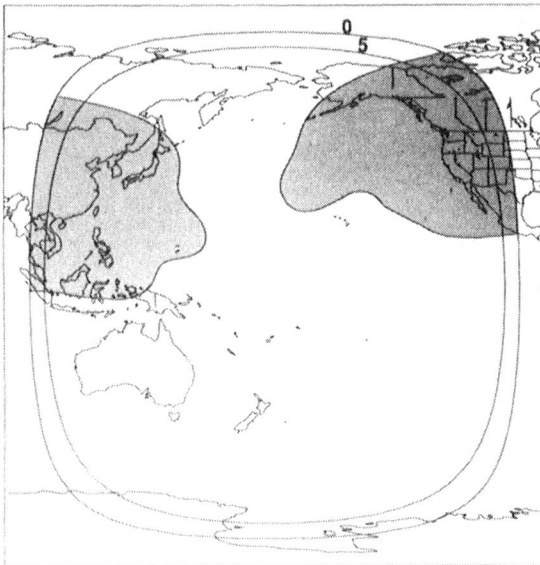

(b)

Figure 2.7 INTELSAT Pacific Ocean region coverage patterns of the east and west hemispheres for the Intelsat V-A satellite located at 180° east longitude (courtesy of COMSAT).

INTELSAT V EAST AND WEST SPOT BEAMS FROM 180.0°E

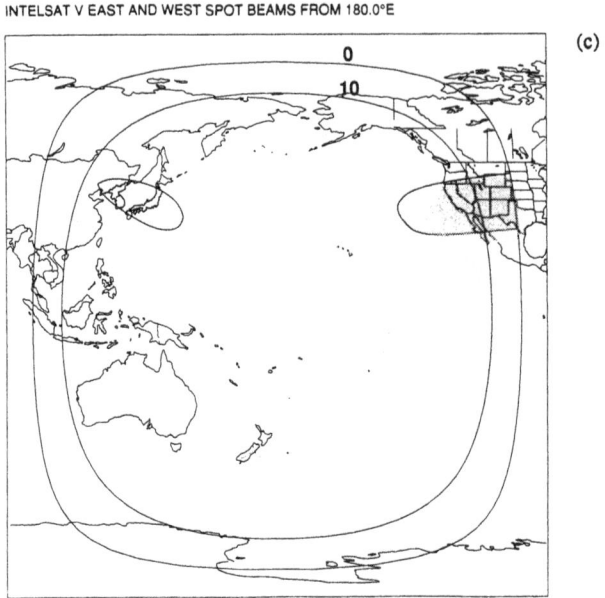

(c)

Figure 2.7 Continued

INTELSAT VI EAST AND WEST HEMI BEAMS FROM 63.0°E

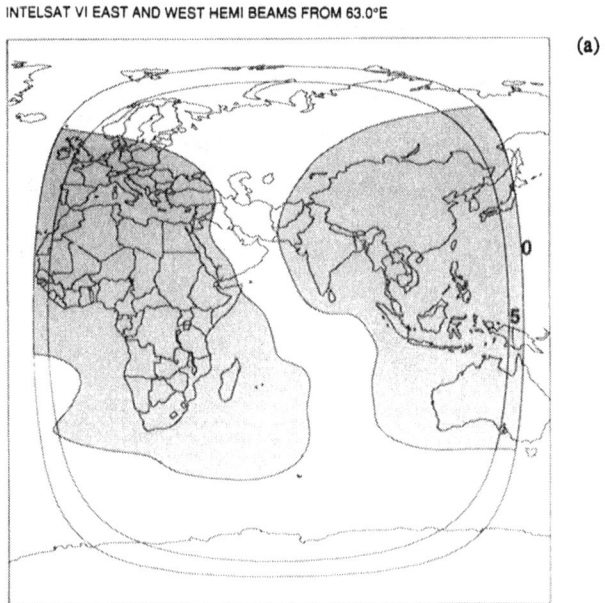

(a)

Figure 2.8 INTELSAT Indian Ocean region coverage patterns for the Intelsat VI satellite located at 63° east longitude (courtesy of COMSAT).

INTELSAT VI ZONE BEAMS FROM 63.0°E

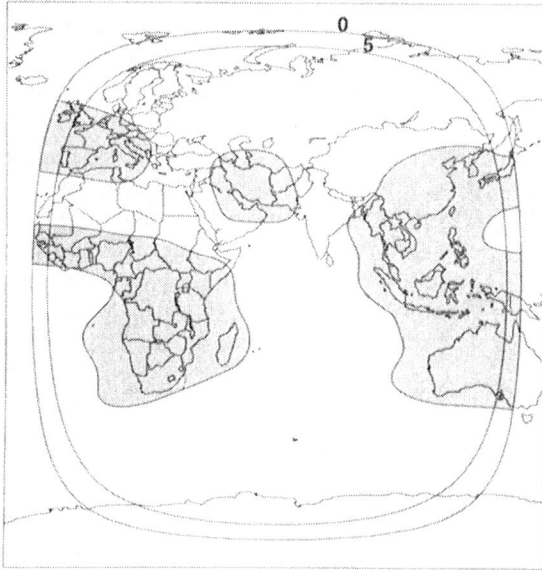

(b)

INTELSAT VI EAST AND WEST SPOT BEAMS FROM 63.0°E

(c)

Figure 2.8 Continued

where a standard-A station is the classic large INTELSAT C-band earth station with an antenna diameter in the range of 15 to 17 m. (The original standard-A antennas were 30 m in diameter, necessitated by the relatively low power of the early generation satellites.) Table 2.2, taken from the INTELSAT technical standards document, lists the standards and identifies the type of service and frequency band of operation.

The INTELSAT earth station standards (IESS) documentation provides actual and potential earth station owners with a common source of technical characteristics to allow INTELSAT to authorize access to the space segment and establishment of communication links with other earth stations [INTELSAT, 1988]. The standards were devised to be compatible with particular transmission approaches, either analog or digital. In 1989, INTELSAT permitted the modulation and access techniques listed in Table 2.3. Generally, analog telephony techniques such as FDM/FM are giving way to digital processing with either time division multiple access (TDMA) or single channel per carrier (SCPC).

Television transmission is almost exclusively done with analog FM because of the low cost of the associated receiving equipment. Various techniques for digitizing the video signal for broadcast television have been tried and will probably be adopted early in the 1990s. INTELSAT participated in a test of digitized high definition television (HDTV) in conjunction with KDD of Japan and COMSAT [Nasoka, 1989]. At 140 Mb/s, the signal was transmitted through a 72 MHz transponder on an Intelsat V satellite over the Pacific Ocean. In comparison, a

Table 2.2 INTELSAT Station Standards

Earth Station Standard	Typical Antenna Size (m)	Type of Service	Frequency Band (GHz)
A	15–17	International voice, data, television, IBS, IDR, INTELNET	6/4
B	10–13	International voice, data, television, IBS, IDR, INTELNET	6/4
C	11–13	International voice, data, television, IBS, IDR, INTELNET	14/11
D1	4.5–5.5	VISTA	6/4
D2	11	VISTA	6/4
E	3.5–4.5	IBS	14/11 and 14/12
E2	5.5–6.5	IBS	14/11 and 14/12
E3	8–10	IBS, IDR	14/11 and 14/12
F1	4.5–5	IBS	6/4
F2	7–8	IBS	6/4
F3	9–10	International voice, data, television, IBS, IDR	6/4
G	All sizes	International Lease Services including INTELNET	6/4, 14/11 and 14/12
Z	All sizes	Domestic Lease Services including INTELNET	6/4, 14/11 and 14/12

Table 2.3 Standard Modulation and Access Methods for INTELSAT Earth Stations

(a)	Frequency Division Multiplex–Frequency Modulation (FDM–FM)
(b)	Companded Frequency Division Multiplex–Frequency Modulation (CFDM–FM)
(c)	Preassigned Single-Channel per Carrier–Quadrature Phase-Shift Keying (SCPC–QPSK)
(d)	Demand-Assigned Single-Channel per Carrier–Quadrature Phase-Shift Keying (SPADE)
(e)	Single-Channel per Carrier–Companded Frequency Modulation (SCPC–CFM) for the VISTA Service
(f)	Frequency Modulated Television with Associated Audio FM Subcarrier (TV–FM)
(g)	Time Division Multiple Access with Digital Speech Interpolation (TDMA–DSI) and without Digital Speech Interpolation (TDMA–DNI)
(h)	Digital Transmission at Intermediate Data Rates (IDR) Using Quadrature Phase-Shift Keying–Frequency Division Multiple Access Carriers (QPSK–FDMA)
(i)	Digital Transmission for INTELSAT Business Services (IBS) Using Quadrature Phase-Shift Keying–Frequency Division Multiple Access Carriers (QPSK–FDMA)

conventional NTSC color television signal would require 45 Mb/s of data throughput in a 36 MHz transponder.

In addition to satisfying INTELSAT technical specifications (which are extensive), all stations are put through a series of exhaustive on-air tests prior to initiation of service. This ensures that INTELSAT's high technical performance standards are met. The document that contains test specifications and procedures is called the Satellite System Operating Guide (SSOG). The testing is conducted in two phases. In the first phase, an earth station is thoroughly checked to ensure that it is capable of accessing the system and meeting all applicable specifications. The second phase occurs just prior to the start of service, when the INTELSAT network operations organization verifies that all things are in order and interference will not occur. Those who complete this gauntlet are rightly proud of their accomplishment. The excellent performance of INTELSAT over the decades testifies to the thoroughness of its procedures.

Important for prospective users of the INTELSAT system is to understand that it serves the signatories, who, in turn, serve the users. In the United States, all services must be coordinated through COMSAT. The international telephone carriers, including AT&T, MCI, and Sprint International, obtain their space segment from COMSAT although the carriers own and operate their earth stations. The same applies to specialized carriers and users who access INTELSAT for IBS. COMSAT maintains an energetic marketing staff on the East Coast because the margin on resale of INTELSAT capacity is COMSAT's primary source of revenue.

To aid users, COMSAT has started a free technical consulting service called Customer Engineering. Its purpose is to help COMSAT customers in the planning, design, and operation of international earth stations. Customer engineers provide this assistance, including training, either at COMSAT headquarters in Washington, D.C., or at the customer's location. This kind of assistance is extremely important due to the experience of COMSAT (which is substantial) and because the complicated rules and procedures of INTELSAT are subject to change. Contact points, service offerings, and other useful information for all foreign signatories and U.S. carriers serving the AOR can be found in [KJH, 1988] and [KJH, 1989]. An overview of the procedures for utilizing INTELSAT for occasional video services is provided in Chapter 7.

2.3.2 INTERSPUTNIK

Established in 1971 as the Soviet and Eastern European counterpart to INTELSAT, the goal of the International Organization of Space Communications (INTERSPUTNIK) was to facilitate the use of existing and planned satellites, facilities, and services [ISD, 1988]. Fourteen socialist nations are members, participating in an organization of similar structure but much smaller size than INTELSAT. Cuba, while not a member, is a user of the system. Members and users own and operate their earth stations and the Soviet government provides the satellite capacity.

From the launch of the first Sputnik in 1957, the Soviets have maintained an ambitious space program, including satellites for communication. Until the late 1970s, Soviet technology had badly lagged that of the West. Their scientists and engineers, however, are highly trained and are recognized for significant accomplishments in communication theory and telecommunication networking. The Soviet Union has built and launched elliptical and geosynchronous satellites for several years, employing them primarily for domestic services across the nation's wide geography and for regional services to its allied nations in Eastern Europe. The first satellites used were of the Molnya type, which travel in inclined elliptical orbits. Two fully steerable antennas are needed to deal with the hand-off from one satellite to the next as they move across the sky. Interestingly, such an elaborate station is located at Fort Dietrich, Maryland, for use in the Washington-Moscow hotline.

The satellites of most interest in the West are those maintained in the geostationary satellite orbit (GSO). The Soviet Union has done a thorough job within the ITU to provide suitable orbit locations for satellites employed by INTERSPUTNIK. While the GSO is literally spotted with positions reserved for Soviet satellites, there are only two or three satellites in continuous use for INTERSPUTNIK services. Statsionar 4 and Statsionar 13 serve the Atlantic and Indian Ocean regions, respectively. The more familiar name *Gorizont* ("horizon" in Russian) is more common than the Statsionar nomenclature used for ITU registration purposes. Home dish owners in North America are able to pick up television signals from the Gorizont 12 (Statsionar 4) satellite transmitting to Cuba. INTERSPUTNIK has not been actively marketing the capacity on its satellites to the West. On a few occasions, Western news organizations have used a Soviet satellite to relay programming from Eastern Europe. During the 1986 Goodwill Games,

Turner Broadcasting System used INTERSPUTNIK to relay one (of many) sporting events in real-time to the United States. Also, the Discovery Channel, a cable television programming service in the United States, obtained permission from the Soviets and the FCC to pick up signals from Gorizont and carry them over their domestic satellite distribution system. INTERSPUTNIK appears to provide an opportunity to users who need temporary or specialized links between the Soviet Union or Eastern Europe and points in the West.

2.3.3 EUTELSAT

The European Telecommunications Satellite Organization provides regional satellite capacity to Europe and is patterned after INTELSAT [ISD, 1988]. The organization is headquartered in Paris, has 26 member countries, and operates as a satellite carrier for the PTTs. Countries in Eastern Europe are being admitted to EUTELSAT, even though the organization was originally established in the west. In other words, to obtain satellite capacity from EUTELSAT, you must work through the PTT of a host European country.

Formed in 1977, EUTELSAT was seen as a vehicle for promoting the development of a European telecommunication satellite industry. This goal has been met because virtually all satellites, launches, and network operations are implemented within the European industrial base. Services on the satellites have evolved from an early expectation to focus on voice traffic to the situation where the bulk of the traffic now is video. A new thrust for VSAT networks is underway, buoyed by the trend toward privatization and liberalization of European telecommunication. EUTELSAT is perhaps the most viable source of satellite capacity beyond the year 2000.

The organizational structure is very similar to that of INTELSAT, although there are three levels instead of four. A permanent staff is led by a Director General, a position first held by A. Carusso (formerly an active participant on INTELSAT's BG for his home nation, Italy). Dr. Carusso stepped down in December 1989, and has been replaced by M. Grenier, formerly of France Telecom. The organization's objective is to meet the international telecommunication requirements of members, meaning traffic between the nations of Europe. Satellite capacity may also be used for domestic communication services. For example, Turkey will lease six transponders on EUTELSAT II (F5) starting in early 1992 for use in a new domestic satellite communication network. This kind of approach simplifies the problem of creating a domestic satellite system (because EUTELSAT takes care of the satellite operation) and Turkey thereby has backup capacity available on a system much larger than any it could afford to launch on its own.

The EUTELSAT system operates at Ku-band, providing voice, data, and video transmission capacity to the PTTs. The satellites are of the EUTELSAT I and II series, manufactured by Aerospatiale and launched by Arianespace, both French companies that subcontract some of the work to other European manufacturers. Four EUTELSAT I satellites were in operation in 1989. This first of the EUTELSAT II series satellites was scheduled for launch in 1990. With five of these newer spacecraft under construction, EUTELSAT will move into an era of higher powered Ku-band transponders.

This improvement in performance was deemed necessary to compete with private European companies offering more power than EUTELSAT for video distribution applications. EUTELSAT is well positioned in video distribution because European television is very fragmented due to differences between countries in language and cultural preferences. Subregions like Scandinavia and even individual countries would require their own set of transponders to deliver their unique brand of programming. Another factor in EUTELSAT's favor is that some domestic satellites operated by European PTTs will probably not be replaced. Increased transponder power will also expand service possibilities for private telecommunication networks in Europe that employ VSATs for data and teleconferencing.

2.3.4 Other Regional Satellite Systems

There are a variety of telecommunication satellites that have been introduced within a limited region of coverage. Two current examples are the Palapa system of Indonesia and the Arabsat system. Other systems are in various stages of planning for the Andean nations of South America, the African nations, and the Pacific Basin. The nations of the Andean Pact signed an agreement in 1988 to pursue Project Condor, a regional satellite system to serve the needs of the members and their PTTs.

What these systems and proposals have in common is that they are backed by one or more governments, giving them political overtones. Generally, as in the case of EUTELSAT, access to the capacity is limited to other governmental PTTs in the particular region. Private network users would only employ the capacity of such a regional network in conjunction with the local PTT or another government agency.

Indonesia created the Palapa system in 1974 primarily to meet its internal needs for telecommunication expansion. Indonesia saw an opportunity to defray construction and operating costs, however, by extending the coverage of the satellites to include the other nations of Southeast Asia. In 1977, after the original two satellites were in orbit and the domestic network was operational, the Indonesian PTT, PERUMTEL, succeeded in leasing some of the transponders to the Philippines and Thailand. The second generation of satellites is in place and is expected to continue well into the 1990s. Again, neighboring nations have leased transponders on Palapa for use in domestic services. The latest addition was Vietnam in 1989. International services (e.g., links between ASEAN member nations) are not provided because Indonesia has decided not to seek INTELSAT's permission, as is required under the INTELSAT Agreement. (A discussion of these procedures can be found in Chapter 9.) Within Indonesia, PERUMTEL subleases capacity to oil and mining companies for communication to remote locations around this archipelago nation.

A different tack was taken by the Arab League nations when they made the commitment to construct a regional satellite system. They formed an organization called Arabsat, which had the charter to organize and construct the system and offer the capacity to member-owners. The project began in 1981 and went into operation in the mid-1980s. Arabsat has experienced more than its share of start-up problems, including technical difficulties and members who were reluctant to use the system. The satellites operate at C-band for standard telecommunication services, and at S-band (2.3 GHz) for video broadcasting. From a technical standpoint, S-band is somewhat more efficient for a wide-area satellite television broadcasting system, and offers the potential of very inexpensive ground receivers. The rest of the world, however, has committed to using Ku-band for broadcasting satellites. Whether Arabsat can offer useful capabilities to the telecommunication manager will depend on the particular circumstances. Obviously, the system is attractive for international communication within the North African and Near Eastern regions.

2.3.5 Transborder Services from Domestic Satellites

Even if the coverage pattern (footprint) of a domestic satellite is intended to focus its transmission within one nation, there will be a usable signal extending into the territory of its neighbors. This was discovered when the first Canadian Anik satellite was put into service. Interestingly, the Canadians were three or four years ahead of the United States in implementing the first domestic satellites. RCA American Communications (now a division of General Electric) obtained limited permission to lease transponder capacity from Telesat Canada to initiate telecommunication services before RCA's own Satcom satellite was launched. This precedent-setting move led to new uses of North American satellites. While not a major business opportunity to satellite operators, this usage of satellite capacity allows networks to extend across borders. This approach could be important in cases where a company had operations in both countries and would want to implement a single satellite network.

Transborder satellite services between the United States and Canada and between the United States and Mexico have been the subject of controversy at various times. Telesat Canada is the sole satellite operator in Canada and must rely on the smaller Canadian market for the bulk of its income. The U.S. market is so large that it can support three or four large satellite operators. In 1990, these included Contel/ASC, Hughes Communications, Inc. (HCI), GE American Communications (Americom), AT&T, and GTE Spacenet. Telesat's position is that U.S. operators have a technical advantage with regard to networks with locations within both countries. Certainly, many large U.S. companies have a minority of sites in Canada. Virtually any U.S. domestic satellite will provide coverage of the complete southern tier of Canada, where 90% of the population and all of the lucrative business are located. Telesat's satellites, however, particularly those with Ku-band transponders, cannot reach very far south into the United States. This technical advantage does well for U.S. operators, but Telesat has some advantages due to its monopoly position in Canada. For example, Telesat must own and operate all transmitting earth stations within its borders, whereas U.S. users have complete freedom to run and own their ground networks. Telesat has taken the step in its next generation satellite to extend the transborder coverage to include the lower 48 states.

The situation with respect to Mexico has its analogies, because the Mexican satellite operator, Secretaria de Comunicaciónes y Transportes (SCT), is a government agency with the responsibility for satellite systems and international telecommunication services. Teléfonos de México, the domestic TA, has been nationalized for decades and now may be privatized by the current government. The Mexican Morelos satellites provide C- and Ku-band transponder capacity. In this case, however, only the Morelos satellites have the desired coverage of their country and U.S. network users are almost forced to employ them. Also, the Morelos satellites only cover a small section of the southwestern United States, making difficult the creation of a large transborder network.

Another complication of using domestic satellites for transborder services is imposed by INTELSAT's Article XIV consultation proceedings. This amounts to a need for the approval of the INTELSAT Executive Organ, BG, and Assembly of Parties that there will be no economic harm to INTELSAT because of lost revenues. Typically, the INTELSAT system would not be used for transborder services, so economic harm should not be an issue. The process is nevertheless laborious and time-consuming. For example, most U.S. domestic satellites provide good footprint coverage of island nations in the Caribbean basin. Permission to transmit video from Aruba to Westar 4, for example, took approximately three years because of the lengthy process involved.

2.3.6 Privately-Owned International Satellites

From the preceding discussion, the reader should expect that international satellites would be the domain of PTTs and domestic satellites could be operated by private companies, where permitted by the local government. In truly American entrepreneurial spirit, a few new companies were started to obtain permission to create privately-owned satellite systems for the provision of international services. The market that appeared to be the most attractive was the Atlantic region, particularly for links between the United States and Western Europe. This required a major change in INTELSAT policy because any such system would definitely "skim the cream" from some of the richest traffic routes. Because of the free-enterprise disposition of the administration of President Ronald Reagan, however, the full support of the U.S. government was given to these private initiatives.

The first private system actually put into operation was Pan American Satellite (PanAmSat), a true startup venture founded by Rene Anselmo, an American entrepreneur with close ties to Mexican media companies. The activities of PanAmSat have, as a minimum, forced INTELSAT to become more competitive. This helped stimulate the creation of international television ventures and overall expansion of the medium. PanAmSat and its supporting company, Alpha Lybracom, are discussed in the following subsection. Orion Satellite Corporation is a second startup company, which plans to serve the Atlantic region. Outside the United States, opportunities for private satellite startup ventures exist where a host nation has the proper disposition. We review the plans of Asiasat, a Hong Kong based company owned by C&W, Hutchinson Cable, and the Chinese government. (See Subsection 2.3.6.2.) Beyond Asiasat, PanAmSat, and Orion, a number of other ventures and startups are being pursued around the world, some of which are reviewed in Subsection 2.3.6.3 to follow.

2.3.6.1 Pan American Satellite

The Society of Satellite Professionals International (SSPI) named Rene Anselmo as the Satellite Entrepreneur of the Year for 1988. Undoubtedly, he deserved this distinction because not only did he conceive PanAmSat, but also funded its construction with his personal fortune and put it into full-scale operation. The first satellite, PAS I, was successfully launched in 1988. Obtaining permission from INTELSAT required a major effort from the FCC and U.S. State Department, but the necessary approvals were obtained from the Assembly of Parties well ahead of the start of service. This process is ongoing because PanAmSat must continue to coordinate with every country to which it hopes to gain access.

PAS I is a hybrid C- and Ku-band satellite with a variety of coverage footprints. From its orbit location east of North America, the satellite can serve the eastern two-thirds of the United States, all of Western Europe, and a portion of South America (primarily the Andean region). The company offers space-segment services only, but operates a teleport facility in Miami, Florida. This teleport can provide "turn-around" services for signals connecting between its satellite and

U.S. domestic satellites. Services are offered on a full-time, part-time, and occasional basis.

Alpha Lybracom Space Communications is aggressively marketing the capacity of PAS I to international broadcasters. PAS I is used to relay CNN and ESPN, the largest U.S. CATV networks, 24 hours a day, to homes, cable television systems, and broadcast television stations throughout Latin America. National television networks such as Televisión Nacional de Chile and Panamericana TV of Peru also use PAS I for their own domestic distribution. Transatlantic television transmissions were provided for C-SPAN through a cooperative agreement with BTI. Full-time services are available for video applications. While video services are attractive to PanAmSat, it is nevertheless pursuing the corporate telecommunication market, primarily in Latin America. This should eventually prove successful because PanAmSat is the only alternative to INTELSAT, except in Brazil. Having C- and Ku-band capacity available, PanAmSat is in a position to satisfy diverse requirements for voice, data, and video transmission between a variety of remote locations. In 1989, PanAmSat was in the process of obtaining the PTT and government approvals for the provision of such services to private companies.

2.3.6.2 Asiasat

Asiasat is a joint venture company set up to establish a satellite communication system to serve Far East Asia. Headquartered in Hong Kong and managed by C&W, Asiasat is well positioned to enter this kind of market. The first satellite was being prepared for launch in 1989 by the manufacturer, Hughes Aircraft Company. Two aspects of this initial launch are very interesting. First, the spacecraft has flown before as Westar VI, a C-band satellite originally purchased by Western Union when it was still in the satellite operating business. (Western Union's other satellites were purchased by Hughes Communications in 1989 and are now part of the Galaxy system.) Westar VI was placed in an unusable orbit by a faulty rocket motor, and was successfully recovered in a history-making operation from the U.S. Space Shuttle. The second interesting point is that Asiasat I will be launched by the Long March rocket, the first commercial launch by the People's Republic of China. The market for Asiasat is primarily that of domestic telecommunication for countries within the satellite footprint. Obviously, the Chinese government PTT and television network would employ the satellite almost immediately after being launched. Thailand has reportedly made a large commitment to capacity on the first satellite. Taiwan, the Republic of Korea, and the Indochina peninsula are good candidates for services. A commercial company to run the system means that services will be competitive and the organization will be willing to make attractive deals to enlist customers.

2.3.6.3 Orion Satellite and Other Proposals

One of the first U.S. applicants for an international satellite system was Orion Satellite Corporation, a startup company founded by John Puente and others. Orion has been working on the legal and financial foundation for the satellite for years. In March 1989, Orion signed a memorandum of agreement with British Aerospace to build and launch two Ku-band satellites. Initial operation could begin as early as 1992, subject to completion of financing for construction. The fact that PanAmSat is already in operation is a problem for Orion; the company, however, is aggressively moving to establish a real system. The principal market is the North Atlantic region, where Orion will offer Ku-band capacity for VSAT applications for domestic and international communication links.

In addition to Orion, there are other potential startup ventures in international satellite communication. PacStar is a planned system that will serve the Pacific Basin. Controlled by Pacific Telecom, PacStar has a license from the government of Papua New Guinea. The economic growth of the region has caused increased traffic flows on the INTELSAT and fiber optic networks. Hence, there is logic behind the system concept. If PacStar, Orion, or Asiasat are particularly successful, we can expect other private satellite systems to appear.

Chapter 3
CIRCUIT-SWITCHED AND PRIVATE LINE NETWORK SERVICES

Network services for voice and data communication are provided over domestic and international facilities by a variety of public and private organizations. Chapter 2 concentrated on the international telecommunication environment from the standpoint of fixed links over terrestrial and satellite systems. In this chapter, we review a broad range of the circuit-switched and private line services available to the telecommunication manager. Readers wishing to focus on business and management aspects may neglect some of the more technical information, such as that dealing with signaling systems.

The North American telecommunication service market, as a frame of reference, is rich in its variety of voice and data service offerings due to the competitive forces at work as well as the innovations of the high technology industrial sector. "Plain old telephone service" (reviewed as MTS below) is the common denominator, representing no particular advantage to a strategic unit. More sophisticated voice services such as WATS and SDN allow flexibility in building private networks. These employ advanced switching and software features within the carriers' digital switches and transmission lines. The data communication service side of the equation, emphasizing the use of packet-switched networks, is discussed in Chapter 4.

Network service innovations tend to defuse over time into the international environment, primarily along the fiber optic pathways between the United States, Canada, Europe, and Japan. As a general rule, strategic units that are not themselves engaged in the telecommunication business will obtain services as discussed below, leaving the long-haul media to the types of carriers discussed in Chapter 2. Facilities tend to be fixed for a period of several years at a time, while the service mix evolves at a much faster rate. This is analogous to information systems, where computer hardware, although fixed after installation, is adapted to different applications by software code. Organizations that extensively use computers tend to employ internal software designers and programmers to tailor the particular applications to the needs of the strategic unit. Similarly, common carriers are able to modify the programming of the switches, which are controlled by very reliable minicomputers.

Will managers in the broadcasting industry find any value in knowledge gained from reading this chapter on circuit switched and private line services and the one that follows, which discusses packet-switched data services? The answer is yes, because any international broadcast operation for video or audio must be planned and coordinated across the same distances that the program material will ultimately travel. Branch offices require voice services so that people can stay in touch with the home office. Also, most network operations are automated to some degree for which reliable data lines will be required. Electronic mail (discussed in Chapter 4) can be used by foreign news bureaus to prepare stories for review by editors at home before the actual broadcast from overseas.

International telecommunication services were originally the domain of the *international record carriers* (IRCs), a breed of common carrier which only did business between countries rather than within them. Principal among them are ITT Worldcom, RCA Globcom, TRT, French Telegraph and Cable (FTCC), and Western Union International (not to be confused with Western Union Corporation, a totally separate company). An IRC obtains transmission capacity from AT&T, the PTTs, and other owners of long-haul facilities and resells that capacity in smaller increments to businesses and government agencies. To break down the capacity into salable increments, the IRC may operate its own switching and multiplexing equipment. At one time, these companies provided a variety of international telex services, leased voice lines and data lines, interconnecting with domestic carriers in virtually every country on the planet. Nearly all of the IRCs have been absorbed by other telecommunication companies. By the year 2000, their subscribers will have been transferred to the main line voice and data services discussed in this book. There will be cases where an IRC that is still in business can supply circuit-switched telephone or data services (e.g., Datel) or private lines for *alternate voice-data* (AVD) service on an economical basis. One can expect that these opportunities will gradually disappear, however, as the strongest players — those who are more committed to the fiber optic and satellite systems of the future — prevail.

The network offerings discussed below can be obtained from domestic TAs and international common carriers. A strategic unit, however, may purchase equipment elements from manufacturers and system integrators and subsequently serve its own needs by operating the equipment. This alternative is reviewed in Chapter 6. Operating your own equipment means that you must solve a wide variety of problems to make the link work (and to keep it working). Among them are circuit interfaces, echo cancellation, telephone signaling systems, synchronization (in the case of digital circuits), and transborder data flow regulation. A network which crosses borders must deal with these variations. The international public switched telephone network and some of the packet networks have devised ways to cross these boundaries and to solve most of the problems that can be encountered.

The more complex and diverse the investment in facilities becomes, the more a strategic unit can be compared to a common carrier. Interestingly, some organizations have invested so much in information technology and international telecommunication that they actually market value-added services to third parties. A good example of this practice, called *reselling,* is Federal Express, the overnight package delivery service, which began in 1988 to sell value-added telecommunication services on its international private network. Network strategies such as reselling are reviewed in our previous book [Elbert, 1989].

3.1 PUBLIC SWITCHED TELEPHONE NETWORKS

Good old-fashioned POTS is not really plain anymore. Switched telephone networks of the past were geared toward voice services. The networks were analog in nature, from the local loop to the switching device, and over the transmission medium between switches. The element of service as well as the standard of measure was, and continues to be, the 300 to 3400 Hz telephone channel, encompassing a usable bandwidth of approximately 3000 Hz (3 kHz). As time passed, people began to use that bandwidth in ways that Alexander Graham Bell could never have considered. One of the early innovations was squeezing several telex channels together, or combining telex and voice into a single channel. During the 1960s, data at speeds of 300, 1200, or 2400 b/s were connected through the telephone network. Modern voice band modems can pass data in two directions at 9600 b/s and in one direction at 19,200 b/s. At these data rates, users routinely transmit fax messages in approximately 20 seconds per page and transfer interactive data almost as effectively as if using a 56 kb/s circuit.

More important to the future of telecommunication is the rapid pace of digitization of the domestic and international telephone networks, providing circuit-switched services at 64 kb/s. The 64 kb/s channel is a standard of measure in wideband digital telecommunication networks, and is known as the DS0 in North America and the B-channel in the ISDN standards. Multiplexing, switching, and routing of DS0 channels is fundamental to the provision of circuit-switched services on the developing international digital backbone. Furthermore, intelligence in the switching nodes and signaling channels can create a private network environment for strategic units.

3.1.1 Message Telephone Service (MTS)

Message telephone service is the basic international circuit-switched medium. Although we use the MTS designation throughout this book, also common is for the normal international switched telephone service offering to be called *international message telephone service* (IMTS). Unsaid, but nevertheless essential, MTS interconnects everywhere, even to Eastern Europe. As an example of its availability, television network news reporters could remain on the air during the June 1989 upheaval in Beijing, China, by using the telephones in their hotel rooms. At the same time, almost all of the city's basic services were disrupted. *International direct distance dialing* (IDDD) is available from the United States to nearly every major city of the world. The PTTs and international carriers value MTS, and have developed the procedures to coordinate usage and, importantly, revenues. Fairly standard among PTTs is to break MTS into three categories: local services, such as within a metropolitan or regional area; long-distance services between cities; and international services. These categories essentially correspond to the organizational boundaries among the North American telecommunication service providers. In a number of key countries, domestic MTS is provided by one company and all international MTS (interconnected through gateways with the domestic portion) by another.

3.1.1.1 International Correspondents

Provision of MTS between countries involves international long-haul links connecting the domestic networks on both ends. The respective MTS providers are called correspondents. For example, consider the relationship between the United States and France. All telephone service for France is provided by the PTT organization known as France Telecom. AT&T and France Telecom, as correspondents, have cooperated along two lines. First, they are in agreement on the sharing of revenue from calls. Second, they are partners in the long-haul link between the United States and France. Calls on the PSTN employ transatlantic cables (TAT-7 or TAT-8) or INTELSAT satellites and earth stations. For a given telephone call, the particular link is essentially selected at random by the international gateway exchanges in the United States. Because local telcos are separate from AT&T, a third company, such as New York Telephone (a subsidiary of NYNEX), is the initial point of access for a subscriber in New York. The local New York exchange must read the sequence of dialing digits, connect to the AT&T gateway, and monitor the call's progress. Important to the viability of international calling, the switch must track the call, account for charges, and pass them along to the subscriber in a monthly bill. Money is transferred by New York Telephone to AT&T, which, in turn, passes a share on to France Telecom.

The process reviewed above is fairly familiar and it is followed for virtually all international calling. In the United States, we have several common carriers that offer international MTS, the largest of which are AT&T, MCI, and US Sprint. From the previous discussion, the reader should realize that for MCI or US Sprint to enter the IDDD market, they had to establish international long-haul links and agreements with foreign TAs. Full-time circuits on cable and satellite systems are needed to support the traffic requirements. At the Global Traffic Meetings of INTELSAT, mentioned in Chapter 2, the three carriers do the necessary "horse trading" with their intended correspondents, producing agreements as to the number of equivalent voice circuits to be installed between the international earth stations. Arrangements for undersea cable circuits are made at the time of construction. Adequate ports subsequently must be provided on the international gateway switches as well.

3.1.1.2 Worldwide Numbering Plan

The CCITT dealt effectively with the rapid expansion of IDDD by adopting an international telephone numbering plan in 1964. It established eleven digits as the maximum recommended length, but a length of twelve was also permitted. As discussed previously, a subscriber must prefix the international number with an access code so that the domestic PSTN exchange will recognize the type of call and route it to the international gateway.

An international number of eleven digits consists of a country code followed by a national number. The first digit of the country code, which can have a maximum length of three digits, identifies the particular world zone, as defined in Table 3.1. A country with a large telephone population would be assigned a short code. For example, the zone and country code for the United States is 1, while the country code for Italy is 39 (which lies in European zone 3). Countries in the same zone have country codes beginning with the same digit. For example, Mexico and Guatemala have country codes of 52 and 502, respectively.

There are restrictions on the numbers to be used in two-digit *versus* three-digit country codes. This is analogous to the N[0/1]X rule for U.S. area codes,

Table 3.1 World Zone Assignments

World Zone	Principal Areas Covered
1	Canada, United States
2	Africa
3, 4	Europe
5	South and Central America, Mexico
6	South Pacific
7	U.S.S.R.
8	North Pacific
9	Far East and Middle East
0	Spare

Note: Specific country code assignments tend to be stable but may be changed by mutual agreement.

wherein the first digit (N) can be any number except 0, the second digit must be either a 0 or a 1, and the third digit (X) may be any number between 0 and 9. This rule allows local switching equipment to separate local from long-distance calls. From the first two digits of the country code, the gateway can determine if the code is two or three digits long. The remaining digits in the eleven-digit international number are available for the national numbering system. Large countries have short country codes (one or two digits) and hence can have the longest national numbers (yielding more combinations for a larger subscriber population).

Suppose a U.S. subscriber wishes to call the number 232-3232 in Naples, Italy. The country code is 39 and in Italy there is also a city code (81 for Naples). The sequence would be as follows: 011-39-81-232-3232. Note the prefix 011, which is used to identify an international call within the United States. A calling party in the United Kingdom wishing to dial the number 555-2345 in Los Angeles, California, would dial: 010-1-213-555-2345. The access number and U.S. prefix are 010 and 1, respectively. Los Angeles is in the 213 area code, which must precede the actual seven-digit number. Note that the complete sequence, less the access code, is eleven digits long, as was the previous example for calling Italy from the United States.

3.1.1.3 International Signaling in Telephone Networks

International direct distance dialing is a working and workable system. The TAs and international carriers have made sure that calls can be placed across the critical interfaces. The public switched telephone network of each country typically evolved in its own way, based on the types of telephone exchange usually obtained from multiple sources. Countries in Europe generally follow one set of signaling and telephone interface standards, those in North America another, and Japan a third. While the 3 kHz voice channel is standard, the techniques for setting up and taking down calls will differ more often than not. The key differences are in the signaling systems used for call processing, including the sending and receiving of the important dialing digits used to route the call.

The following discussion of signaling systems is somewhat technical, and it is important for understanding how domestic and international PSTN facilities can be interconnected. The discussion, however, can be skipped by those readers more concerned with business and management issues. An excellent summary of signaling systems, covering nearly 30 years of evolution, can be found in [Welch, 1979]. Telephone signaling has developed slowly over the decades because of the enormous established base of telephone exchanges in each country. The two basic types of signaling are *supervisory* and *interregister*. Supervisory signaling directs the process of initiation and taking down of the call. On the originating end, a signal is transmitted to "seize" the line (i.e., to inform the other end that there is a demand for service on a particular line). The other end acknowledges the seize and holds the line open for the subsequent interregister signaling, which conveys the city or area code and local number.

On most telephone networks, supervisory signals are energized through a separate pair of wires on transmission equipment, called the "ear" (E) and "mouth" (M) leads. The E lead accepts signaling input from the originating switch or transmission line, and the M lead transmits the signaling to the next link in the chain or the terminating switch. With conventional transmission equipment, supervisory signaling is carried at a frequency of 3825 Hz, which is outside of the voice frequency band ("out-of-band"), yet follows the same path as the associated voice channel. The tone cannot be heard by subscribers because it is blocked by the 300–3400 Hz filters in the transmission system. Some older North American transmission systems use "in-band" supervisory signaling, also called *single-frequency* (SF) signaling, with an audible tone of 2.6 kHz. In this case, a narrow "notch" filter is used to remove the tone from the voice frequency bandwidth. AT&T discovered that telephone pirates, called "phone freaks," made free calls by using "blue boxes" that generated SF and interregister signals (discussed below) to fool long-distance switches into allowing nonpaying calls through the network. Phone freaks now are largely out of business by virtue of SS-6 and SS-7, which do not recognize the tones coming from blue boxes.

During the 1960s and 1970s, the most common form of interregister signaling used the voice channel to pass the dialing digits, also called *address information.* By the in-band approach, each digit (0 through 12 or 0 through 16, depending on the system) is coded into a pair of audible tones, there being seven or eight different tones to be employed. You can listen to the combinations when you touch the key pad of a touch-tone telephone. After the digits are received and interpreted by the switch at the other end, the call is completed to the distant party. When the party answers, the distant switch returns an appropriate supervisory signal to confirm the connection. Now, the switching equipment can track the duration of the call for billing purposes. The final step occurs when one party hangs up, causing a "clear" supervisory signal to be transmitted. The circuit connection is taken down by the switches so that another call can be connected. Billing information is also recorded.

The predominant in-band signaling systems in North America and Europe are called R1 and R2, respectively, and have been adopted as international standards by the CCITT. The R2 system is also used in many countries of Asia and Africa. Variants of these systems have been developed to make them compatible with satellite links. This was required because the digits are transmitted in what is called a "compelled" mode, wherein the sending register holds the tone until a response has been received from the distant register. The tone is disabled, which must also be acknowledged by the distant register, giving two waiting periods per digit. If a satellite link were introduced, this "handshaking" would add unacceptable time delays to set up a call. *Semicompelled signaling* was devised, wherein tone pulses of fixed duration are used. The delay for call setup is reduced because the sending equipment only waits once per digit for acknowledgment from the receiving end before sending the next tone pulse.

International trunks over transoceanic cables imposed additional complexity on the problem of interconnecting telephone exchanges. A form of semicompelled tone signaling, called CCITT signaling system number 5 (SS-5), resulted from a joint development effort of AT&T Bell Laboratories and the British Post Office. This system operates effectively over transoceanic cables equipped with *time-assignment speech interpolation* (TASI) equipment, and via satellite circuits as well. TASI is a circuit multiplying technique that detects an active speaker at the input to the cable and switches the speech into an available channel. With proper synchronization and control of speech clipping, TASI can double the number of conversations that a cable can accommodate. The original work on TASI was performed at Bell Telephone Laboratories in the United States to increase utilization of voice channels on analog frequency division multiplex cable systems [Bullington, 1959]. The signaling digits must be held long enough to prevent the TASI equipment from interrupting the call setup process. Signaling system number 5 continues to be used internationally between gateway exchanges in many countries. Like R1 and R2, SS-5 is a system that uses in-band interregister signaling.

More advanced signaling systems use an independent data path called a *common signaling channel;* the resulting approach is called, not surprisingly, *common channel signaling.* AT&T introduced Common Channel Interexchange Signaling (CCIS) to combine the supervisory and interregister functions on a dedicated data network backbone [Bell, 1983]. Later, CCIS was adopted by the CCITT as signaling system number 6 (SS-6), the precursor of the more advanced version called SS-7. Under this concept, the supervisory and interregister signaling information is stripped from the telephone circuit at the originating exchange and transmitted as a message over an entirely separate data communication link, which runs in parallel with the voice network. The message format takes advantage of the principles of packet switching and processing, which can be made extremely reliable. Error detection techniques such as *cyclic redundancy checks* and *forward error correction* allow the data communication network to yield messaging reliabilities of greater than 99.9999999%. The developers of SS-6 thought that this type of reliability was necessary because the distant exchange would not be in direct contact with the originating exchange when completing the call. Also, SS-6 does not interfere with the operation of TASI and is not seriously affected by satellite propagation delay.

Due to the dominance of AT&T's network and technology, the most prevalent signaling system in North America is SS-6 as of 1990. The signaling channel operates at 4800 b/s within the United States, and is capable of supporting 4000 voice channels of traffic. Internationally, the rate and channel-support capacity are 2400 and 2000 b/s, respectively. The number of voice channels is determined by the allowable delay for call setup, which usually must be less than 10 s. Because the signaling channel services many voice channels, the error rate performance must be much better than that of a normal voice band data channel. Consequently, error detection and control techniques are included to reduce the chance of an undetected error down to only one in 10^{10} bits.

In addition to normal call servicing, the packet switching and database access features make SS-6 a powerful business tool for AT&T and the telcos. One of the first innovations was credit card calling from payphones, where the holder of a calling card would be able to enter his or her own credit card number and to obtain instant credit for any type of call. The number is entered through the key pad, where it is passed through the common channel signaling network to a central database for a credit check. Upon verification, the connection is completed and billing information is subsequently returned to the subscriber's home account. As discussed in the section on SDN, SS-6 gave AT&T a significant advantage over rivals MCI and US Sprint in the creation of innovative private network services. This advantage has already been overcome as all carriers upgrade to SS-7.

The common channel signaling approach is part and parcel of SS-7, the datagram message structure of which is based on the OSI model. Key to the future of digital networks, SS-7 is also the signaling system for ISDN. The physical channels transmit synchronous data at the elevated rate of 64 kb/s, raising the number of telephone channels within a given group by several multiples. At the link level, SS-7 employs *high-level datalink control* (HDLC), a very robust protocol employing the "look back N" technique. The higher level protocols of SS-7 are used for reliable routing of the specific content of call-processing messages. Other features are possible because SS-7 is potentially a very versatile message communication medium. SS-7 includes ample capacity for new features for use by subscribers as well as by the TAs, but this capacity must be activated by all players for its potential to be realized.

Finally, consider the fact that the national PSTNs differ as to the structure of the signaling systems in use. How do the TAs interconnect with one another to implement an efficient IDDD capability? The answer lies in the gateway concept, as discussed in the next section. Ultimately, SS-7 will preclude the need for the gateway because any domestic switch can conceivably be connected to a switch in another country. Gateways will still be needed where dissimilar systems must be connected, a condition that will persist well into the next century.

3.1.1.4 International Gateways in the PSTN

Gateway exchanges perform the important conversion from the domestic signaling system to a common international one. On the domestic PSTN, subscribers indicate that a call is to be routed to the gateway with an access code of some type. In the United States, the access code is 011 for IDDD. The gateway then acts as a point of domestic termination for the call, and international routing is established by an entirely different system. The major international carriers can afford the investment in trunks and switching programming to accommodate variations between countries. Also, the standards for international circuits provide a convenient conversion between the national and international signaling systems.

The gateway concept fits well with the topology of international long-haul transmission systems. INTELSAT earth stations are typically directly connected to gateways, allowing traffic to be aggregated for connection to the satellite links. Likewise, cable heads at the landing points of transoceanic fiber optic links are extended to international gateways. Traffic channels are aggregated into large bundles at the gateway so that long-haul media are efficiently loaded.

Because MTS is still provided on an analog basis, users can connect virtually any device over the PSTN. There are certain issues concerning differences in the domestic transmission plans of countries. Regarding the audio level (loudness) at the telephone channel interface, there is considerable disparity among countries and regions of the world. The basic unit of measure is the decibel above (or below) one milliwatt (dBm), with systems rated as to whether the input and output levels at the receive end are the same or differ by a fixed amount of dB. The dBm output of a channel is an absolute level, which can be measured with a power meter, whereas a dB difference is effectively the ratio of two levels based on a logarithmic scale. If two dissimilar links are connected, the audio volume will be too low to one listener and it will be too high to the listener on the other end. There is also the possibility that elevated levels will cause audio feedback. Expedients such as external *resistive pads* and *automatic gain control* (AGC) can provide immediate cures, but the long-term solution to this problem is for the world to adopt a common transmission plan. Another important facet is *echo return loss* and the use of *echo cancellers* on domestic and international long links. Echo is a serious impairment to voice and data communication on telephone circuits. Fortunately, echo can be almost entirely eliminated with modern digital echo cancellers. Outmoded analog echo suppressors (based on analog switching devices) should be avoided as they would detract from speech quality and were very inflexible when passing the full range of services now common in the PSTN. Conservative design suggests that an echo canceller should be used on both ends of a circuit of greater than approximately 500 miles in length, particularly if added time delay from digital multiplexing and switching systems is encountered along the route.

The advent of ISDN with signaling system number 7 has its advantages and disadvantages. The digital bearer circuit (B-channel) at 64 kb/s will be usable for a variety of services in addition to voice. From the analog domain into digital transmission, however, there is a need for coordinated timing. The fact that countries may synchronize their domestic networks to independent clocks makes timing problematical for critical data communication applications, which is why error recovery is so important in information systems. Voice communication can withstand the effect of differential timing because of the relative length of conversation and the tolerance of users to infrequent clipping or static bursts. Telephone customers are accustomed to occasional dropouts and disconnections, realizing that redialing is a simple remedy.

In the next few paragraphs, we review the domestic MTS primarily in countries outside of North America. This is in keeping with the focus of the book. Readers wishing to learn more about North American MTS and expanded calling services can review [Elbert, 1989], [Noll, 1986] and [Briere, 1990].

3.1.1.5 MTS Offerings of PTTs

Basic message telephone service is the business foundation for most PTTs because it is how they obtain the bulk of their revenue. In addition, the public service aspect of the PTT's operation is often measured through the quality of domestic telephone service. This is an important factor because there is evidence that economic development is correlated to some extent with the diffusion of the telephone network. The *telephone density,* which is the ratio of the number of telephone lines divided by the total population, is a recognized measure of the state of overall economic development of a country. As shown in Figure 3.1, this measure ranges from a low of one per hundred in Africa to twenty-two per hundred in Europe to a high of fifty per hundred in North America [Luhan, 1989]. An interesting situation exists in Asia, where the density of 1.34 per 100 is nearly as low as that of Africa, yet the density in the city state of Singapore is, like Japan, among the highest in

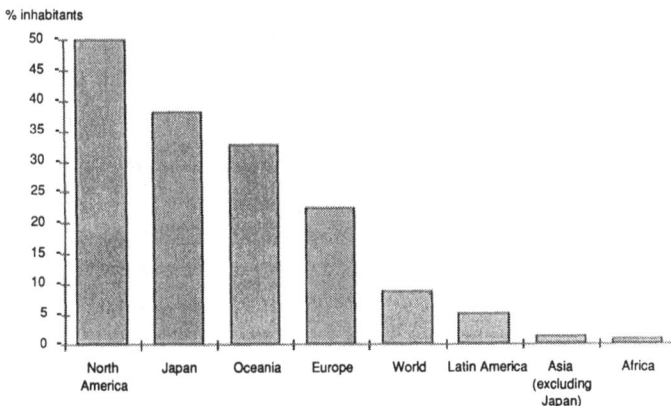

Figure 3.1 World telephone densities at the start of 1987 (courtesy of the International Telecommunication Union).

the world. In some developing countries, MTS may be considered to be a luxury, except possibly for the government and businesses. On this basis, telephone service can be expensive, both in terms of the cost of installation (including the instrument) and fees for usage. Service fees are divided into four elements: basic line or access fee; local calling charges; toll charges for domestic long-distance calling; and tolls for international calls.

An excellent overview of the similarities and differences among the rates for telephone services within Europe can be found in [Tarifica, 1988], which is produced by Logica Consultancy Ltd, London. Although centered on Western Europe, the information is still useful as being a consistent set of data across a broad group of countries. The PSTN tariff structure therein is fairly uniform, but the actual charges vary. A typical example of this variation is shown in Figure 3.2 for a multinational company located in the capital city of the respective country. The following were used to prepare the bar graph, which is for the year 1988. There are a total of 2000 minutes of telephone calls per month per line, and the breakdown of the type of call is as follows:

30% local calls (5 km);
10% domestic long-distance (trunk) calls;
30% international calls to adjacent country;
30% transatlantic.

The totals show a significant variation among countries, Spain being the most expensive, approximately twice the cost of Sweden, the least costly. The next lowest cost is in the United Kingdom, where both privatization and limited competitive entry are environmental factors. Interestingly, Sweden, a country with

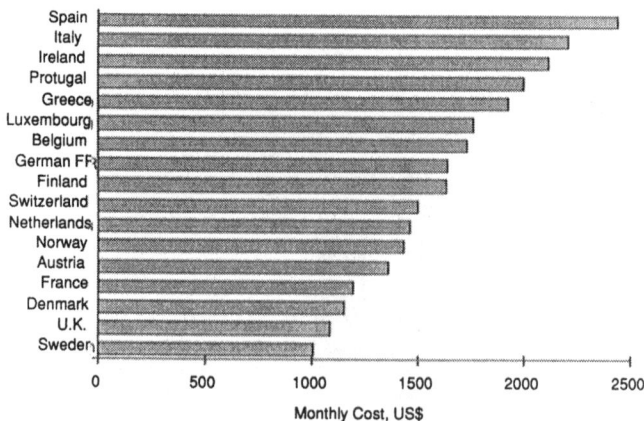

Figure 3.2 Monthly costs for switched telephone service for a hypothetical multinational company in the respective capital city in 1988, in US dollars (courtesy of Logica Consultancy).

neither factor applicable, still has the lowest charges for telephone service. This probably relates to the policy of the Swedish PTT to provide basic service at reasonable cost and the efficiency of its all-digital network. According to Logica, the domestic tariff is determined by the PTT in its sole discretion without regard to international convention; international tariffs, however, follow the D-Series Recommendations of the CCITT, which relate to tariff structures.

As discussed in Chapter 2, we are unsafe to generalize about how individual PTTs operate. What we attempted to do in that chapter was to develop some typical profiles of PTTs and to place them into certain categories. An important development is a trend toward privatization of PTTs and the introduction of a limited amount of domestic competition. These two trends are good for businesses because the service providers tend to offer more varied services, often at more attractive prices. Regardless of how diverse is the domestic telecommunication network, industrialized countries have telephone networks that provide MTS with typically excellent quality and reliability.

3.1.2 WATS and Toll-Free Calling

Wide Area Telecommunication Service (WATS) is geared to the business customer who wishes to facilitate telephone contact with distant offices, customers, suppliers, and other outside organizations. Consequently, WATS is tailored to the needs of strategic units that use the public telephone network in the daily exercise of their business. AT&T first introduced WATS to encourage use of the public switched telephone network. Later, rival long-distance companies emulated the WATS system to win and hold business customers who found WATS to have important benefits, one of which is reduced cost. Also, inbound WATS with its familiar 800 area code is the key to having consumers use the telephone to place orders for the widest variety of products and services. Customers appreciate being able to call a business free of charge. "Dial our 800 number, toll free" has become a familiar marketing lure.

WATS in the United States is offered by long-distance and local carriers in two forms: outbound and inbound. Outbound WATS is a discounted calling package to provide bulk rates. The traditional form of outbound WATS takes the subscriber's location as the center of a series of concentric circles, which establish the distance bands used for pricing. The particular WATS line, which is a dedicated access line to the carrier's switch, can be used to place calls to any point within a specific band. Calls are charged only on the basis of the duration of the call because the distance is already set. Subscribers were able to save 10 to 30% on the cost of long-distance calls. Since the mid-1980s, the long-distance carriers offered a wide variety of bulk calling packages, such as AT&T's ProAmerica Plan, MCI's Prism, and US Sprint's version of WATS, which allows a given access line to be used for all bands. With these innovative cost-saving initiatives, stimulated by the highly competitive business market in the United States, outbound WATS, *per se,* is no longer considered to be a unique offering. Rather, the long-distance carriers package the bulk deals for virtually every customer. The ultimate case is typified by AT&T's Tariff 12 offering, where network services and the calling package are customized for a large business user.

Inbound WATS, with its provision of toll-free calling, is the format that will no doubt endure and expand on the international level. As discussed previously, the 800 prefix identifies an inbound WATS call, whereupon the local exchange routes the call to the appropriate carrier. The carrier then uses a table to look up the correct area code and local number for the particular call to reach the correct party. To do this, the conversion of telephone numbers must be performed for each WATS call. AT&T was able to introduce WATS when SS-6 was installed. In the use of SS-6, the look-up process is performed when the request message reaches the centralized database via the common signaling channel. Without this separate communication path, there is no convenient way to perform the conversion of the called number from, for example, 800-234-5678 to an arbitrary number such as 212-433-2211.

Domestic WATS has appeared in Europe and Japan, and international WATS will soon be a mainstay for some businesses. British Telecom offers the 0800 toll-free calling service, the equivalent of inbound WATS in North America. A prospective customer can reach a business toll-free by dialing 0800 followed by a unique six-digit number. The signaling system routes this special number to a centralized database where the actual city code and local number are determined through a look-up process. On the continent, where the service is typically called "green number," there is either no charge or a reduced charge to the caller. *Numero Vert* in France, *Numero Verde* in Italy, and Green Number in Switzerland are all toll-free. The 020 service in Sweden and the 130 service in West Germany are toll-reduced, these numbers being the prefixes that callers use to identify this type of call.

AT&T has an international inbound WATS called 800 Service–Overseas, which provides toll-free calling to the United States from approximately 35 countries. A U.S. customer signs up for the service and then obtains an 800 Service–Overseas access line for each foreign country. The cost of using the service to the U.S. customer consists of three components: (a) a usage-sensitive rate for the particular foreign country; (b) a country access charge; and (c) a charge for installation and any additional services. The costs for usage are tiered, meaning that the rate per hour of calling declines as the total monthly hours increase, with the steps occurring at 10, 25, and 35 hours.

The advanced forms of CCITT signaling system number 7, such as ANSI No. 7, have the ability to process WATS calls and much more. Therefore, as more countries introduce SS-7, services like WATS can become more widespread. The

European PTTs have agreed to use a form of SS-7 called Telephone User Part–Plus (TUP+), which currently does not provide some of the message-handling features necessary to implement number-conversion services such as inbound WATS and SDN (reviewed below). As implemented, TUP+ is really intended to deal with the processing and management of circuit-switched operations at 64 kb/s, the B-channel of ISDN. Perhaps the limitations of TUP+ will be corrected soon after SS-7 is in widespread operation in Europe, allowing advanced services for private switched networks to be more universally provided.

3.1.3 Software Defined Networks

The flexibility and sophistication of modern circuit-switched digital networks have given rise to a private networking capability called software defined network (SDN) or, equivalently, virtual private network (VPN). As an innovation of AT&T, SDN allows companies with several locations to build private telephone networks without leased lines. Instead of having dedicated trunks between PBXs and CENTREX installations, the customer employs the switched telephone network to make all connections. While only available in the United States in 1989, SDN is being extended to Europe and Asia on a limited basis. The key feature of SDN is that a user can reach another telephone extension at any distant location on the network by using simplified dialing. In many installations, this is done with seven digits or even less (depending on the number of stations in the network). SDN is clearly another application of common channel signaling and a central database because the network must take the abbreviated dialed number and convert it into a standard ten-digit telephone number, complete with area code, exchange, and line. Therefore, AT&T was able to implement SDN by using SS-6, which had already been deployed around the country. Call setup time (also called postdialing delay) must be held to a minimum, which is why SS-7 is preferred.

The advantages of SDN are both economic and operational. For many organizations, there are locations that have comparatively small calling needs into the private network. To connect these locations into the network with dedicated private lines would be prohibitively expensive; hence, they tended to be serviced by normal MTS or WATS. The larger sites could properly use leased lines because of the ability to load them with sufficient quantities of calls to justify the fixed expense. Therefore, two classes of users evolved: those "on-net" and those "off-net." Locations on-net could reach each other by abbreviated dialing. To reach an off-net location from one on-net, the caller had to dial as many as 12 digits. These consist of an access code (typically 8) to reach the PSTN, the digit 1 to indicate a long-distance call, and the final 10 digits of the called party. Off-net locations therefore had a kind of second-class status. (Speed dialers and advanced PBX features now allow abbreviated dialing independent of the PSTN signaling system.) With SDN, all locations employ the carrier's switched network, but can reach any other location in the organization with seven digits. Because of the large aggregation of business that this offers the carrier, the pricing of SDN is competitive with the old way of doing things. An important advantage of using the carrier's network is that it tends to be more reliable than leased lines, because the PSTN automatically reroutes calls in the event of line problems or congestion.

From an operational standpoint, the work load on the customer is greatly reduced because the carrier takes care of all facilities' maintenance and administration. A monthly bill (typically on computer tape or available to an on-line computer terminal) contains all call detail information. Also, the SDN can be reconfigured by the customer through a terminal located on the customer's premises and connected to the carrier's servicing node.

US Sprint and MCI are both using SS-7, and have responded to the potential market with SDN offerings of their own. There are subtle differences, but all three carriers are in a position to continue to imitate one another as innovations are made. MCI's offering is called Vnet while that of US Sprint uses the VPN service name. In [Briere, 1990], many of the detailed characteristics of SDN, Vnet, and VPN are analyzed and compared. All employ the data networking and database management capabilities of the digital switched telephone networks of these carriers. One advantage to this competitive approach is that each carrier has total control of its network and can introduce any desired changes in the signaling system and database structure. In the international environment, changes occur slowly and innovation is hampered by the complexity of dealing among multiple service providers.

At present, international implementations of SS-7 do not have the capability to support SDN features. To employ SDN will be possible in certain special circumstances. Certainly, an SDN in the United States can be used to reach an international private line, allowing abbreviated on-net calling to a specific foreign location. The SDN simply reaches the U.S. access point to the private line and thereby can be connected overseas. Transfer to the PSTN at the distant end, however, is typically not permitted. On May 1, 1989, AT&T announced a new service called International City Service Center (ICCS) and opened it to users of their SDN through more than 500 SDN serving nodes. Others must use private lines to access ICCS. The rates are very competitive with MTS for overseas calling and are distance-insensitive within the distant country. Customers for both SDN and ICCS will have the added convenience of seeing ICCS charges included on a single bill.

An interesting opportunity in international telecommunication is developing in the United Kingdom and Japan, as a result of the activities in private fiber optic cables. As discussed in Chapter 2, Cable and Wireless, US Sprint, Pacific Telecom, and International Digital Communications (of Japan) are completing fiber optic linkage among the United Kingdom, the United States, and Japan. The employment of a common telephone switching architecture from Northern Telecom will make possible the offering of the complete set of on-net and off-net calling services available in the United States. Certainly, SDN could very feasibly be included. C&W's Mercury Communications subsidiary is already a local and long-distance carrier in the United Kingdom and will be able to access subscribers of US Sprint in the United States. On the Japanese side, there are organizations offering domestic leased-line and switched telephone services that use privately owned facilities. These Type 1 carriers can interconnect IDC to render international SDN services to Japanese companies.

3.1.4 Facsimile Networks

The facsimile machine has become pervasive throughout the world. Hard-copy reproduction of the printed page and graphic images simplifies communication for a multitude of business purposes. Once the fax machines became sufficiently inexpensive for every business with a need, critical mass was rapidly established. Operating at speeds up to 9600 b/s, the fax machines transfer a typed page in 20 s over the PSTN. Other features have appeared through a variety of digital processing and communication devices and services. In Chapter 4, we discuss the use of PDNs and E-mail services to deliver documents by fax machines. These services provide other attractive features, such as the ability to use store-and-forward messaging facilities like multiple addressees. Personal computers can be equipped with fax boards (called PC fax) to allow users to transmit graphic and textual data to distant fax machines and to receive fax messages as well. A discussion of fax machine technology, including details of the Group 3 and Group 4 standards, can be found in Chapter 5.

The growth of fax has encouraged operators of the PSTNs of the world to offer special connections for fax machines. In the United States, MCI was the first to make such a major move with the introduction, in November 1988, of MCI Fax. This network features store-and-forward service to allow fax messages to be held and then sent to several recipients. An inexpensive fax machine, costing less than $1000, can now work like a machine costing twice that amount by using MCI Fax's networking features. MCI claims better transmission quality than the PSTN because of maximum usage of digital facilities. Access is through a dedicated line to the MCI exchange, providing operation at up to 9600 b/s (as in any Group 3 link). Because of the digital network capability, for MCI to convert these access lines to 56 kb/s will be relatively easy, thereby allowing the use of digital Group 4 machines with their higher resolution and faster page delivery rate.

PTTs in industrialized countries are also beginning to provide these capabilities to fax users. In some countries, the PTT supplies the fax machine along with the line. A fax transmission is inefficient in terms of the quantity of information per thousand bits of data applied to the network. This is why a circuit-switched network is the most effective approach. A fax could be processed as a message, however, and handed off through a packet-switched network as if it were any kind of digital data. The inefficiency has been reduced over the years by use of compression algorithms. Evolution of fax networks is likely to continue for several years before a standard architecture is established.

3.2 ANALOG AND DIGITAL PRIVATE LINE SERVICES

Domestic and international private lines have generally been the means of implementing private telecommunication networks. Many private line voice applications within the United States are being replaced by SDN services. Data communication links, however, are still very dependent on private lines, primarily under the telco's Dataphone Digital Service (DDS) offering. For international voice applications, however, private lines are still very important because of the expense of heavy MTS usage of the PSTN and circuit quality problems on the domestic networks of many countries. Intercontinental private lines (East-West) may be less attractive for voice applications because of the reduced overlap of working hours when the circuit is actively used.

There is an important constraint on the application of private lines that PTTs around the world tend to recognize: the information being carried is private and only intended for the internal use of the particular business. Any circuit must terminate on the customer's premises as opposed to being connected to the PSTN or other domestic network. Resale of telecommunication service is usually forbidden. One reason for this restriction is that the PTT would not want a private leased line to be used in the telecommunication business in competition with the PTT. Also, by precluding connection with the public network at either end, the user cannot employ private lines and switching facilities to bypass the PSTN or the PTT's PDN.

Aside from issues of quality and availability of private lines in different regions, users are most concerned about the wide variability of tariff charges. The range of tariff rates for leased lines are shown in Figures 3.3 and 3.4 for domestic and international private leased lines for the European countries. The premium on price of international private lines is clearly substantial. Of course, an international private line must transit the local country's network, pass through an international long-haul cable or satellite link, and then employ the domestic network at the distant end. At the low end of the cost scale are countries where competition has been introduced (cost-based pricing). The most flagrant of the PTTs have tariffs that charge, literally, what the traffic will bear (value-based pricing). In general, the more competitive is the local telecommunication environment, the closer the tariff structure relates to cost. Competitive environments also motivate carriers to find the least cost approach to service provision.

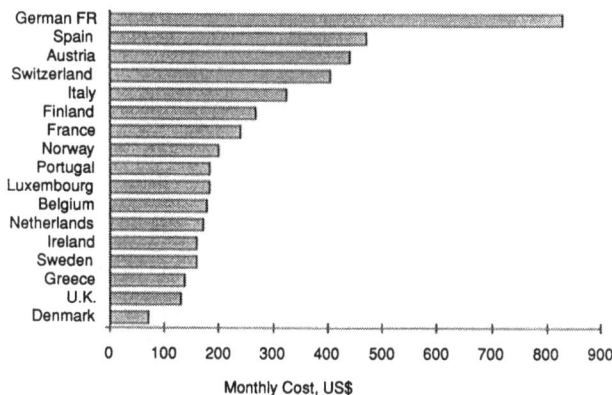

Figure 3.3 Monthly costs for national private line service for a hypothetical application in the referenced country, in US dollars (courtesy of Logica Consultancy).

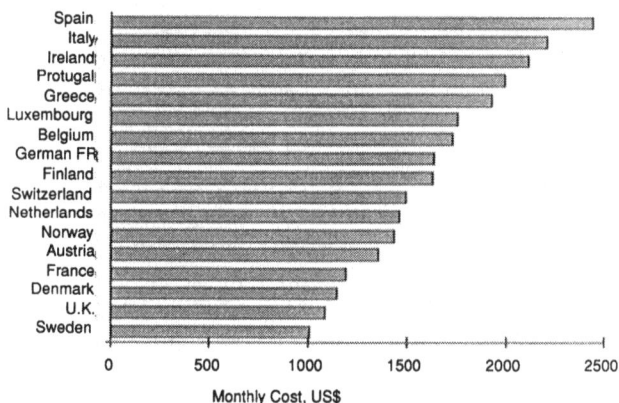

Figure 3.4 Monthly costs for international private line service for a hypothetical application in the referenced country, in US dollars (courtesy of Logica Consultancy).

Although analog and digital private lines are essential to most international private networks, the PTTs and privatized carriers prefer to transfer all business services to the PSTNs and PDNs that they operate. Evolution of separate packet networks and analog voice networks into the true digital infrastructure and ISDN would be welcomed by users. Therefore, the amount of private line traffic may peak and perhaps wane when PTTs deploy their intelligent networks. For the time being, however, private lines must coexist with the public switched networks. In the following sections, we discuss typical offerings of service providers in the way of analog and digital private lines. For convenience, currently available switched digital services at 56 and 64 kb/s and T1 rates are included because these are primarily aimed at private network developers for such applications as teleconferencing and high-speed data link restoral.

3.2.1 Analog Private Lines

An analog private line is usually nothing more than a normal telephone circuit established as a fixed connection between two locations. They use the same transmission facilities as the PSTN, but bypass the carriers' switching equipment because individual calls are not processed. There is a charge for initially making the connection and then another charge for each month of service. Because of the special routing from premises to exchange to long-haul facility (which is usually done by telephone technicians), a private line must be ordered well in advance of need. This can take from weeks to months. If adequate local-loop or long-haul facilities do not exist, the customer is faced with substantial extra charges for special construction. In addition to paying more money, the time to implement the service could then increase to a substantial fraction of a year. Because many of these circuits are used for private trunks and data transmission, separate wire pairs are provided for transmitting and receiving (i.e., a four-wire interface).

Applying private lines in a global network can be a complex and frustrating process. In the United States, private lines are offered by the RBHCs and the long-distance companies. Specialized carriers are involved in the private line business, including a U.S. subsidiary of C&W; Communications Transmission, Inc. (CTI), based in Austin Texas; and Williams Telecommunications, a fiber optic network subsidiary of the largest gas pipeline company in the United States. The private line market is therefore very competitive within the United States. A similar situation exists in the United Kingdom, where BT and C&W compete for the lucrative business customer, and in Japan, where significant competition is permitted by the MPT. The situation on the European continent is very different because there is only one supplier of private lines in each country, namely the PTT.

Nevertheless, private lines represent a good moneymaker to the PTTs that have the facilities and personnel to support the added workload. To aid PTTs in setting rates for the expected range of private line requirements, the CCITT has recommended the relative prices through tariff multiplier coefficients. These are summarized in Table 3.2 for analog private lines applied for a variety of purposes. The baseline measure is the standard four-wire telephone leased line with a bandwidth ranging from 300 to 3400 Hz. As discussed previously, this bandwidth can support a variety of voice band devices. The coefficient of one simply indicates that it is the baseline for setting the rest of the leased line rates; the CCITT will not recommend absolute prices. This particular rate would correspond to the cost of using a dial-up circuit on the PSTN for a period of approximately 9000 minutes, which is 20% of the time available in one month.

Table 3.2 Analog Private Line Circuit Tariff Multiplier Coefficients Recommended by the CCITT

Circuit Type	Coefficient	Equivalent Minutes per Month on the PSTN
All uses-telephone	1.00	9000
Alternate voice-data (AVD) telephone	0.833*	7500
Voice-only or facsimile-only telephone	0.667*	6000
48 kHz wideband	6.67	60,000
50 baud telegraph	0.25	2250
100 baud telegraph	0.30	2700
200 baud telegraph	0.40	3600

* CEPT members offer a circuit for these uses, having a coefficient of 0.75.

To complete an end-to-end circuit, a user must obtain a *half-circuit* from each PTT or international carrier. A half-circuit is defined as service from the international gateway in a country to the midpoint of the circuit. Conceptually, the midpoint of a transatlantic cable is in the middle of the ocean and that of a satellite circuit is 22,300 miles above the equator at an imaginary point within the satellite itself. A terrestrial circuit could traverse a third country, giving rise to the need to consider the share for the particular TA. Costs will differ among countries because of the respective private line rate structures, even if a common transmission facility is employed. The complexity of building a private network in this way has led to the concept called "one-stop shopping," where a major international carrier provides the end-to-end service by making all of the arrangements with the other involved TAs and carriers. One-stop shopping is being introduced in Europe and is already available between the United Kingdom and the United States and Canada. The lead service provider also takes care of collecting and distributing charges, providing the user with a single bill for all private line services.

The baseline "all uses–telephone" circuit shown in Table 3.2 is established on a full-time basis; employs an echo suppressor or (preferably) an echo canceller; and has reasonably good frequency response and dynamic characteristics. In terms of data capacity, the baseline would support up to 9600 b/s with modems capable of self-equalization. Nonequalizing modems could operate at up to 4800 b/s. Circuit equalization, according to CCITT Recommendation M.1020, could be obtained from the PTT at additional cost. Importantly, the "all uses–telephone" circuit connection does not go through a cable TASI system and has the full bandwidth of 300–3400 Hz available to it. Very often, however, this type of circuit goes over a satellite link to obtain this bandwidth and performance. A satellite-derived private line can give excellent circuit quality, provided that there is an echo canceller on each end. The quarter-second one-way time delay should be acceptable for most voice applications, and the circuit could support asynchronous and synchronous data transmission with an appropriate link-level protocol [Elbert, 1989].

The remaining circuit types are specialized in their capacity or service capabilities, and are discussed later in this section. Each coefficient is a multiplying factor to be applied to the monthly lease rate specified by the PTT for the baseline "all uses–telephone" type of circuit.

Although PTTs generally follow the recommendations of the CCITT worldwide and those of CEPT in Europe, there are subtle differences in the quality of facilities and details of the technical specifications that are followed. Michael Thomas of Northern Telecom has reviewed the situation in Europe and determined that a telecommunication manager had better be prepared for technical problems that can prevent a link or network from working properly or even working at all [Thomas, 1989]. Among the things to consider are signaling standards (discussed previously), synchronization on digital transmission systems, loudness on analog voice lines, propagation time (terrestrial and satellite links), and PCM quantizing distortion (Europe uses *A-law* and the United States uses μ-*law*). Any of these technical issues can be handled; the concern is that any item overlooked can produce a serious roadblock. Telecommunication managers or their technical staffs must be available at the proper time to ask the right questions and to specify the additional elements needed to implement the private line network. Of course, the alternative would be to hire an integrating contractor, which could be an international carrier or a specialized company with experience in the international telecommunication environment. Thomas points out that the switches manufactured by NTI can be equipped for any signaling system in the world, and, in fact, PBXs used in international private networks may have trunk lines set up for the complete range of cases (e.g., R1, R2, SS-5, SS-6, and SS-7).

3.2.1.1 Voice-Only or Facsimile-Only

Private lines classified as *voice-only* should be used solely for telephone tie lines; their use for data transmission should be restricted to relatively low-speed applications. The other important use is for facsimile communication with equipment meeting CCITT Group 2 or Group 3 Recommendations. This is because the circuits may have fixed routing within the countries at both ends, but can pass over analog transoceanic cables with limited bandwidth and TASI. The PTTs have the right to specify the use of the circuit and conceivably can require inspection of customers' facilities. The fact that the cost coefficient indicated in Table 3.2 is the lowest for voice band circuits makes this type relatively attractive and therefore of interest to telecommunication managers. PTTs in Europe, however, are moving to eliminate the cost break of voice-only–facsimile-only circuits relative to the more expensive alternate voice-data circuit type.

Any application using the voice-only private line classification must tolerate the potential for reduced bandwidth and TASI operation. Voice communication and fax operating at 4800 b/s or less should not experience difficulties. The key is that each transmission ought to be able to verify that the circuit is working all along the line, a normal property of voice or fax calling. Also, the speech detection equipment in TASI should see a constant analog signal within the circuit bandwidth; otherwise it would disconnect the transmission until such energy were present. When the connection is made again, it will most likely be via a different circuit of the cable link. This is why the transmission must be continuous to avoid a change in the circuit parameters during the call.

3.2.1.2 Alternate Voice-Data (AVD)

The AVD type of circuit can connect between private telephone switches at both ends or modems for use in data communication. Historically, AVDs were offered in the United States by the IRCs such as ITT Worldcom and RCA Globcom. Today, the AVD classification has less meaning than it once did. European PTTs such as DB Telekom differentiate only between two-wire and four-wire private lines, where the latter are charged at twice the rate of the former. As previously mentioned, there is also an extra charge for equalization of four-wire circuits. Such equalization would not be required for modems with built-in adaptive equalization (e.g., Recommendation V.32).

3.2.1.3 48 kHz Wideband

A much broader analog bandwidth can be obtained on domestic and international leased lines for transmission of high-speed data, wideband audio (7.5 and 15 kHz), and multiple-channel voice with customer-owned FDM equipment. The 48 kHz bandwidth, equivalent to the standard 12-channel FDM group, is the most common offering. PTTs have adopted the rule of charging 10 times the rate of a single voice channel for this class of service. By using the V.35 modem interface standard, a customer can transmit 56 or 64 kb/s of data over a 48 kHz private line circuit. Alternatively, a number of high-quality audio channels can be combined, where each typically occupies 7.5 or 15 kHz, depending on the application. The analog interface frequency range is 12 to 60 kHz, which precisely matches that of a standard FDM group.

3.2.1.4 Telegraph

As given in Table 3.2, there are a variety of narrow-band analog private line services available to support low-speed data applications. With rates of 50 to 300 b/s, users can connect teletypewriter machines, personal computers, and low-speed serial printers. The coefficient of less than one indicates a significant savings over the cost of an AVD circuit. These speeds, however, are so low by modern standards that there are few worthy applications in data processing. Examples are simple point-to-point teletypewriter messaging and a host application like credit verification.

3.2.2 Digital Private Lines

Domestic and international digital private lines at rates of 56 kb/s up to 2.048 Mb/s are available in developed countries. These services are differentiated from analog circuits because there is a direct digital baseband connection, eliminating conventional modems. Instead, a *digital service unit* (DSU) provides the interface. Transmission at the higher end of the rate spectrum requires a more elaborate device called a *channel service unit* (CSU), which incorporates the interface features of the DSU and can also be commanded by the international common carrier or TA into various test modes. Those who employ T1 services in North America are very familiar with the CSU because one of these devices is needed on the termination of each T1 circuit.

The public networks in the United States, the United Kingdom, West Germany, France, Scandinavia, Japan, and other industrialized countries are effectively digitized, simplifying the installation of end-to-end digital private lines. Between these countries we find digital fiber optic cables and satellite links, which contribute to the possibilities. Another important trend is the introduction of ISDN for handling domestic and international traffic. From a technical standpoint, passing user information in digital form allows the domestic and international carriers to combine the functions of switching and transmission. The user often would not notice whether the digital data passed over a fixed connection, or was switched and multiplexed with the data of other users and a variety of other services.

3.2.2.1 Low and Medium Speed Data Transmission (DDS)

For more than a decade, users in the United States have been able to obtain dedicated digital private lines from AT&T and local telcos under the DDS offering (which originally stood for Dataphone Digital Service). DDS is provided over the analog and digital transmission facilities of the carriers, using TDM equipment. Within the United States, available rates include 4.8 kb/s, 9.6 kb/s, 19.2 kb/s, and 56 kb/s.

A common application in host-based computer networks is the *multidrop line,* where the circuit is routed as a "daisy chain" from a communication controller to a series of remote sites. With appropriate software, a central computer system could address the individual sites along the way as if they were individually connected to the host computer. This technique was pioneered by IBM and is the way in which the model 3274 cluster controller routes data traffic between a quantity of remote terminals or small processors. The controller polls the remotes sequentially in time using the series channel provided by the multidrop line. Because of the criticality of the combined operation of this kind of subnetwork, high-quality digital private lines are preferred. Often crucial is that such an international DDS circuit be obtained, regardless of the cost, so that a distant branch operation in a foreign country can be brought into the corporate information environment. An attractive alternative would be to employ an established PDN, as discussed in the next chapter.

The DDS class of service can be used for a variety of point-to-point applications that cannot otherwise be replicated in a foreign country. A strategic unit can use several DDS circuits as a backbone for a private packet-switched network. For example, an international airline might install computer messaging nodes in various key cities throughout the world for use in E-mail and administrative transactions. Each node would consist of a packet switch and a computing device, such as a minicomputer, to service the location and to hold the local mailboxes. More information on E-mail is contained in Chapter 4. Unless there were a requirement for rapid downloading of large blocks of data, this international packet-switched network could rely on low-speed DDS circuits operating at 9600 b/s. The 56 kb/s type of circuit could support a higher volume of packet traffic, such as would exist on a major path between the United States and the European hub. At the higher end, newspaper page facsimile transmission requires at least 56 kb/s to transmit the page images with adequate speed and resolution. More common is to find that newspapers employ T1 transmission to shorten the amount of time required to transmit the entire paper.

The telcos and long-haul carriers historically have charged more for DDS than for analog circuits with comparable capacity. This recognized the improved circuit quality and higher cost of providing the service because some of the facilities were separate from those used for telephone. With competition among carriers in the United States, the prices of DDS services have declined to levels that compete with analog. Also, analog plant is almost obsolete from an economic perspective as the carriers race to complete their respective digital networks. A digital end-to-end circuit over a digital network eliminates the conversion to analog format (which may occur several times on the hybrid transmission facilities still prevalent). This provides a cleaner path and allows the carrier to be more efficient when combining user channels.

Pricing of digital private lines was still changing rapidly as of 1989. U.S. carriers are offering 56 kb/s circuits for essentially the same price as analog voice lines (local connections are extra and often cost as much as the long-haul section). European TAs are still defining their pricing structure, and service is not available in some countries. The United Kingdom, France, BeNeLux, West Germany, and

Sweden are delivering service at prices that are considerably higher than for voice. The relative pricing is over a wide range, similar to that indicated in Figures 3.3 and 3.4. Readers requiring specific information are advised to gather current data by contacting the relevant TA or international carrier. Recent surveys have shown that the privatized carriers are particularly helpful in providing such information, but a trend toward a market economy among other TAs in digital private lines probably is soon to come.

From a technical standpoint, a properly configured DDS circuit will provide superior service compared to an analog circuit with modems. This is because modems introduce distortion into the data signal, and the analog circuit will add noise and distortion as well. A digital circuit from end to end can be precisely controlled and distortion may be removed through data regeneration at various intermediate points along the route. This is understood by many U.S. telecommunication managers, who also realize that DDS is difficult to obtain on global networks. In fact, many small-to-medium sized communities in North America are not serviced by DDS. Organizations that must move quickly to implement networks of this quality have needed to consider VSAT satellite networking technology, which is indifferent to the physical location of circuit terminating points (see [Elbert, 1989] and [Morgan, 1988]). The time will come, however, when the majority of public networks are digitized and high-quality digital private lines are readily available.

There is an evolving standard of 64 kb/s, known as the B-channel of ISDN. The first introduction of the B-channel is for use in the PSTN. To have such a channel established by software and maintained in a continuously operating state for an indefinite period, however, is a relatively simple matter. This technique has been adopted in West Germany by DB Telekom. Users requiring extremely high reliability should consider the fact that hard-wired connections would be more reliable because they could still function, even if a main telephone exchange were experiencing difficulties. Tariff structures are evolving to provide part-time and full-time circuits at the 64 kb/s rate. Multiples of 64 kb/s would be available on a special basis. A new classification of digital private line, called the *fractional T1,* was introduced in the United States in 1989 by AT&T. Data channels are derived from the T1 by using conventional TDM equipment, supported by the use of DACS (reviewed in Chapter 6). The capacity is provided by AT&T's digital backbone network, unlike DDS, which uses dedicated facilities. Importantly for telecommunication managers, AT&T has priced fractional T1 below DDS. More information on T1 services is provided in the next section.

There will be customer requirements that are difficult to meet with the digital networks and ISDN services over the terrestrial networks. Satellite communication continues to offer the medium for delivering virtually any data speed or bandwidth requirement. In particular, INTELSAT Business Service is a means for accessing the international satellite system by way of specialized digital facilities. This is reviewed later in the present chapter.

3.2.2.2 *Multiplexed (T1 and E1) Channels*

The comments for DDS can be extended to apply to what are commonly known in North America as T1 services. The T1 is defined to be 1.544 Mb/s and is obtained by combining 24 individual DS0 channels, each operating at 64 kb/s. A somewhat different standard has been adopted in Europe, designated as the E1 for the purposes of this book (the designation of CEPT-1 is sometimes used as well). The European E1 standard of 2.048 Mb/s is not directly compatible with the North American T1. As reviewed in detail elsewhere [Elbert, 1989], the megabit bandwidth of T1 links permits a variety of services to be integrated within a private backbone network. This provides an economy of scale and the opportunity to adapt flexibly to the changing needs of the strategic unit.

Before delving into the matter of international T1 services, we will review some of the important technical details about the T1 channel. The frame structure of the T1 was devised to match the needs of voice communication in the 300 to 3400 Hz range. Adequate voice quality is obtained if samples of the time waveform are transmitted at a rate of 8000 per second, resulting in a frame duration of 125 μs. The analog to digital conversion process converts each voice sample into an eight-bit binary code format called *pulse code modulation* (PCM). A complete T1 frame consists of 192 bits for the 24 channels plus a 193rd bit, included for frame synchronization.

Early implementations of T1 circuits earmarked one of the eight bits per DS0 channel for telephone signaling and synchronization, effectively reducing the data throughput to 56 kb/s. Another failing of data communication is that older T1 circuits have protection circuits that preclude passing less than three 1s out of every 24 successive bits; this is called the *1s density* restriction. Many data transmission applications require a clear channel that can transmit any bit sequence. A special encoding scheme, called B8ZS, which stands for *bipolar transmission with eight zero substitution,* can be used to preprocess a clear T1, thus satisfying the 1s density restriction.

AT&T came up with a concept called the *extended super frame* (ESF) as a more complete solution to the previously recounted difficulties with the traditional T1 format. Implementation of ESF requires that the 193rd bit be converted into an 8 kb/s data channel, which is used to handle the appropriate signaling and synchronization functions. This leaves the full 64 kb/s available for data throughput on each DS0. Deployment of ESF, however, is still proceeding slowly in the United States as of 1989, and other events may well take precedence. In particular, the eventual solution under ISDN is to limit the throughput of the T1 to 23 B-channels at 64 kb/s, leaving the 24th channel (the D-channel) for signaling and messaging support of the network. As discussed previously, ISDN signaling and messaging is by way of the packet-switched SS-7 network, using a data rate of 64 kb/s. These characteristics compose the *primary rate interface* for North America, also known as 23 B + D.

In Europe, CEPT has established its E1 multiplexed channel at 2.048 Mb/s, containing 30 traffic channels at 64 kb/s each, one signaling channel, additional bits for synchronization, and other support functions. An aggregate throughput rate of 1.920 Mb/s is available within and between European countries that have digitized networks and transborder links. In terms of the ISDN primary rate interface, the CCITT recommendation characterizes the European standard as 30 B + D. This expressly includes the 64 kb/s signaling channel for SS-7 plus another 64 kb/s for framing, synchronization, and other purposes.

Because of the difference in transmission rates, the North American T1 and European E1 cannot be directly connected to one another. North American equipment manufacturers have responded to this problem by delivering both interfaces for the necessary switching and multiplexing equipment, allowing 64 kb/s channels to be routed in groups of 24 or 30. The problem remains, however, if a user simply wants to connect a 2.048 Mb/s service on one end to a 1.544 Mb/s service on the other. The T1 is at a lower rate and therefore can be used on both ends of the link, but at least six bearer channels will be lost from the connected E1 if the process is not done on an individual DS0 basis.

The PTTs and international carriers have found a solution at the aggregate level of approximately 140 Mb/s. Allowing for synchronization and framing overhead, a multiplexed stream at this rate can support either 84 T1s or 64 European E1s. As a consequence of this agreement, the long-haul fiber links will utilize 140 Mb/s as a building block of capacity.

As a concrete example, consider the arrangement of North American and European digital hierarchies presented in Figure 3.5. This is the type of scheme employed on fiber optic cable systems to combine the channel requirements of a variety of users on both ends. The digital hierarchy allows groupings to be connected at 64 kb/s, 6 Mb/s (4 North American T1s or 3 European E1s), and 140 Mb/s (84 North American T1s or 64 European E1s). The PTAT cable, in particular, operates at 420 Mb/s per fiber pair, which is a factor of three still greater than the new ISDN high-level standard of 140 Mb/s. Clearly, the place where the systems come together conveniently is at the highest traffic level. Economies of scale exist at the higher levels of the hierarchy, making such fiber links best suited for the international carriers and possibly the major resellers of private lines. This also corresponds to early plans for broadband ISDN using 140 MB/s as a basic transfer rate.

The utility of T1 channels in private backbone networks has been verified in the United States and holds potential in the international context. Intelligent multiplexers, discussed in Chapter 6, are the key ingredients to the T1 network. By offering the ability to assign bandwidth (time slots) dynamically in response to network requirements, the intelligent multiplexer becomes an automatic manager of resources. The network adapts to service demands and allows the telecommunication manager to focus attention on strategic applications rather than day-to-day traffic dimensioning.

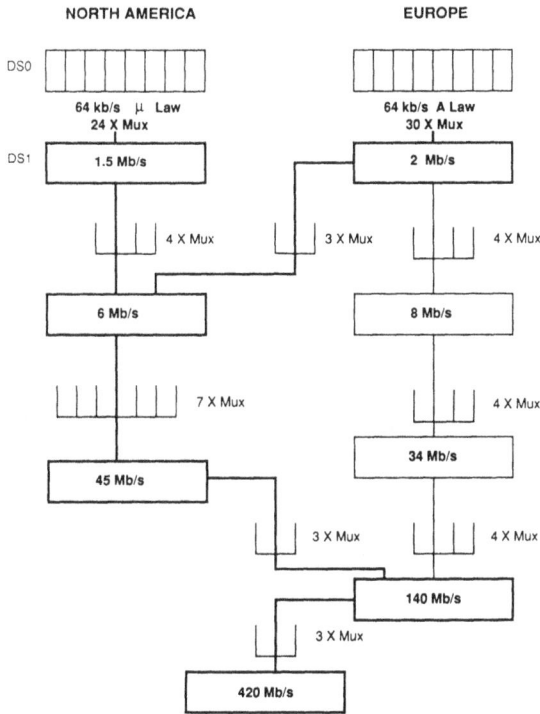

Figure 3.5 The digital hierarchy used by the PTAT-1 fiber optic cable system to integrate North American and European standards (courtesy of PTAT Systems, Inc.).

On a more basic level, T1 private lines can be directly connected to digital PBXs. The requirements for private traffic, of course, must justify this much capacity. Modern digital PBXs already operate on 64 kb/s channels and have standardized interfaces for either the T1 or E1 channels. An international T1 link via fiber optic cable or IBS with echo cancellation would deliver very high voice quality to users accustomed to ordinary international telephone service.

Fractional T1 private line service appeared in the United States in 1989, giving users the chance to commit to a smaller unit of point-to-point bandwidth. The service is available from the leading long-distance carriers, in increments of 64 kb/s. A strategic unit can use a 384 kb/s fractional T1 to add a smaller location to an overall digital backbone network. Fortunately, the T1 multiplexer manufacturers are offering the necessary fractional interface trunk cards to avail users of the reduced domestic rates. For international connections, fractional T1 service is offered by common carriers and several PTTs through the INTELSAT Business Service. The domestic "final mile" must still be provided by the local telco or PTT. Unfortunately, users may be faced with taking a full T1 or E1 to extend from the fractional T1 service point because of inflexibility in the local-loop plant. Efforts are underway by U.S. local telcos to resolve this difficulty on at least one side of the international connection.

3.2.2.3 Circuit-Switched 56 or 64 kb/s

The interest of business customers in circuit-switched 56 and 64 kb/s service on international circuits has motivated a number of international carriers to introduce such capabilities on fiber optic cable and satellite links. Modern digital telephone exchanges are capable of performing that function, as discussed previously in this chapter. Switching capabilities have been made available at 56 and 64 kb/s via public networks in several countries. The fundamental issue is whether the facilities are the same as those used by the PSTN (which is entirely feasible) or separate. When AT&T first introduced switched 56 kb/s service in the United States, it was forced to use separate facilities because its telephone network was primarily analog in nature. Today, PSTN operators such as BT, DB Telekom, and NTT support 56 or 64 kb/s data services as if they were for voice. The key, however, is to provide the right interface and to deal with some of the idiosyncrasies of the PSTN.

In 1988, US Sprint began to give its largest customers the opportunity to access their network at the DS0 level. The important point is that instead of paying a premium for switched 56 kb/s services, the user could establish circuit-switched data connections as if they were normal telephone. The price savings relative to AT&T's switched 56 kb/s are potentially substantial. For the user, the particular challenge is the final mile from the long-haul carrier's switch to the customer's premises. This could mean that a dedicated 56 or 64 kb/s access line would be required from the local telco. Such lines could be difficult to obtain and, when they are available, might be very costly. Alternatively, a user may be able to justify a T1 access line because several circuits and services can be aggregated.

As was mentioned in Chapter 2, DB Telekom operates a circuit switched data network covering a range of interface speeds from 2400 b/s up to 64 kb/s. This network is separate from the PSTN in West Germany. Networks of this type, including AT&T's switched 56 kb/s network, are probably soon to be obsolete because they cannot achieve the economy of scale of the PSTN.

International switched 56 kb/s service was introduced in 1989 for connections between the United States and Japan, and the United Kingdom and France. The respective TAs (i.e., NTT, BTI, and France Telecom) offer the service in conjunction with AT&T. Initially, transmission is provided over INTELSAT satellites and the gateway earth stations of these carriers. The fiber optic cables connecting these countries with the United States are being brought about to assure customers of the highest quality of service possible. Ultimately, switched 56 kb/s will coalesce with domestic and international ISDN basic rate services, which are discussed in the next section.

3.2.2.4 ISDN Basic Rate Services

The promise of truly economical public circuit-switched data service is to be realized with full ISDN development, where the *basic rate interface* (BRI) carries two 64 kb/s circuit switched lines plus the 16 kb/s packet-switched data line. These channels are derived from a single wire pair and are separated at the subscriber location by a multiplexer. In addition to the economy of scale due to aggregating voice and data traffic on a common network, the ISDN solution puts these capabilities in everyone's hands. This would achieve the critical mass of subscribers needed to make viable a whole new range of services. Switched 64 kb/s makes possible the utilization of the PSTN for many of the business data communication applications that rely on private lines. Also, Group 4 fax could be extended to essentially every business location and even the home. Another promising application is compressed video teleconferencing, which has been limited due to the lack of availability and high cost of switched 56 or 64 kb/s service in the United States and elsewhere.

In the countries where ISDN is available at the BRI, tariffs are generally geared toward the business customer. The monthly access charge is typically at least twice that of a conventional analog telephone line. Two countries where the PTTs have made a substantial investment in ISDN BRI service are France and Singapore. France Telecom and Singapore Telecom have already connected business customers to ISDN-compatible switches, allowing users to engage in voice, data, and Group 4 facsimile networking.

Aside from the port equipment on the telephone exchange, the actual line is not appreciably different from the standard loop. There must be a multiplex device on the customer's premises, however, to connect terminal devices (i.e., to achieve 2 B + D service). Similarly, the cost per minute of usage is also in the range of two to three times that of a conventional analog circuit. This seems not to be justified because the amount of channel capacity being used on the loop or by trunks is the same as in the analog case. In fact, ISDN does not require analog-to-digital (A/D) conversion on the network because it is performed within the terminal device, instead of within the telephone exchange.

3.2.2.5 Switched T1

Recognizing that there were applications where users would prefer to use T1 channels on an occasional basis, AT&T introduced their switched T1 service in 1985. The facilities to provide this capability existed in portions of AT&T's evolving digital network. Switching is primarily performed by using *digital access and cross-connect systems* (DACS), which are discussed in Chapter 6. As in the case of switched 56 kb/s service, the key requirement is that a dedicated access line be installed between the customer's premises and AT&T's T1 facilities.

Switched T1 service is most appropriate for graphic and video communication applications, because this type of information usually is transmitted for a limited period of time. High-quality laser reproductions of newspaper pages can be sent in approximately one minute per page at the T1 rate. National editions of the *Wall Street Journal* and The *New York Times* are examples of the use of high-speed digital lines for remote printing of newspapers. In 1989, these two papers were still using satellite networks rather than terrestrial paths. Another good use is for two-way video teleconferencing which is digitally compressed down to the T1 rate. The circuit could be established on either a reservation or dial-up basis. Teleconferencing was slow to gain acceptance because T1 lines typically needed to be ordered months in advance and used on a permanent basis. Switched T1 service appears to offer an attractive alternative.

Looking at international networking in 1989, no carrier offers switched T1 services between the United States and overseas. Likewise, switched T1 service, either at 1.544 or 2.048 Mb/s, is not available within Europe or Japan. A customer can obtain international T1 service for a limited service period, in the range of weeks or months, but an "on-call" or dial-up service does not appear to be in the offing. Users wishing to engage in occasional international video teleconferencing can employ the carrier's meeting room facilities. The carrier, therefore, has already invested in the necessary equipment and the associated full-time links, rendering service for an hourly charge. The disadvantage of this approach is that participants must travel from their offices to the carrier's meeting room location.

3.2.2.6 DS3 Point-to-Point

Looking back at digital private line services at very high data rates, the long-haul carriers in the United States are offering point-to-point connections at 45 Mb/s (DS3). This amount of capacity could be used by a major corporation or government agency for very thick routes along a private network backbone. High-speed computer-to-computer links are appearing (sometimes called a "channel" connection). Also, for a major corporation or government agency, the DS3 supports 28 point-to-point T1s. Manufacturers of T1 multiplexers (discussed in Chapter 6) provide direct access ports for DS3, eliminating an external M1–3 TDM multiplexer.

The really attractive application for DS3 is for the transmission of broadcast-quality video signals for the commercial television networks. Bellcore, several of the RBHCs, and long-haul carriers like MCI and Williams Telecommunications are cooperating in field trials of DS3 transmission between the studio and television stations. Full deployment of DS3 for video distribution is many years away because the fiber infrastructure in the United States does not extend to enough locations. Also, there is the basic question of the reliability of running vital network programming over a single-line facility. A more practical application is for backhaul of programming from news bureaus and sports stadiums to the network studio. A switching capability can prove very attractive because television networks usually only require the connection for a few hours at a time. More advanced technologies like broadband ISDN and fast packet switching would need to be implemented before switched DS3 services could be available to enough locations to be of value.

Very high-speed private lines can be obtained between the United States and other locations, and among other locations, using fiber optic cables and international satellites. Both classes of facilities can carry data rates of 45, 60, 120, and 140 Mb/s, for example. In a 1989 demonstration by KDD and COMSAT, a digitized HDTV signal was transmitted from the United States to Japan at 140 Mb/s. While the need for this type of connection is limited at this time, an organization will be reassured to know that extremely high-speed digital service can be made available. In the next section, the INTELSAT Business Service will be discussed. IBS is particularly attractive for the large-bandwidth, high-data-rate applications.

3.3 INTELSAT BUSINESS SERVICE

In Chapter 2, we introduced INTELSAT as the operator of the most all-encompassing satellite system in the world. The primary role of INTELSAT is to meet the international public telecommunication needs of its members through the provision of satellite capacity. In addition, members are encouraged to employ INTELSAT satellite capacity to solve difficult domestic telecommunication problems, like connecting remote islands and traversing uninhabitable territory. The expansion of domestic, regional, and now private satellite systems has encouraged INTELSAT to go beyond its basic charter, resulting in the creation of a variety of new service offerings. Among the more important to telecommunication managers is the INTELSAT Business Service.

IBS effectively is a way for commercial customers to obtain private line services via the international satellite system, possibly even bypassing the normal international gateway. As in satellite communication, in general, earth stations and satellite transponders can provide wide bandwidths and clear signals, making the links very suitable for very high-speed digital transmission. Prior to the advent of IBS, users could only access the satellites through the routine PSTN and private line services of the international carriers, the IRCs, and the PTTs. With IBS, the user can obtain direct access to a wideband digital service with the sophisticated digital multiplexing and switching capabilities of a private telecommunication network. A listing of the countries in the AOR for which IBS service is available is presented in Table 3.3, arranged by the year that access was initiated. In preparing this section on IBS, we appreciate the assistance provided by KJH Communications and their excellent publications on IBS and other digital offerings of international service providers ([KJH, 1987] and [KJH, 1989]).

Table 3.3 INTELSAT Business Service (IBS) — New Service Points by Year

1985	1986	1987	1988	1989
United Kingdom	Antigua	Australia	Botswana	Argentina
United States	Belgium	Bermuda	Dominican Republic	Austria
	France	Canada	Italy	Bahamas
	Mexico	Germany, West	Korea, Republic of	Colombia
	Netherlands	Hong Kong	Malaysia	Costa Rica
	Switzerland	Ireland	New Zealand	Egypt
		Japan	New Mariana Islands	Israel
		Singapore	Norway	Portugal
		Sweden	Spain	South Africa
		Turk and Caicos	U.S.S.R.	Venezuela
		Islands		

Following the lead of INTELSAT, a private digital communication service called Satellite Multiservice System (SMS) was introduced by EUTELSAT within Western Europe. SMS is employed in exactly the same manner as IBS, except that members of EUTELSAT use the satellites operated by that particular consortium. Therefore, a user requiring wideband digital services from the Netherlands can go to PTT Telecom to obtain IBS service over the Atlantic and SMS service to Italy. As of 1989, only two transponders on one EUTELSAT were committed to SMS; the service may become more popular, however, as the environment becomes more open.

The international carriers and domestic telecommunication authorities, notably the PTTs, are INTELSAT's sanctioned users. Private companies and normal telephone subscribers access this international backbone network via the sanctioned users, some of whom also direct the activities of INTELSAT. The system works very efficiently as an international cooperative, but many of the specific and specialized needs of commercial users are not always met. Under traditional procedures, private lines via INTELSAT are connected through the country's gateway earth stations. These stations operate at C-band, the microwave frequency band approximately between 4 and 6 GHz, which is shared equally with terrestrial line-of-sight microwave systems. Consequently, the gateway earth stations are located out in the countryside, behind high terrain to shield their receivers from terrestrial microwave stations. The counterpart of protecting microwave station receivers from earth station transmitters is also of concern.

The inconvenience and cost of using C-band earth stations as well as the greater interest in digital communication led INTELSAT to focus on Ku-band (which lies approximately between 10 and 14 GHz) when it introduced IBS in 1983. Unlike C-band, portions of Ku-band are not shared with terrestrial microwave services. Service is further enhanced because the satellite transmissions at Ku-band are at a significantly higher power level so that earth station receiving antennas can be made smaller. Because of these two factors, Ku-band antennas can be located within cities. Bandwidth within the satellite transponders is sold in increments of 64 kb/s, the recognized international standard unit of capacity. Service at C-band is also provided because INTELSAT Ku-band coverage is available only in certain regions of the world (see Figure 2.6).

Like other INTELSAT service offerings, IBS capacity is provided through its signatories. They, in turn, either supply end-to-end services directly to the end user or they work through independent carriers who, in turn, offer a complete end-to-end service. As an example of the former, DB Telekom offers the complete package including earth stations as it is the signatory, PTT, and sole IBS service provider within the Federal Republic of Germany. The latter case is typified by COMSAT World Systems, which is the U.S. signatory that offers space segment to the United States. The IBS service providers are shown in Table 3.4. To become an IBS service provider in the United States, a company must apply to the FCC for a common carrier license. COMSAT must then approve the technical design of the facilities, as discussed below, and the newly recognized IBS carrier will then be able to file a tariff.

Because of the attractiveness of access by Ku-band earth stations and the availability of digital point-to-point links, IBS has grown rapidly since its inception in 1983. A total of 7300 equivalent 64 kb/s circuits worldwide existed in 1989, as compared to 14 in 1984. Although the users may employ this increment of capacity, the actual links over the INTELSAT network primarily operate at the T1 or E1 rate. This allows the earth station operators to economize on equipment, because 64 kb/s and 2 Mb/s satellite modems cost about the same. Of course, the IBS carrier must have reasonable confidence that it can resell sufficient DS0 channels to make the link pay.

Table 3.4 IBS Carriers in the United States All Obtain Space Segments
from COMSAT World Systems

AT & T
Overseas Telecommunications, Inc.
GTE Spacenet
Hughes Communications, Inc.
TRT Telecommunications, Inc.
Contel/ASC
IDB Communications Group, Inc.
MCI International
COMSAT International Communications, Inc.
Houston International Teleport/STARS
Worldcom
FTCC/National Gateway Telecom
Vitacom/EDS
GE American Communications, Inc.
Carnival Cruise Lines

As discussed in Chapter 2, INTELSAT specifies various sizes of earth stations, such as the standards A and E stations used for international telephone service and IBS service, respectively. INTELSAT Standards E1, E2, E3, F1, F2, and F3 were designed for IBS installations from community gateways (in or near cities) and user premises. The E series are relatively small, being 5 to 7 m in diameter, and operate at Ku-band. C-band stations employ the F series of standards, and may be in the 7 to 12 m range. As explained previously, to place a Ku-band station within a city environment is easier because much of the frequency band is free from terrestrial microwave interference. C-band is still useful, however, where Ku-band is not available (due to the limited coverage of the satellites' Ku-band spot beams), or local rainfall statistics are so severe that Ku operations are not sufficiently reliable to meet circuit availability requirements. Performance of Ku-band can be improved through Super IBS, which is discussed later in this chapter.

When IBS was originally introduced, many potential users were hopeful that customer premises earth stations would become readily available. The terrestrial access line (tail circuit) between community antenna to customer site is eliminated, improving the effectiveness of the overall service to the user. There are exceptions, but the cost of a dedicated antenna has been concluded to be prohibitive when viewed against the amount of service that a single customer can generate. Consequently, most IBS installations are maintained by carriers who aggregate customer requirements. The station may still be located in relative proximity to densely populated areas, minimizing the length of tail circuits.

The satellite and earth station densities are high in the AOR, with the result that IBS activity is high in the region. Because five of the seven INTELSAT AOR satellites provide IBS services, a community earth station can require several antennas to reach a multiplicity of country destinations. Teleport installations are called "antenna farms" because of the number of dishes usually in place. The full-time nature of IBS and SMS prevents the service provider from moving one dish from satellite to satellite to serve different customers. In contrast, this is the practice in the occasional video service market in the United States, where customers may only require one or two hours of transmission on a given satellite. Once a service provider has sufficient business directed toward a specific satellite, it can usually justify "locking down" the particular dish for an indefinite period of time.

There are many locations where the only satellite coverage available is at C-band. To allow interconnection between IBS stations operating in different frequency bands, INTELSAT offers a capability called "cross-strapping." A connection is set up within the satellite repeater to allow a C-band uplink to be connected to a Ku-band downlink, and *vice versa*. This capability exists on Intelsat V and Intelsat VI model satellites, both operated in the AOR. Cross-strapped and noncross-strapped transponders exist on the same satellite, so a particular Ku-band earth station antenna can communicate with distant C-band and Ku-band IBS earth stations.

INTELSAT has adopted a set of standards for the digital communication services that are provided. As presented in Chapter 2, the standards and specifications are rigorously maintained and enforced, which promotes the quality and reliability of the service. Keep in mind that these standards are followed by earth station operators located in virtually every country of the world. The earth station standards (E and F) have already been covered. Transmissions between earth stations meet performance objectives that generally follow CCITT and CCIR Recommendations regarding international digital transmission in general and ISDN in particular. In this sense, INTELSAT was the first international telecommunication organization to introduce ISDN on a global basis. We are, however, considering the quality of the link in terms of *bit error rate* (BER), measured by the number of bits received in error divided by the total number of bits, and by availability (i.e., the percent of the time that the link is workable).

The are two service quality objectives: Basic IBS and Super IBS. These are defined as follows [KJH, 1988]:

Basic IBS is offered at both C- and Ku-band and provides a high degree of reliability and channel availability. Basic IBS is ISDN compatible at C-band; at Ku-band, quality (BER) is 10^{-8} or better under clear sky and 10^{-6} under degraded conditions for 99% of the time.

Super IBS is designed to offer higher channel availability at Ku-band. Super IBS is ISDN-compatible and provides 10^{-8} or better under clear sky and 10^{-3} under degraded conditions for 99.96% of the time.

Satellite communication links are engineered with extra power "margin" to guard against rainfall-induced signal loss (rain attenuation). C-band links require a relatively small margin, which is why ISDN compatibility can be maintained. Rain attenuation at Ku-band is two to four times worse simply because Ku-band wavelengths are more readily absorbed by water droplets. The only way to overcome this problem is to increase power on the satellite link. Because power is a valuable resource and must be shared among signals in the same transponder, however, the relative cost of the service increases. Note that the availability figure for Super IBS is significantly better than for Basic IBS, amounting to only three hours of outage per year instead of more than 800 hours per year. Super IBS, however, only guarantees that the BER will increase to and exceed 0.001 during outages, which is a relatively poor level, adequate only for voice service. Ultimately, for Ku-band operation, Super IBS may only be needed where rainfall is considerable, and Basic IBS will be satisfactory for dryer regions. In the tropical regions where thunderstorm activity is extremely high, IBS links will operate best at C-band.

With regard to IBS and SMS within Europe, customer requirements are individually too small to justify dedicated customer premises earth station installations. Carriers in the United States have installed city gateway IBS stations which are located close to downtown areas. Tail circuits are relatively short and can be obtained from the local telco or bypass carrier. In New York, one of these bypass carriers is New York Teleport, which has installed a fiber optic cable system ringing Manhattan Island and extending to IBS facilities operated by IDB on Staten Island. On the European side, BTI and Mercury in the United Kingdom and PTTs on the continent have installed community earth stations which are accessible by customers in areas of high concentration. KDD in Japan, C&W in Hong Kong, and Singapore Telecom all provide IBS via their gateway earth stations. Although this approach is less than optimum, IBS has nevertheless turned into a success for users who wish high-quality digital communication links between foreign offices and the home base.

3.4 HYBRID NETWORKS

The development of an international digital infrastructure is changing the way telecommunication networks are assembled for use by strategic units. For example, on May 31, 1989, General Electric Company of the United States entered into simultaneous five-year contracts with AT&T, British Telecom International, and France Telecom for a private global telecommunication network [Foley, 1989]. The three international carriers are acting as partners in a network that will service all of GE's domestic sites with locations in twenty-five countries on six continents. The network will be centrally managed and fully connected with GE's domestic network. Earlier in the year, GE signed a landmark deal with AT&T for a bulk package deal to be filed under Tariff 12, a mechanism for implementing customized private networks for major U.S. customers. This package would most certainly include SDN, digital private lines, and switched 56 kb/s and T1 services. The international part of the network would use fiber optic cable and satellite communication capacity, carrying voice, data, and digitized video traffic for GE and its subsidiaries. AT&T, BTI, and France Telecom as lead carriers should greatly facilitate the challenge of coordinating all the pieces in Europe, Asia, and other parts of the world.

Chapter 4
PACKET-SWITCHED DATA
NETWORK SERVICES

Data communication is the most rapidly expanding segment of the telecommunication industry, and its importance has spread to the international market as well. The terrestrial and satellite long-haul infrastructure provides the point-to-point pathways for global services. In data communication, users employ one of the established networking forms to interconnect information processing systems. Among the approaches are circuit-switched and private line services, discussed in Chapter 3, and packet-switched data communication services, reviewed in detail in this chapter. A strategic unit may create an international data networking environment by obtaining private lines and installing its own node equipment, or engage one or more international service providers. We focus on the *public data network* (PDN), the E-mail service, and the videotex service.

The most flexible data communication services generally operate through packet-switched networks employing the type of tiered architecture found in the seven-layer model Open Systems Interconnection [Elbert, 1989]. The PDN is a common carrier that provides a convenient transport medium for all varieties of user data, circumventing the PSTN or private line service. Because means are provided to convert from one data format or speed to another, the PDN can greatly expand the utility of a corporate information environment. The PDN represents a foundation for higher levels of data network services such as E-mail and videotex. E-mail has developed from its simple beginnings as an expedient for computer network operators. By the year 2000, there is little doubt that E-mail will be one of the dominant data communication applications. Videotex is perhaps a promising information medium, offering content as well as convenience. The simplified framework in Figure 4.1 shows the relationship among X.25, X.400, and higher level applications such as videotex.

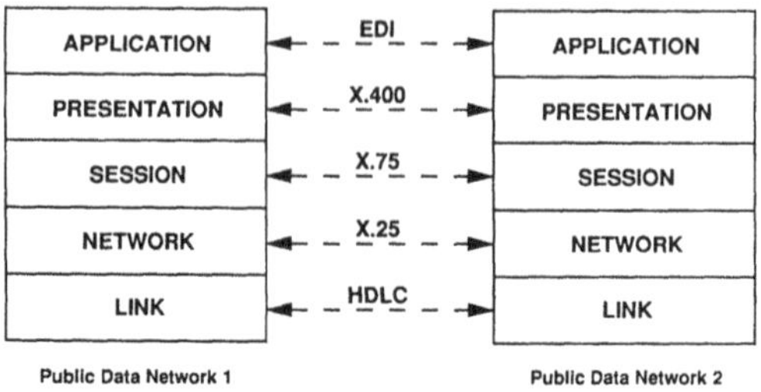

APPLICATION	← – EDI – →	APPLICATION
PRESENTATION	← – X.400 – →	PRESENTATION
SESSION	← – X.75 – →	SESSION
NETWORK	← – X.25 – →	NETWORK
LINK	← – HDLC – →	LINK

Public Data Network 1   Public Data Network 2

Figure 4.1 Interconnection and application of packet-switched networks at various levels of the layered architecture, using the Open Systems Interconnection model.

The packet-switching concept is very powerful in the way that it gathers traffic from low-speed devices and routes it over an elaborate network of private lines. Reliability is also excellent, provided that there are multiple paths between nodes. A network of this type is self-healing, meaning that the loss of a node or link can be overcome by the automatic rerouting features built into the network control system. On a routine basis, long messages are automatically split into multiple packets with procedures implemented to account for the individual packets. Garbled or lost packets are detected at the receiving end and replacements are automatically requested. The user would not even know that this process was underway, except possibly for an increase in the response time.

4.1 PUBLIC DATA NETWORK CONCEPTS

The PDN primarily evolved from the computer time-sharing and information service industries, which require low- and medium-speed data communication links on demand. Another general classification is the *value-added network* (VAN), which encompasses PDNs, but may be applied to a variety of specialized telecommunication networks. To be successful for the intended applications, the PDN must be convenient to access from any business location and must be extremely reliable. An unreliable network will cause the information to be disrupted before or during delivery. Even momentary interference with the smooth interaction of user with information source-destination could cause an entire session to be abandoned. This is analogous to what happens when a computer is "hung-up" due to a software problem; it may need to be "rebooted" or undergo initial program load. In applications such as electronic funds transfer, errors in transmission can have serious financial consequences. Because any communication link will experience technical problems at one time or another, the PDN has automatic means to reroute or repeat portions of the data without the user being aware that corrective action was taken.

Another important PDN feature is called *protocol conversion*, wherein a variety of different terminal types can be connected to the network and still access a host computer or another terminal (which itself may be incompatible with the first terminal). Protocol conversion is an important value-added service because organizations employ a wide variety of computer systems and due to differences in the character of individual data networks. For example, a remote user could have a personal computer operating in the asynchronous mode, as is common with most PC software. The host computer using synchronous X.25 software could not talk to the remote PC without the PDN performing the conversion. In 1989, AT&T and the RBHCs were still restricted from providing protocol conversion services over their respective PDNs, but may interface with other vendors that are not so restricted.

The PDN plays an important role in linking different networks. In the international environment, the domestic PDN of one country may not be directly compatible with that in another. Because there is not one universal PDN, the concept of the gateway between operating PDNs is vital in the international environment. Owing to international standards such as X.75, one PDN can generally connect with another as well as a private packet-switched network.

In this section, we review the implementations and services that employ the packet-switching capabilities of domestic and international PDNs. The basic arrangement of the PDN is illustrated in Figure 4.2, where each circle labeled N is a packet-switched node in the network. We emphasize the most common network structure, that of CCITT Recommendation X.25. For a detailed discussion of X.25, see [Schwartz, 1987]. Other packet network structures, such as TCP/IP and SNA, are very comparable as to arrangement and often use X.25 as a packet transfer

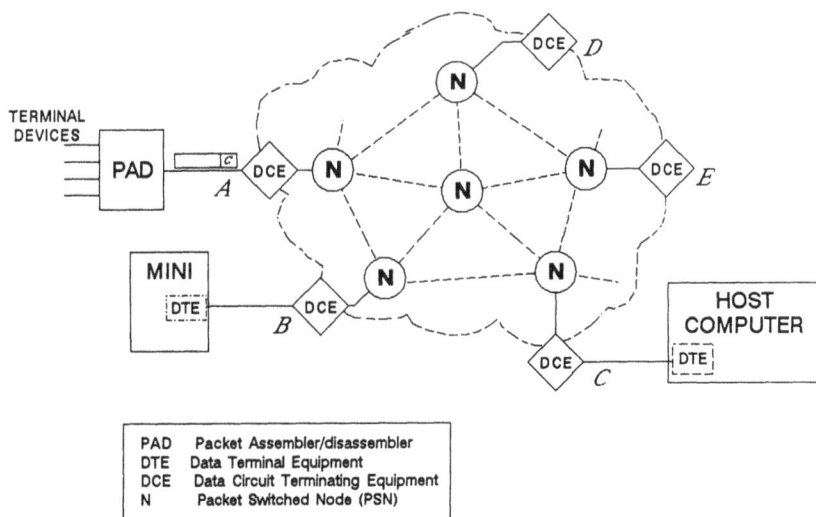

Figure 4.2 Packet-switching terminology.

medium. The X.25 standard is embedded in the OSI layered architecture and has generally been adopted in PDNs throughout the world. Between packet-switched nodes are full-time data links, which pass the packets as synchronous data at rates of 9600 b/s, 56, or 64 kb/s, or even T1 or E1. The operator of the PDN must analyze the expected packet loading of these full-time circuits and then select the appropriate link speeds. The data link protocol between nodes should be reliable to minimize the frequency of lost packets. If transmission is unreliable, the network compensates by requesting retransmission, increasing the time required to convey the complete message. The links employ synchronous transmission with error detection and *look-back N* error recovery (in particular, *modulo 8* or *modulo 128*). The two most common synchronous data link protocols are *synchronous data link control* (SDLC), used by IBM in the Systems Network Architecture, and *high-level data link control* (HDLC), used in X.25 and other OSI-compatible network approaches.

Access to the network is through *data circuit terminating equipment* (DCE), which is the point of presence at which users can employ the capability of the PDN. To reach a DCE point, users may have a dedicated connection over a physical circuit or private line, or they can reach it through the PSTN. Dedicated circuits may be digital (DDS) or analog and typically provide the most reliable service. For dial-up connections and analog private lines, the PDN operator provides the modem ahead of the DCE while the user is responsible for his or her end. Most information networks are aimed at terminal-to-host applications, supporting personal computers or video display terminals. These end devices need the resources of a minicomputer or mainframe host computer which would be connected to a dedicated DCE access port somewhere in the network. All that the user needs to know is the name designation or address of the host and the network properly routes the data. Also, terminals conceivably could connect to one another and hold telex-like conversations.

The terminals and hosts require an interface called *data terminal equipment* (DTE), the element which creates data packets from the local input and recovers information from packets coming from the network. A host normally contains software to perform this function while a simple terminal may employ a separate device called a *packet assembler-disassembler* (PAD). A user of a PDN may or may not need a separate PAD, depending on the inherent processing power of the terminal device or host. Personal computers and intelligent modems now contain software PADs to provide a direct DTE capability.

Before the advent of modern packet-switching technology, many of these host-remote data communication applications used the telex network. This was expedient because the service is available worldwide, even within the Third World and Eastern Europe. Telex, however, has major disadvantages: it is character-oriented, using the limited set of the Baudot code; and the device speed is very slow, at less than 100 b/s. Due to the presence of impulse noise and other impairments, data transfer over the telex network is generally unreliable.

Circuit-switched data networks, such as Datel in most European countries, are upgraded versions of telex, which handle speeds of transmission ranging from 300 to 9600 b/s. Some networks provide users with switched 64 kb/s services. Software-driven controllers can convert from one character format and protocol to another. Generally, the domestic and international telex networks (and possibly even Datel) are unsatisfactory for the kinds of information services and computer processing needs that exist in the 1990s.

As discussed in [Caswell, 1988], the basic purpose of packet switching is for multiple low-speed terminals to share the same data channel (which can be a leased telephone line, a DDS circuit operating at 56 kb/s, or even a T1 line) in such a way that each user thinks that he or she is the only one using the channel. The user data stream, which may be continuous or intermittent, is distributed by the DTE or PAD into blocks called *packets*. To each packet is added the addresses of the sender and the destination. Using Caswell's analogy of a childhood game, the packets are handed off between nodes as if they were hot potatoes. There is a distinction between packet-switched networks that transmit the packets completely independently of one another even if they are part of the same message or session (called *datagram service*) as opposed to those which route all packets in a given session stream in precisely the same way (called *virtual circuit service*). These aspects are discussed elsewhere in summary form [Elbert, 1989] and in technical detail [Schwartz, 1987]. Caswell makes the interesting point that the packet approach was conceived by the defense sector as a way to transmit computer information over a network that would function even if a particular switching center were destroyed by a nuclear attack. Although this thought is disturbing, to know that PDNs are very reliable is nevertheless reassuring.

The digital trunks which connect packet switches in a PDN or private packet network are typically provided over terrestrial facilities. This reduces the propagation time of the information bits, a significant contributor to overall packet transmission time, to an irreducible minimum. Network operators, however, find necessary the use of satellite links for many of these trunks simply because high-speed terrestrial lines are not always available. If the speed of the satellite trunk is high enough (e.g., 56 kb/s), and the protocol is properly "tuned" to provide appropriate error recovery (modulo 128, typically), the end-to-end delay of the satellite link can be made acceptable [Steinbruck, 1989]. With this in mind, some operators of PDNs and E-mail services are adopting satellite communication for hard-to-reach places, where DDS terrestrial lines are difficult to obtain. GE Information Systems, discussed later in this chapter, has even made a commitment to use VSATs for domestic services to the remote branch offices of large customers.

In the remaining paragraphs of this section, we identify the PDNs that provide international packet-switching data transport services. These are divided into three categories: domestic PDNs (including those provided by commercial companies and PTTs); transit PDN services of international carriers and IRCs; and the true international PDN. This discussion of the PDN environment is basically a snapshot as of 1990, and the reader should expect significant changes throughout the ensuing decade in the providers of PSNs and the capabilities of the service offerings. By the year 2000, many of these will have become obsolete (like telex is becoming) as more sophisticated networking capabilities become clearer. For the moment, however, the PDNs have remarkable abilities and provide reliable services at reasonable prices.

4.2 COMMERCIAL PDN COMPANIES

The commercial PDN companies are typically based in the United States, having started in the 1960s as offshoots of computer time-sharing and as a way to bypass the PSTN. Telex service, although useful as an interactive record communication medium, was often inadequate in terms of line speed and too unreliable, even on an occasional basis. Local telephone service with voice-band modems proved useful to reach the customer's premises, so time-sharing providers installed local access numbers to reduce the cost to subscribers for the services. Meanwhile, packet-switching technology was developed by research activities of the U.S. government in the late 1960s and early 1970s, particularly by the Defense Research Projects Agency (DARPA). The technology was spun off and developed by the companies discussed below. In addition to PDNs, major users have implemented private packet-switched networks for internal data communication. The equipment that can be used, such as packet-switched nodes, is discussed in Chapter 6.

There are a number of features that most commercial PDNs have in common. Of course, the PDN provides dial-up access for terminal devices that use an asynchronous protocol. Full-time connections are also allowed, although the customer must pay for the use of a dedicated port on the PDN's closest node. Another feature is the permanent virtual circuit, wherein the user makes fixed connections to two nodes with the PDN maintaining the same routing of packets. In this way, a user gets the benefit of a leased data line without paying for a full-time circuit from the long-haul carrier. Due to the sophistication of the programming of the PDN, a feature called *closed user groups* is available to provide a private network environment for an organization. A member of a closed user group may only initiate calls to, and receive calls from, another member of the group. This allows the PDN to achieve an economy of scale, reducing the incremental cost per user group, while giving a private customer exactly the service that is wanted.

BT Tymnet and SprintNet are the leading pure PDNs in the United States, representing over 85% of the market [DATAPRO, 1987]. Also reviewed below is Infonet, a PDN that established an international focus. Another PDN worthy of note is AT&T's ACCUNET Packet Service (APS), a shared packet network arrangement providing only X.25-compatible services. The APS network plays an important role in the internal operation of AT&T's various networks; capacity on APS, however, is provided to users and other PDNs, allowing AT&T to recover operating costs. The ability of X.25 to allow various PDN businesses, international connections, and private packet networks to coexist is particularly significant, and is the reason that the protocol set has evolved so quickly to become a worldwide standard.

4.2.1 BT Tymnet

The Tymnet PDN was created in the early 1970s by Tymshare, Inc., a leading computer time-sharing company. It was purchased by McDonnell Douglas, and in 1989 was subsequently sold to British Telecom International. The operations of Tymnet and Dialcom (discussed later in the chapter) have subsequently been merged. The developers of Tymnet pioneered a number of important commercial developments in packet switching, not the least of which was the implementation of the first such public data network in the world.

The network can be accessed in 234 U.S. cities and 30 countries, and consists of over 1000 packet-switched nodes, each of which is interconnected with at least two others. This provides the alternate routing capability that gives packet switching its robustness. Packet assembly and disassembly are done with a proprietary structure, using virtual circuit routing. Tymnet is interconnected with the Canadian Datapac PDN and networks in Europe via U.S. international record carriers. Furthermore, Tymnet can be accessed through the domestic and international telex networks.

The packet-switched nodes in the Tymnet PDN are actually designed and manufactured by BT Tymnet. Consequently, it is a significant supplier of this equipment to private companies and common carriers. Interconnecting Tymnet nodes are leased lines, satellite links, and private microwave links. The network is used as a public utility, serving over 10,000 public access ports.

Users are basically charged for service in two ways. First, there is a charge per hour of connection time, analogous to telex service. The second charge is for the packets actually sent over the network. Users are charged for blocks of 1000 characters (kilocharacters), monthly bills being usage-sensitive (this means that the more kilocharacters you send, the lower the rate per kilocharacter). Also, there are peak-time and off-time rates to encourage use during nights and nonworking days. The combination of the two charges is typically less than using the PSTN would cost and possibly even the telex network, depending on the device speed. Users can employ Tymnet to reach information services on distant hosts, a capability common to many PDNs and E-mail networks. A third-party host provider using Tymnet (such as a financial information service) would pay a fee per user and per access port connected to the PSTN.

Tymnet is a flexible PDN in that the network can allow devices of dissimilar speeds and different character formats to communicate with one another. These two facilities come under the general category of protocol conversion. For example, a terminal that uses ASCII characters and transmits them at 2400 b/s in the asynchronous mode can be connected to a synchronous host operating at 9600 b/s. The PDN node at the terminal end converts the input data to Tymnet's format and transmits to the distant node, where the information is converted to the format required by the host. The general arrangement and operation of Tymnet is very representative of other U.S. PDNs and those in different countries as well. Tymnet also provides X.75 gateways into external X.25 PDNs and private networks.

4.2.2 SprintNet (Telenet)

The SprintNet PDN was started by BBN, the supplier of packet switches for the original ARPANET. SprintNet has grown impressively to become one of the most sophisticated packet PDNs in the world. The company was acquired by GTE in 1979; subsequently, GTE transferred SprintNet to US Sprint, where it was combined with United Telecommunications' Uninet PDN. In July 1989, SprintNet announced that it would be merged and fully integrated into US Sprint so that all sales, administration, and research and development (R&D) activities would be conducted as one entity.

SprintNet provides a broad range of data communication services and also is a manufacturer of packet-switching equipment. The packet switches employ X.25 in the virtual circuit mode, distributing user data into packets that are 128 characters in length. In 1986, there were 800 host computers and 100,000 terminals connected to the network. The traffic on the network for the year amounted to 16 million domestic calls and more than a million international calls [Vignault, 1987]. To facilitate interconnection with PDNs in other countries, SprintNet has built on the X.25 protocol and interfaces using the CCITT Recommendation X.75 on internetworking. The first such interconnection was demonstrated with Datapac of Canada, a service of Telecom Canada, in 1980 [Unsoy, 1981].

The predecessor Telenet became familiar to many home computer hobbyists as the primary means of accessing the CompuServe recreational time-sharing service. To reach CompuServe, you would dial a local number on the PSTN and use a 300, 1200, or 2400 b/s modem. The local SprintNet node would answer your call with a simple prompt, to which you would respond with a series of abbreviations and identification codes to demonstrate that you were a legitimate CompuServe customer. From that point, you were connected to the CompuServe host computer as if you had dialed it directly by using long-distance. One cannot tell that the information is being carried over a packet-switched network and being combined with the information of thousands of other users. Connection time charges for SprintNet are billed by way of CompuServe. These are, in fact, the primary features of the PDN service, making it readily usable and attractive as an alternative to the PSTN. Also, as in the case of BT Tymnet, SprintNet nodes can perform protocol conversion so that most types of terminals, personal computers, and modems can be used to access a distant host computer.

Host computers are connected to the SprintNet PDN through dedicated access facilities, which individually would be a leased line from a local telco and, if appropriate, a long-haul carrier. The most efficient type of access is with the X.25 protocol, because no conversion is necessary on the SprintNet side of the interface. Users with terminals could connect to the closest access point on a dial-up or dedicated basis with any of a variety of protocols and speeds. For example, SprintNet can support asynchronous ASCII terminals; synchronous devices using SDLC, HDLC, and X.25; and with a range of line speeds from 110 baud to 56 kb/s.

One of SprintNet's strong points is the advanced stage of development of SprintMail, their E-mail offering. SprintMail is discussed in more detail in Section 4.4.3.5 below. Because of their support of international standards activities, SprintNet has made their PDN and E-mail networks easy to connect to those of other carriers and with domestic networks in different countries.

4.2.3 Infonet

The Computer Sciences Corporation (CSC) introduced the Infonet PDN in 1968 to support the data communication needs of its time-sharing and software customers. Importantly, Infonet developed into an entity of its own with particular strengths as an international backbone between users in the United States and Europe. Infonet is a self-contained marketing, engineering, and operating company, and it claims to be the worldwide leader in providing a comprehensive standards-based, value-added, international data transport medium. The organization has focused on reliability, quality, and customer support, three features which are potentially very valuable for global data networking in a world filled with dissimilar domestic networks and service providers.

In 1988, CSC set a new strategic direction for Infonet by selling pieces of the network enterprise to PTTs. As of 1989, ownership is divided among: France Telecom, DB Telekom, Singapore Telecom, Telecom Australia, PTT Nederland, RTT Belgium, Telefónica of Spain, and Teleinvest of Sweden. CSC retains a minority interest. Each investor elects a member to Infonet's board of directors. Interestingly, the foreign government PTTs derive a controlling interest in what is a U.S. private corporation. Infonet is currently headquartered in El Segundo, California, as is CSC.

Infonet provides packet-switched data communication services in a manner similar to the U.S. domestic leaders, BT Tymnet and SprintNet. Packet-switched nodes employ the X.25 network protocol and are connected primarily by leased digital private lines operating at 56 kb/s. The full range of user terminal devices and host computer systems can be connected to the network, with nodes performing whatever protocol or speed conversion is necessary. Many users employ Infonet to extend their IBM SNA information environments to diverse overseas locations. The access points support a variety of IBM devices that connect through the following protocols: SNA/SDLC, 3270 BSC, and 2780/3780 remote job entry (RJE). Of course, Infonet supports CCITT X.25 and X.75 and asynchronous telex and teletypewriter attachments.

Infonet provides dedicated access lines for customer host computers. Occasional access to Infonet is primarily via the U.S. domestic telephone network, with points-of-presence in approximately 100 cities. What is more important is that Infonet has a substantial global presence, with direct access in 21 countries via dedicated data lines. This connectivity is increased to a total of nearly 100 countries through the use of IRC services. Furthermore, there are X.75 bridge connections to PDNs in 18 countries of the ensemble. Infonet's international backbone can be reached via the domestic PSTN in a particular country, a dedicated access line, or the domestic PDN.

In 32 countries, Infonet maintains a dedicated support staff to assist with installation and operation problems. This means that a user on the other end of a customer's network, for example, in Hong Kong, can call a local telephone number and immediately speak with a local Infonet technical-support person. Most problems, however, are solved through an international 800-number customer-support hotline, which connects a user with the Infonet Support Center in El Segundo. From there, a 24-hour customer-support person can "talk" the user through a solution (which has been successful in 80% of the cases) or enter a trouble order into Infonet's international network management information system. This is an administrative network management system with connections to every customer-support location.

Some customers currently use the network for what are effectively EDI applications. Infonet helped Liz Claiborne create an on-line information network for E-mail and EDI services between the New York area and six overseas locations: Hong Kong, the Republic of Korea, Taiwan, Singapore, the Philippines, and Shanghai. In 1990, Infonet introduced this capability by offering E-mail and EDI applications that adhered to recognized international standards. These topics are reviewed later in the chapter in services available from other organizations.

According to a 1987 Link Resources estimate, Infonet has nearly twice the international service market share of its closest rival, SprintNet. Although primarily a PDN organization, Infonet provides other services, such as network consulting and facilities management. The company's goal is to serve customers with "one-stop shopping" for international information networks.

4.2.4 ACCUNET Packet Service (APS)

We introduced APS as the PDN transport service of AT&T. As discussed, AT&T is restricted (as of 1990) to interfacing only with users having the X.25 and X.75

protocol sets. Networking and messaging capabilities are not new to AT&T, as Bell Laboratories developed UNIX, the computer operating system that serves as an excellent platform for distributed processing. APS is well established, and it has been used by AT&T and a number of third-party service providers, including SprintNet and Control Data Corporation. AT&T Mail, an E-mail service, reviewed later in the chapter, is employing the network as a backbone. An important role of APS is as the long-haul medium to extend the PDN offerings of the RBHCs, which are restricted by the U.S. Federal District Court from directly offering long-distance services.

Customers can reach APS either with dedicated DDS or voice-grade access lines or through a dial-up service called REDI-ACCESS. The former follows the X.25 interface, where users need to supply the DTE function or to incorporate a PAD. Line speeds on the dedicated access circuit include 2400, 4800, 9600, and 56,000 b/s. REDI-ACCESS is provided in cooperation with Control Data, offering the protocol conversion services. Through REDI-ACCESS, the APS PDN appears to users to operate basically in the same way as other full-featured PDNs in North America.

The principal operating mode of APS is the virtual circuit. This allows call setup and clearing functions on a per-call basis, which is appropriate where destinations vary. A permanent virtual circuit can be created for applications where a full-time connection is required. This eliminates the need to initiate a call each time that information is to be conveyed via the circuit. Packet-switched nodes are supplied by AT&T, based on the 3B2 minicomputer. There were 21 of these PSNs in operation in 1990. The network employs CCITT Recommendation X.25.

Billing for APS is set up for each customer where there is a minimum charge, regardless of the amount of usage. This considers the location and quantity of access points and associated access lines. Above the minimum, the customer pays for the number of virtual call requests and for the quantity of data routed over the network. The cost per kilocharacter, the unit of measure, is higher during the business day and lower at night and on nonbusiness days.

AT&T has arrangements with 40 countries for the transfer of packets from the United States to points in Canada and overseas. For this type of access, the available port rates are up to and including 9600 b/s, as of October 1989 [DA-TAPRO, 1989].

4.2.5 GE Information Services (GEIS)

GE Information Services (GEIS), the telecommunication and information service organization established by General Electric Company, is a leading provider of PDN and associated enhanced services. GEIS is a worldwide operation, having grown rapidly over two decades. The offerings of GEIS in the area of E-mail and EDI are covered later in this chapter. Here, we focus strictly on the data transport services of the GEIS PDN itself. The service name of MARK*NET has been adopted for the PDN capabilities offered by GEIS.

Domestically, GEIS provides asynchronous dial-up service from 650 U.S. and Canadian cities. Leased line connections would be required for host access, for a variety of protocols including IBM SNA/SDLC and X.25. GE has supplied the bulk of the packet-switching node equipment and can perform a variety of protocol conversion services. Equipment such as data multiplexers and PSNs can be installed on the customer's premises if required for the particular application.

GE maintains a full-time staff and major processing node in Amsterdam, the Netherlands. This facility has the capability to monitor the European and transatlantic networks. There are leased trunk lines and nodes deployed in some 30 countries, which are also interconnected with 70 PDNs worldwide. At other points-of-presence, GEIS representatives are available to assist customers with technical and other problems.

Network services are charged in the typical way, with customers paying for connection time and for kilocharacters transmitted or received. Dedicated access lines for hosts and other applications like point-of-sales terminals are charged on a monthly basis.

GEIS has long been a major force in teleprocessing and information services. Before the personal computer was invented, GE was the principal supplier of engineering and scientific computing services. As with many time-sharing operations, GEIS introduced a data transport capability to allow companies to have extra service usage (and to generate additional connection time charges for GE!). The worldwide extent of the network has become attractive to strategic users needing a well integrated data communication medium. In this context, GEIS is a viable competitor to INFONET.

4.3 PDN OFFERINGS OF THE PTTs

While the first PDNs to use packet switching were implemented in the United States, other nations were quick to adopt the technology. NTIA determined that by 1988, PDNs were operating in more than 100 countries, many with 100 or more nodes. Canada, being in close proximity to the United States, implemented its Datapac PDN. In Europe, Spain was a leader with its Datex-P packet-switching network. Other European nations have made an early commitment to data circuit switching with Datel, giving them an advantage with this advanced version of telex. France and the United Kingdom, however, quickly introduced packet switching when trends became clear. In the following paragraphs, we briefly present examples of some of the more significant PDN activities in the industrialized world to give the reader an idea of the approaches taken in different countries. More detailed information on PDNs in Europe can be found in [Tarifica, 1988]; other comments regarding the PTT posture toward data communication are contained in Chapters 2 and 9.

4.3.1 Canada's Datapac

The Canadian PDN, Datapac, is operated by Telecom Canada, one of the two terrestrial long-haul carriers in Canada. Telecom Canada is not a government agency or PTT; rather, it is a cooperative comprising most of the Canadian local telcos, many of which are stock companies. The Datapac network began service in 1977 and provided the foundation for the development of packet-switching products by Northern Telecom. Although users may access the network with the virtual circuit mode of X.25, packets are actually routed as independent datagrams. The network architecture ensures that packets are all transmitted correctly and arrive at the distant port without loss and in the proper sequence [Schwartz, 1987]. A user who accesses the Datapac network will "see" the same end-to-end virtual circuit connection as in SprintNet. Between the nodes are fixed 56 kb/s data circuits maintained by Telecom Canada as part of its long-haul network.

The rates for using Datapac are volume- and distance-sensitive; there are additional charges for installation and special features. The access charge is dependent on the service provided, and usage charges are based on the number of packets transmitted [NTIA, 1983].

4.3.2 TRANSPAC

France Telecom claims that TRANSPAC, its domestic PDN, is the world's largest and most successful packet-switched data network. The network celebrated its tenth anniversary in 1988. In 1987 alone, TRANSPAC experienced an annual growth rate of 38%. A total of 12,000 dedicated customers use the network, with 60,000 direct accesses as well as 3120 accesses via the PSTN and telex networks. Approximately 50% of the packet traffic on TRANSPAC is generated by the Teletel videotex service [FT, 1988].

As in SprintNet, the TRANSPAC packet-switched architecture uses the virtual circuit mode of CCITT Recommendation X.25. Each network node has at least two permanent data lines connected, which operate at 72 kb/s. This arrangement ensures that there is diverse routing for any access point and destination. France Telecom announced in 1988 that access to TRANSPAC is being expanded through dial-up facilities. Traditionally, only the X.25 interface was supported. The new access facilities include dial-up asynchronous access with error correction or with V.32 modems. An international gateway from TRANSPAC using Recommendation X.75 is available through France Telecom. This is in addition to the international data transport facilities provided in France by Infonet (discussed in Section 4.2.3).

Expansion and improvements in the capabilities of TRANSPAC are common occurrences. France Telecom is pursuing a growth policy by expanding capabilities in E-mail via this packet-switched network. As an example, the TRANSPAC network is the backbone for a growing EDI market within France. In 1987, a project was begun to supply the French automobile industry association, Galia, with an EDI value-added network, which will use the Atlas 400 electronic messaging system.

As France moves toward ISDN, TRANSPAC will be adapted to the new environment. ISDN access became available in 1989 at speeds up to 64 kb/s on the circuit-switched B-channel. Subsequently, users are expected to be able to use the packet-switched D-channel at 9600 b/s. France Telecom's continued support of digital networking technology is evident from the innovations in TRANSPAC's PDN, providing for its evolution into the new realm of ISDN.

4.3.3 PDNs in Japan

At the foundation of Japanese networking is the major telecommunication provider, Nippon Telegraph and Telephone. Since 1981, NTT has offered packet-switching services to business customers, particularly financial institutions. As mentioned in Chapter 1, NTT is able to provide the full range of value-added services (unlike AT&T) including data processing and database operation. The introduction of Type II carriers in Japan (discussed below) stimulated NTT to expand its PDN services, and it now allows access by personal computers via the PSTN. Packet nodes are interconnected with high-speed synchronous links, facilitated by NTT's all-digital Intelligent Network System (INS), NTT's blueprint for ISDN in Japan. This new PDN feature allows users to access at 4800 b/s, which is four times the rate previously permitted.

International packet-switched network services are provided by KDD as a part of their established position as a Japanese overseas carrier. Under the trade name VENUS-P, KDD served 14,000 business and individual users in 1988 with packet-switched data transfer at 9600 b/s on international lines to 43 regions of the world. Users can be provided with electronic mailboxes for storage of E-mail messages, and access is provided to U.S.-based public databases and information services. Charges are assessed separately for connection time and volume of data transmitted. KDD provides full-time access lines to the customer's premises via NTT's network at a variety of speeds between 1200 b/s and 64 kb/s, for which there are separate charges for initial connection and each month of access.

Public data networks have been established since deregulation opened the market in 1985, primarily through joint ventures of Japanese and U.S. companies.

As many experienced international business people know, this is usually the most successful way of doing business in Japan. The U.S. company provides the expertise while the Japanese company provides both capital and entrée into the complex Japanese market [Hills, 1989].

Under deregulation, MPT allows two classifications of telecommunication service companies that can compete with NTT. The Type I carrier owns and operates its own facilities; foreign ownership of Type I carriers is prohibited. Value-added networks are in the Type II carrier category, which comprises organizations that must lease their transmission capacity from NTT or a Type I carrier. Foreign ownership of Type II carriers is permitted. Under a further classification, General Type II carriers are set up to serve the internal data communication needs of one company or a related group of companies, while Special Type II carriers can provide a variety of services to third parties, both domestically and internationally. As a Type II carrier, a PDN would install its own nodal equipment and obtain leased lines from NTT or a Type I carrier for transmission.

An example of a Type II PDN is the Fujitsu Enhanced Information and Communication Service (FENICS). Based on packet-switching principles, FENICS provides a wide range of services from basic data communication to more specific applications. In particular, Fujitsu has been focusing on the financial community, where more sophisticated applications are desired [Morita, 1989]. As with U.S. PDNs, FENICS can be reached by dial-up through the PSTN or via dedicated access lines. The network can provide real-time connections with protocol and speed conversion. Customers may connect host computers to the network and offer information services to other users. In addition, E-mail services are offered through a store-and-forward capability.

4.3.4 Telecom Singapore's Telepac

As the single supplier of data communication services in Singapore, the Telecommunication Authority has done an excellent job of creating an advanced PDN for the internal and international needs of business customers. The Telepac network, based on X.25 networking principles, branches out from Singapore to almost 40 countries and interconnects with every major PDN operation in the world. Users can connect their host computers and terminals on a dedicated basis or by using dial-up facilities. Access to the PDN can be from telex terminals at 50 b/s; asynchronous modems at 300, 1200, and 2400 b/s; and synchronous modems at speeds between 2400 and 9600 b/s.

The features available to individual and corporate subscribers encompass virtually all of those familiar to PDN customers in the United States. Among these features are permanent virtual circuit, closed user groups, and availability of E-mail services. A host connected to the PDN would incur charges for either a dial-up or dedicated port, plus a charge for each network user identification code. Users pay for connection time plus a fee per kilosegment (thousand segments), where a segment is equal to a maximum of 64 characters.

One of the attractive features of using the Telepac PDN is that Telecom Singapore has established an international networking hub in Asia. The excellent connectivity throughout Asia and to North America and Europe presents an interesting (and potentially attractive) opportunity to a strategic unit that requires a global networking capability using modern data communication techniques. Singapore has always had an international business focus in areas ranging from manufacturing to shipping and, more recently, telecommunication. Because of its business climate and attractiveness, Singapore Telecom has been able to invest in an excellent infrastructure and can be expected to continue to do so.

4.4 E-MAIL

Electronic Mail (E-mail) is a computer messaging application that grew fairly rapidly during the 1980s. According to NTIA, the number of users of in-house electronic messaging systems reached 5.6 million in 1987, an increase of 40% over 1986. The number of subscribers to public E-mail services had grown an average of 20% per year between 1982 and 1987, and reached 1.3 million at the end of the period. In terms of total volume, the combination of public and private messages is still small in comparison to regular mail and telephone calling [NTIA, 1988]. With this kind of trend, telecommunication managers are not surprisingly seeking global interconnection for their E-mail systems and services. As the eventual replacement for telex and telegram, E-mail has been proved valuable to multinational corporations, airlines and other transportation companies, media organizations, and international cooperative activities.

In his thorough book on the subject, Steven Caswell defines E-mail as "the generic name for noninteractive communication of text, data, image or voice messages between a sender and designated recipients by systems utilizing telecommunications links" [Caswell, 1988]. This definition might have been considered too broad when that book was first published. The type of information that can be conveyed via E-mail, however, is becoming more encompassing, and inclusion of image and voice mail with text and data is a reality. Figure 4.3 illustrates examples of the information "packages" that E-mail systems are capable of delivering.

The key feature of E-mail is that it is noninteractive, meaning that the originator compiles the message that he or she applies to the network for transport to one or more recipients. The sending party does not enter into a real-time dialog with the recipient because the passage of the message occurs over some period of time. Receipt could be acknowledged at one or more points in the process,

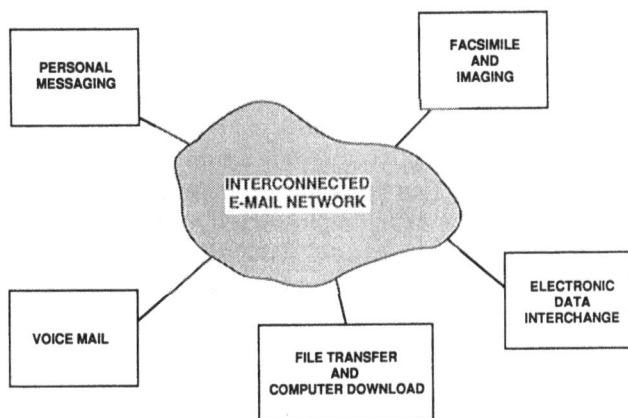

Figure 4.3 Messaging services typically available on a fully interconnected E-mail network.

depending on the particulars of the E-mail system. As an apparent contradiction, some E-mail services actually permit end-to-end connections for interactive sessions with information services and software packages for remote job entry or time-sharing.

In this section, we will not discuss the fundamentals and structure of general E-mail systems (this is done very well by Caswell), but we will instead focus on the international dimensions of the technology and services offered. Fundamentally, E-mail began as an internal function for users of time-sharing computer networks. A number of large organizations tailored system software for their unique needs, such as for internal correspondence (memoranda, personnel forms, work orders, *et cetera*) and to allow system programmers to access any device in the underlying computer network. A fundamental problem arose when a company wished to allow users on one network to exchange messages with users on another (either within the same company or to other organizations). Consequently, the proprietary E-mail networks were incompatible.

Incompatibility of E-mail networks can exist at each layer of the data communication architecture. There are problems in physically connecting one network to the other (the *physical layer*). Even if physical connection is possible, the binary coding and point-to-point protocols on the links may not work with each other (the *link layer*). Moving up the hierarchy, the data may be packetized and, if so, the routing algorithms may be incompatible. Packets may not reach destinations in the other network, or they may not be presented in the correct order (the *network layer*). The previous issues are resolved if a gateway such as X.75 is used, as discussed previously. Packets are delivered to the distant network node or computer device, but lo and behold, the message itself is meaningless to the mailbox or terminal because the address of the recipient is not coded properly and other information about the communication procedure is missing (the *transport* and *session layers*).

Assuming that all of the networking processes are correct and operation is validated, we still do not have an acceptable E-mail service connecting the two systems. The final layer to be considered is the actual application. Most of the effort to rationalize E-mail focuses on the X.400 *message handling systems* (MHS), the recognized E-mail application being implemented on top of most of the X.25 PDNs of the world. Interestingly, whereas X.400 is an OSI application, often the case is that X.400 is used on top of other networking schemes such as SNA and TCP/IP [Elbert, 1989]. In this way, X.400 is useful as a link between incompatible data networking systems for E-mail and file transfer functions.

Regarding message handling systems other than X.400, IBM has delivered literally billions of dollars of hardware and software, much of which supports its proprietary office automation systems, called PROFS and DISOSS. Within these applications can be found excellent E-mail service capabilities. Although proprietary in nature, this installed base of IBM systems carries a substantial fraction of corporate data message traffic.

Strategic units such as international banks and airlines often have private data communication networks interconnected between IBM computer devices in several countries, allowing users to send E-mail messages to one another. The IBM information network, discussed later in this chapter, enhances this connectivity by allowing E-mail to cross over between private and public networks. Another important feature of IBM networking is a built-in messaging capability called Systems Network Architecture Distribution Services (SNADS). (The SNADS approach to E-mail is discussed in the Section 4.2.2.) Through the facilities of SNADS, users can send messages containing almost any kind of data, ranging from letters to large batch-processing files. The SNA network hands off the message, as in any effective E-mail system, so that it is routed from node to node until arrival at its destination. SNA employs packet switching so that a large file is divided and sent with high accuracy. Connection from IBM's messaging systems to X.400 has not been ignored. Software suppliers and X.400 service providers are encouraging IBM's customers to employ bridges, which are appearing on the market. Among other bridging services is Western Union's OfficeAccess software, presented in Section 4.4.3.6.

4.4.1 X.400 Message Handling System

The CCITT has established a set of standards and protocols under Recommendation X.400 for exchange among message handling (E-mail) systems. As indicated previously, E-mail is an application in the sense of being at the highest levels of the OSI layered architecture. Both private and public E-mail systems are embracing the X.400 protocols to bridge incompatible systems, as bodes well for the achievement of truly global messaging. In particular, even if lower layers of OSI do not become universal, X.400 can still be employed by users with appropriate conversion software and gateways to proprietary data networks. In Chapter 6, we review the types and manufacturers of data network equipment, including packet switches. Where the demand exists to justify such expenditure, a user can invest in a private data network, including a significant capability for E-mail. As is reassuring, the X.400 MHS is to be available for users to build their proprietary systems and to interconnect with the public and private systems of others.

The general arrangement of the X.400 message handling system specification is shown in Figure 4.4. In X.400 terminology, the term "domain" is used to refer to an individual E-mail service or network. A public X.400 service is called an *administrative domain* (ADMD) and a private E-mail system is called a *private domain* (PRDM). The typical user will employ a text editor or data formatting program of some type to prepare the information for delivery. The user then employs the *user agent* (UA) service function of X.400 to assemble and address an "envelope" containing the information. Next, the UA enters into a dialog to deposit the envelope into the network through an element called the *message transfer agent* (MTA). Routing of the message is then conducted between MTAs around the network, until the destination is reached and the delivery to the distant UA occurs. As most X.400 systems will use the X.25 standard for networking, the message will be divided into packets and routed like any other data via one or more packet-switched networks (public or private).

Recommendations within X.400 provide three fundamental data transfer protocols, designated as P1, P2, and P3. Much like predefined dialogs or scripts, the protocols provide the "canned" instructions for exchanging the properly formatted messages. The P1 protocol is used by the MTAs to communicate with each other. Therefore, P1 is much like any other networking protocol, but in this case it

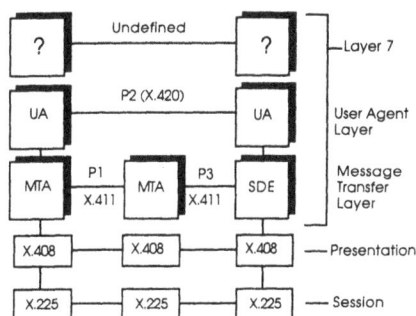

Figure 4.4 X.400 message transfer protocols.

understands only the transfer of information between X.400 defined nodes. Message transfer between a local UA and its associated MTA is with protocol P3. The purpose of protocol P2 is to transfer interpersonal messaging application information between two UAs via the network. The routing and control information of protocols P1 and P3 are appended to that of P2, which allows the UA-to-UA dialogue to occur transparently. Therefore, P2 is actually at a higher level of the X.400 application layer of OSI.

A particular E-mail system, whether OSI-compliant or not, has a directory that identifies every user with an associated address identification. The identification code of a user is the address of his or her electronic mailbox or personal computer, which is connected to the network and capable of holding messages. Sending messages between different domains is tricky because a user in one domain probably does not have complete address information for a recipient in another. The sending domain may not know if the person is even an E-mail subscriber. The purpose of Recommendation X.500 is to set forth protocols and procedures for locating users on distant domains and for maintaining directories around the world. As discussed in [Brabner, 1989], X.500 refers to "directory" in the singular because of the intent that there ought to be one interconnected directory system of global scale.

A search of the directories on different domains is done through a hierarchy, where the first piece of information needed is the distant domain identification (analogous to the country code and area code in the PSTN). X.500 has its own signaling protocols (separate from X.400) to locate the proper address of an intended recipient. Directory request messages can flood the E-mail networks of the world if care is not taken in selecting the criteria for the requests. If the address cannot be located within an agreed time frame, the effort is terminated and an appropriate message is returned to the user. The techniques for searching the directories of the interconnected domains must be very powerful to meet the needs of a practical system.

Several PDNs and public E-mail services have introduced X.400 to allow users to send messages among public and private systems. An important demonstration of eight interconnected X.400 E-mail services was conducted in June 1989 in San Francisco, California, under the sponsorship of the Aerospace Industries Association (AIA), an organization which represents major U.S. aerospace manufacturers. When all major services are connected in this manner and workable directory search capabilities are really available, many observers expect that "critical mass" will be reached. The concept of critical mass in networks, developed by Professor David Allen of the Sloan School of Management at the Massachusetts Institute of Technology, states that once a certain quantity of users are able to communicate with one another, the network then begins to expand rapidly. This rapid growth occurs because of the attraction to new users of being able to use the network to receive or exchange information with many more other users. In 1990, E-mail had not yet reached this point, but, as critical mass is achieved, X.400 will become a business necessity much as the telephone and fax machine are today. To paraphrase Mark Winther of Link Resources, a telecommunication market research firm based in New York: the more places to which people can send messages, the more messages they will send. To have this happen in the industrialized world by the year 1994 would not be surprising. In particular, Ovum Ltd, a London-based market research firm, estimates that about 500,000 personal computers and 150,000 large systems will communicate using X.400.

Recognizing this eventuality, the PDNs, E-mail services, and manufacturers of packet-switching equipment are actively participating in international standards committees. The X.400 Application Program Interface Committee was formed in 1989, and includes such leaders in data communication as Digital Equipment Corporation, Sun Microsystems, Sprint International, British Telecom/Dialcom, Retix (a developer of OSI-compatible products), and 3Com Corporation (a leading manufacturer of local area networks). To reinforce these efforts, the U.S. government and the AIA have said that they will only use in-house E-mail systems and PDNs that support X.400. The federal government is requiring that the two winning contractors for the giant FTS-2000 network, AT&T and US Sprint, interconnect their respective E-mail networks by using an approved X.400 protocol. Absent such a requirement, the two long-distance rivals might not have found the fortitude to work together in this way. On the international side, Sprint International announced in 1989 that it has agreements for SprintMail E-mail connectivity with Belgium, Austria, and Sweden. This represents the beginning because of the importance to users and E-mail providers of reaching the critical mass stage.

4.4.2 IBM Message Distribution Systems

There is considerable momentum behind what could be called the "X.400 movement" in the world. When discussing critical mass in E-mail, however, we should not forget the large installed base of IBM mainframe computer and minicomputer systems that support more than 50% of all of the West's data processing and data communication applications. We introduced the various capabilities under IBM's SNA environment, including the PROFS and DISOSS office automation systems and SNADS store-and-forward messaging service. Systems Network Architecture is a layered architecture which is similar in concept to the Open Systems Interconnection model. Because of maturity and heavy financial backing from the largest computer manufacturer in the world, IBM computers are easily networked and the software products are very well integrated. An organization that wants to install an automation system for virtually any type of business can ask IBM to create a working environment. Also, SNA environments of different strategic units can be interconnected to provide advanced applications such as *electronic funds transfer* (EFT) and EDI services for *just-in-time* (JIT) manufacturing. The point is that SNA behaves as if it already has achieved critical mass.

Systems Network Architecture has within it the powerful messaging capability of SNADS. This is an SNA transaction-layer service to transmit documents and files on an asynchronous basis between peer nodes. In one of the upgrades to SNA, IBM made the important step of creating in the network what is called a *logical unit 6.2* (LU 6.2) element, which can directly exchange information with any other LU 6.2, bypassing a common mainframe computer [Sundstrom, 1987]. The SNADS system software handles large or small messages that a user can create at one node in the network. The message can be applied to the SNA network, where it travels effectively from node to node. This is in contrast to the normal mode of SNA networking, where a synchronously timed session is established for the duration of data transfer and response. In SNADS, there is no session, nor even a virtual circuit, to control the passage of the message from the originating node to the final destination. An intermediary node accepts the message and can retain it in memory until an open link is available to the next node along the route. In SNADS, confirmation that the complete message has been received without error is an important attribute.

To use the power of SNADS, an application program of some type is required. Therefore, applications like EFT and E-mail employ SNADS as an underlying messaging protocol. The application system hands off the user inputs and responses to SNADS, which, in turn, provides the transportation throughout the network. This is different from lower layers of SNA, which, like X.25, use the virtual circuit mode to route packets of data from originating node to destination node. SNADS takes care of important functions such as maintaining the directory of addresses, where there is a conversion from user names to these addresses, and allows messages to be carried to multiple addressees. The convenience of using E-mail under DISOSS, for example, is provided by the software that interfaces with SNADS or another transaction type of structure. Sophisticated users can, however, write their own interfaces to SNADS and do other things that cannot be performed by "canned" software.

IBM recognized that, although the SNA environment provided almost all that internal systems could need, there are many cases for which users on a dedicated network would need to exchange data with users within domains of external E-mail systems. A variety of companies offer software for IBM computers to allow users to work between SNA and non-SNA networks, employing an X.400-compatible bridge. Some examples are included in the following paragraphs. Of particular note is the IBM Information Network, which is a worldwide service, allowing customers to share applications and to create international E-mail systems. This network is reviewed in Section 4.4.3.8.

4.4.3 Public E-Mail Offerings

A comprehensive review of public E-mail offerings is contained elsewhere [Caswell, 1988]. In the paragraphs that follow, we identify some of the primary players in this particular market. Our international focus causes us to emphasize those networks which are achieving a global presence. There are perhaps three classifications of providers, based on their overall business strategy. We have the leading PDNs, such as Infonet, which view E-mail as a step toward vertical integration, providing more specific user-oriented service. This strategy tends to increase loading on their respective networks and potentially increases revenue from more diversified sources. A second category contains major carriers, both domestic and international. AT&T and MCI are in this category as is British Telecom International. Through the previously mentioned merger, US Sprint is in the E-mail business both as a carrier and by virtue of its ownership of the SprintNet PDN. The last category of E-mail service organization is the computer company or service bureau. Computer giant IBM provides E-mail and teleprocessing services through the IBM Information Network. IBM's E-mail capabilities derive from its dominant position in computer networking with SNA. GE no longer manufactures computers, but its information systems subsidiary continues to be very strong in its markets. Also mentioned is Western Union, a company that has undergone several transformations and is now focusing on business network services. System integrators, such as Electronic Data Systems and Computer Sciences Corporation, are likely candidates to make strong showings in E-mail, either as builders of such systems for users or as providers of E-mail and other messaging services.

Before entering into the discussion, we will give a word about how E-mail services are priced. The tradition of on-line services like CompuServe encouraged the first E-mail services to charge for connection time. The user pays the same rate per hour, regardless of the amount of E-mail that is originated or received. Another approach was to charge for the number of characters or blocks of characters, as practiced by PDNs. A combination of connection time and actual usage is a further mode of pricing. MCI introduced the system of charging a user for an entire transaction (i.e., a flat fee for a message up to a certain size). Multiple billing units would be incurred for large messages. In dial-up situations, the E-mail host would automatically disconnect an inactive user to reduce idle time and free facilities for new callers. All told, Caswell estimates for dial-up access with a PC that the typical monthly bill (in 1988 dollars) across the various suppliers may be in the range of $25 to $100 per month. This is because there is a wide range of pricing and service features.

We now consider some of the leading E-mail services. There is considerable variety to the capabilities of these services, including on-line database access and interactive communication, but all offer the capability of store-and-forward messaging. Users would be advised to examine the options carefully, in light of the specific usage that is contemplated. Because of the complexity of this type of analysis, the only practical approach could be to evaluate more than one E-mail service under typical application conditions. An important factor, almost independently of local charges, could be the convenience of connection to other countries. Even if not the lowest in price, a well connected E-mail service that satisfies X.400 and has presence in foreign countries of interest may provide vital benefits to a strategic unit.

4.4.3.1 AT&T Mail

AT&T entered the E-mail business in 1986 with a service that is carried by the APS X.25 packet-switched network described previously. According to Caswell, one of AT&T Mail's strengths is its off-net delivery capabilities via telex or laser-printed hard copies, which are delivered either by the U.S. Postal Service or an overnight courier service. AT&T Mail is easy to use with PCs (particularly IBM-compatible and Apple Macintosh computers) and UNIX computers because of the Access Family proprietary software packages, which provide a convenient electronic "desk" environment for the user. The message may be prepared off line with a text editor, or the customer can use an on-line message preparation system. Users connect to AT&T Mail by using a modem and the PSTN.

Pricing for AT&T's E-mail service is based on transactions, where there is a fixed charge for each message unit. There is a monthly membership fee and no connection time charge. A cost advantage for short messages occurs at 400 characters with the next increment extending all the way up to 7500 characters. Use of printers and the different hard-copy delivery systems add significantly to the price of sending the message. There is no charge to receive a message directly from the user's mailbox by way of the network. Electronic delivery without a computer terminal is permitted through a voice synthesis service called AT&T Mail Talk, which will read the correspondence over the telephone.

AT&T mail should have a very large user base available through the FTS-2000 government network. In addition to its own subscriber base, the FTS-2000 program includes Dialcom as AT&T's subcontractor to provide access to a variety of proprietary databases. Dialcom, discussed in the next subsection, employs an X.400 interface with AT&T mail so that AT&T's previous 35,000 corporate mailboxes can exchange messages with Dialcom's 170,000 business users. AT&T has reached agreement with PTTs in Australia, Finland, Sweden, and the Republic of Korea for similar X.400 message exchange. Considering all of these developments, AT&T Mail would appear to be worthy of consideration as either a gateway provider or even the source of most E-mail capabilities for a major strategic unit. This will depend on the company maintaining its rate of growth, as it still considerably lags behind leaders such as US Sprint and BT Tymnet.

4.4.3.2 BTI Dialcom

Purchased by British Telecom from ITT in 1986, Dialcom is a mature E-mail service with an extensive international presence. Dialcom has two particular competitive advantages: first, it has licensed its services in some 20 countries as well as the United States; and, second, it offers several unique on-line databases, which can be reached from any access point. In countries where Dialcom has a presence, users can send E-mail for a fixed rate and they are not taxed with charges for use of the domestic PDN (unless they use it to reach Dialcom's local access point). There is a fixed charge per hour of connection time plus a charge per kilocharacter.

There are extensive information-retrieval capabilities to allow users to reach Dialcom's own databases plus those of third-party vendors. This access can be very powerful to companies that conduct market research from a variety of locations worldwide. Among the information services that can be reached include the Dow Jones News/Retrieval Service, FEDNEWS federal government news, Bureau of National Affairs' BNA on-line tax news and analysis, and the DIALOG bibliographic database. The Electronic News Clipping Service allows a business subscriber to inform the system of the types of news story that are of interest. Subsequently, the appropriate stories are selected for the subscriber, much like an in-house clipping service.

In addition to its international presence, Dialcom is a leading provider of E-mail and information services to the U.S. federal government. As was mentioned previously, Dialcom is a subcontractor to AT&T for the FTS-2000 government network. Dialcom provides an E-mail capability, called FTS Mail, with access to private and government databases. As of 1990, 62 federal agencies were already taking E-mail and information services from Dialcom.

Regarding international standards activity, Dialcom has such an international presence that it is designated as an administrative domain for X.400 message traffic. As an example, the Sheraton Hotels division of ITT has been using Dialcom's X.400 service to allow worldwide staff to keep each other informed on developments. By using the file transfer capability, Sheraton developed an internal financial reporting system, which permits managers of individual hotels to consolidate, balance, and transmit crucial data before it becomes outdated. (This would be precisely the kind of system that Harold Genneen, former chairman of the board of ITT, should have greatly appreciated, as it would have reduced the amount of time that he spent flying between corporate locations to receive this type of information.)

Dialcom and the Overseas Telecommunications Commission (OTC), the Australian government's monopoly international carrier, have a joint venture to provide international E-mail services between Australia and the United States, using an X.400 gateway. The facility will allow users to send facsimile and value-added business services. Telecom Singapore has licensed Dialcom's E-mail system and markets it under the name of Telebox. Users of Telebox within the island city-state of Singapore can exchange messages with subscribers of Dialcom-compatible E-mail systems in other countries. These are examples of how Dialcom is working to enhance its reputation as an international E-mail network.

4.4.3.3 GEIS QUIK-COMM

The QUICK-COMM E-mail service has evolved from a sturdy foundation at a time when time-sharing is moving to the background due to the wide availability to technical people of personal computers and engineering workstations. GE has emphasized its capabilities in EDI and offers specialized services to the automobile industry. GEIS is expending great effort to serve as a gateway among private corporate and government E-mail systems, many of which employ IBM's proprietary PROFS office automation system.

The QUICK-COMM service has an international flavor because many of GEIS's traditional users work for multinational corporations. Half of its 250 customers use the system for international message communication. In Europe, QUICK-COMM has direct access within Denmark, Finland, France, West Germany, Italy, Luxembourg, the Netherlands, Norway, Sweden, and the United Kingdom [Vignault, 1987]. The service offers telex access, yet allows the transfer of large data files used in heavy data processing applications, the kind of flexibility that makes modern E-mail such an attractive messaging application. GEIS has altered its pricing structure from connection-time-oriented to transaction-oriented in response to the general trend among competitors in international E-mail service. GEIS's traditional computer time-sharing services are available through QUICK-COMM, which will be attractive to an organization that has a standardized set of engineering or scientific applications that must be applied, regardless of location. The emphasis on EDI targeted to particular industries is also a strong selling point for QUICK-COMM as an international E-mail offering. With the kind of support needed to maintain and expand QUICK-COMM on a worldwide basis, to have a major player such as GE providing backup is desirable.

4.4.3.4 MCI Mail

When MCI began using television advertising in 1983 to introduce their new service, MCI Mail, many of us wondered if it was really viable. This author had already experimented with CompuServe in 1979, when it was a challenge for an individual just to get a microcomputer to work with a modem. MCI, however, accomplished a great deal by creating the first user-friendly E-mail environment, and making it particularly effective through connection with telex and with the U.S. Postal Service for hard-copy delivery. As mentioned previously, MCI was the first to employ transaction-based pricing; with MCI Mail, users pay for each message unit of 500 characters or less. An initial subscriber base was achieved by giving users of Dow Jones News/Retrieval Service an MCI Mail user identification code and mailbox. During its evolution to become a leading E-mail service for private individual users, MCI Mail has added features such as bulletin boards, databases, business forms, and access to private E-mail systems. Gateways are provided to university and research center messaging environments such as INTERNET and BITNET, making MCI Mail particularly popular with the scientific community.

The user orientation of MCI Mail has been maintained as MCI has sought to broaden its base among business customers. Software packages have been developed by Lotus Development for use on the IBM PC and compatible computers to allow subscribers to exchange virtually any type of computer file or application output via the network. There is a software package for the Apple Macintosh, which maintains the graphic interface (icons, mouse, and pull-down menus) to simplify the user interface. MCI maintains Apple Laserwriters at a number of locations to print text and graphics remotely for subscribers, offering overnight mail or messenger delivery.

Regarding the international environment, MCI Mail is available to telex subscribers throughout the world. In 1988, MCI Mail could be directly accessed in 57 countries [Elbert, 1989]. SprintNet SprintMail subscribers can exchange E-mail messages with MCI Mail users by the X.400 MHS, giving MCI Mail users access to approximately 300,000 mailboxes worldwide. Another 400,000 mailboxes were introduced through a relationship with CompuServe, the leading U.S. videotex service.

4.4.3.5 *SprintMail*

US Sprint has made a substantial commitment to its SprintMail service, which in 1990, represented the largest integrated E-mail structure in North America. Many corporations use SprintMail for internal E-mail services due to extended coverage of the SprintNet PDN and convenient access to telex services. SprintMail provides private messaging as well as office automation systems. Access to the public networks in addition to other private and public E-mail domains is controlled to provide data security, an important function in these days of concern about the transmittal of computer viruses. Reliability is enhanced through the use of the parallel processing capabilities of the Tandem Computer hosts wherein the mail boxes reside.

In addition to basic E-mail, a variety of information services are available to subscribers. SprintMail Gateway is the connection to financial news, travel information, and current events. There are electronic bulletin boards and facilities for creating and using forms. The latter allows an organization to reduce paperwork and exchange formatted information more conveniently than with conventional text messages. Messages can be prepared off line with PC SprintMail, a personal computer program with simplified text editing, and the capability of conveniently sending and receiving messages.

SprintNet is expanding the scope of SprintMail to be global in nature. As discussed previously, SprintMail incorporates the X.400 MHS to allow messages to be passed to outside E-mail systems. This facilitates connection with foreign PTT E-mail systems in Europe and Asia. Under the brand name of SprintMail 400, the interconnection capability will include X.500 directory services to enable users to search a directory and to locate addresses on other public or private messaging systems. The X.500 standard is still under development as of 1990, as discussed previously.

As of 1988, SprintMail direct access was available in the following countries: Australia, Belgium, Canada, Chile, Italy, Japan, Mexico, Norway, the Netherlands, Taiwan, Saudi Arabia, Sweden, and the United Kingdom. The SprintNet PDN is used extensively by the U.S. government as a result of parent US Sprint's responsibility for 40% of the massive FTS-2000 network. In 1989, SprintNet and AT&T interconnected their respective E-mail networks by using X.400. Connection is possible between SprintMail and the office automation E-mail systems of Hewlett-Packard (HP Desk), Wang Laboratories (Wang Office/X.400), Xerox Corporation (Xerox Network Systems X.400 gateway), and SoftSwitch. According to Caswell, SoftSwitch, of King of Prussia, Pennsylvania, has developed a gateway product that can link more than eight different systems without requiring the twenty-eight unique gateways that are theoretically possible. The product, called DAD, translates the format of one system into a standard "envelope" which is conveyed over the gateway to the other where the appropriate format is produced. Importantly, one of the formats that DAD handles is IBM's SNADS, discussed earlier in this chapter. The advantages of using DAD are (a) users need not wait for X.400 to be universally adopted or critical mass to be achieved, and (b) the cost of using a separate X.400 service for bridging purposes is avoided.

SprintMail is a mature service carried by the SprintNet PDN and accessible from other networks including telex, MCI Mail, AT&T Mail, and X.400 gateways to foreign countries. Pricing is based on connection time and number of characters transmitted (as in the case of the X.25 PDN). There are also additional charges for extra addressees.

4.4.3.6 WU EasyLink and OfficeAccess

During the 1980s, Western Union underwent tremendous change as it switched from being a generalized telecommunication common carrier to focusing on business network services. In 1989, in fact, the company basically rid itself of its heavy investment in long-haul transmission facilities. The telex network was seen as one of the elements of strength that Western Union could count on for a considerable time; as stated numerous times, however, telex is fast becoming an outmoded service in the United States and internationally as well. Recognizing this fact, Western Union has found ways to improve upon the shortcomings of telex yet protect the customer base that continues to use telex in everyday business.

In the 1990s, Western Union has a clear focus on the business data customer. During its evolution, Western Union introduced EasyLink in the mid-1980s to allow subscribers to reach mailboxes and telex switches through the PSTN. Telex users quickly found the service to be economical, because inexpensive local dial-up telephone calls replaced the costly dedicated access circuit. The inconvenience of old-fashioned mechanical telex machines is also avoided. Western Union makes available a software package for the IBM or compatible PC that simplifies use of the EasyLink network. As of 1988, EasyLink had approximately 30% of the domestic market in the United States, primarily because of the preexisting telex subscriber base. This amounted to approximately 175,000 mailboxes, transmitting over 9,000,000 messages a month [NTIA, 1988]. Users can send Mailgrams through the U.S. Postal Service for next-day delivery of hard copies. A variety of information services and a thousand on-line databases are accessible via the network, with charges aggregated on the monthly EasyLink statement. The pricing policy of Western Union is to collect an hourly fee for connection to the service, plus extras for third-party information services.

A program called Western Union OfficeAccess was introduced in 1989 to simplify the connection from office automation systems and EasyLink's E-mail capabilities. The type of software package is typical of a generic approach for integrating E-mail environments without resorting to a common standard. OfficeAccess would allow users of IBM PROFS, DEC VAX, and Wang Office systems to send and receive messages to external addressees as if they were internal to the particular system. Thus, it provides a bridge between private E-mail systems and public networks, particularly telex and EasyLink. The proprietary software takes the message and assembles it to employ the EasyLink network without user involvement. Therefore, the OfficeAccess software, running on the user's internal computer system, has been preprogrammed with the necessary directory information for the external users. In addition to EasyLink subscribers, users of OfficeAccess can send telex messages anywhere in the world. Other kinds of messaging include facsimile output and delivery by the U.S. Postal Service through the mailgram and telegram services of Western Union. The charges for Office-Access are separated into an initial software license per system, a monthly subscription charge, which increases with the number of internal mailboxes, and charges for pages actually sent over the EasyLink network.

4.4.3.7 Tymnet OnTyme

As mentioned previously, Tymnet had been acquired by BTI in 1989 and integration with Dialcom is anticipated. The following discussion relates to the OnTyme E-mail service as it existed prior to integration. OnTyme is geared toward Tymnet's existing customer base and gives access to personal computers and host computers for messaging purposes. Pricing is similar to the PDN, based on connection time and the quantity of characters transmitted. The service includes good form capabilities, with the option of allowing the information to be passed over the network to a host on a transparent basis. This capability had not appeared for SprintMail as of 1990.

4.4.3.8 IBM Information Network

The services of the IBM Information Network (IIN) extend to more than 270 U.S. cities and some 125 cities outside of the United States, with full SNA connectivity in 21 countries and PTT-based X.25 connections in another 55 countries. We provide a fairly detailed description of IIN and its services owing to the extensive deployment of the network on a worldwide basis and the importance of IBM computing services in general. Inclusion of this information in the section on E-mail can be considered somewhat arbitrary as IIN encompasses nearly all data networking capabilities. The following discussion is based on information supplied by IBM's IIN staff.

To reach the closest IIN node, users can employ dedicated access lines or dial-up service via the PSTN. The standard interfaces available include SNA/SDLC, *binary-synchronous communication* (BSC), and asynchronous device support. Protocol and speed conversion are also provided, where leased-line data rates can range between 300 b/s and 56 kb/s. As a consequence of this versatility, users can employ IIN resources from a private SNA network (complete with mainframes), a minicomputer such as the AS/400 business computer series, an IBM PC or compatible, or a 3270 IBM terminal. Based on an extensive SNA computing and networking base, IIN offers a broad range of software and services that complement IBM hardware and software products. This, in essence, is the particular advantage of IIN and probably the main motivation for IBM to undertake such an extensive endeavor. To comfort customers, there will always be external connectivity, as IBM provides an X.400 network interconnection between IIN and the E-mail networks of MCI and SprintNet.

The SNA network that supports IIN uses IBM processors and communication controllers located in 20 U.S. cities and in Europe. The interconnecting data lines are all on terrestrial links to minimize propagation delay. Multiple routes between nodes, however, ensure reliable transport of data because a given link can be disabled at any given time. The network is managed from a central location by using a combination of NetView software products plus other software developed especially by IBM for operation of IIN as a worldwide customer service business. For example, the system keeps an inventory of customer attachments and requires customers to notify about any changes in attachment or operating environment. The software arranges for charges to be assigned on the basis of a variety of criteria, some of which are chosen by the customer. Participants maintain control over authorization and sharing of charges, which include message transmission, message storage, network traffic, and dial access.

Begun in 1982 as a domestic extension of IBM's internal nationwide computing network, the facilities of IIN allow IBM's customers to link their terminals, minicomputers and mainframes, and internal SNA networks to remote counterparts in other organizations. Furthermore, customers could use IIN to obtain a variety of services that would not be available on dedicated systems. Examples include information databases, use of IBM software that is too costly or complex to justify on a dedicated basis, and availability of computing power for temporary peak loading and backup. The network is available on a full-time basis and incorporates such reliability features as parallel links, multiple routes, redundant processors and controllers, and enhanced recovery features. Those familiar with SNA will recognize these as characteristics inherent in a fully developed SNA environment. Most importantly, any user, even one having a single node, could connect into a much more elaborate and powerful network than would be justified on a permanent basis. Users can therefore share in an economy of scale because of the size and extent of IIN.

The main divisions of IIN services are Network Services, Information Exchange, Electronic Data Interchange, and International Information Services. These services allow users in separate SNA networks to exchange a variety of information in either real-time or message format. The messages may be standardized, as in EDI, or they may take any form or length, as in file transfer and document transfer. The other main use of IIN services is to extend the reach of a given customer's network because the IIN is so extensive yet so well integrated with the customer's (IBM-manufactured) computer system.

IIN Network Services provides high-speed data communication between a wide variety of remotely located terminals, processors, and PCs, both inside an organization and within other organizations. Network Services enables users to link two SNA networks on an interactive basis, a powerful capability when organizations wish to automate a combined activity. This capability is particularly relevant when the operations of one company or division are merged into another, typically very wrenching for computer-support activities. Indeed, one of the supposed benefits of a merger is increased efficiency of operation, which usually means that administrative functions can be combined. Although IBM systems may be used in the two strategic units, the IIN provides sophisticated conversion routines to perform a manipulation of otherwise incompatible data. The Network Services tools would allow the separate SNA environments to employ each other's application programs and databases as if they were a single network. The problem of consistent addressing of devices and users across an enterprise's boundaries is accomplished by assigning a unique enterprise code to each customer network.

The Information Exchange product is probably the most powerful aspect of IIN, wherein customers can employ software and information databases that are otherwise unavailable to them. The IIN, in effect, is an enterprise network into which any customer can connect. Use is on a "pay as you go" basis, so customers need not make a long-term commitment to a particular class of computing system or software package. Once the service arrangements are made with IBM, any user on a customer's SNA network or anyone with a terminal connected to an IIN node can establish a session with an application on the Information Exchange. The purpose of the service is to send messages to the mailboxes at distant locations, in a manner similar to E-mail. We reviewed the capability of SNADS, which is an underlying IBM transaction system suitable for this function. IIN simplifies the process by helping the user collect the information, arrange it into the proper message format, locate the distant mailbox by the various directory services, and send the message. In addition, a user's mailbox can be checked for any received information, either automatically or on call. Software is available to allow personal computers to exchange files, such as EDI data, spreadsheets, and word processing documents. The expense of using the network is minimized because the PC software can format the information off line before actually sending it.

Electronic Data Interchange is an application for which IIN is particularly well suited. Employing the store-and-forward system of the Information Exchange service, EDI services relay electronic versions of standard business documents to reach any interconnected customer network. Two of the EDI formats implemented are ANSI X12 and UCS, reviewed in the next section. As with the other services, a simple PC is all that is needed to access the EDI facilities of IIN. The PC approach allows a company to apply EDI for selected customers or suppliers without a major investment in dedicated EDI software and computer hardware. In addition to standard document formats, the system can exchange unformatted E-mail messages among different users, such as memoranda, letters, and reports.

The remaining IIN service, International Information Services, is a store-and-forward electronic mail capability to reach IBM customers anywhere on the worldwide network. The service enables the creation, storage, retrieval, and amendment of documents, memoranda, and short messages for distribution to one or more registered users. The trade name for this service is Screenmail, which combines a variety of existing IBM software capabilities, such as DISOSS, with a unique set of user-friendly interfaces. The services can be accessed through a customer's in-house network via PROFS, DISOSS, and the appropriate productivity peripherals for the IBM personal and business computers, such as the System/

36, System/38, and AS/400. Security is provided through multiple levels of identification and passwords, and directory information can be restricted as well. This is particularly important because of the number of private customer networks that can be reached through IIN. Obviously, separate companies that wished to communicate with each other by using Screenmail, EDI, and other IIN services would want to maintain tight control of access to outsiders.

IIN has particular importance in international telecommunication management because of the extent of the IBM customer base. A company with a good SNA network in the United States may find that IIN is the most practical way to extend familiar capabilities to Canada and Europe, the areas where IIN has the greatest presence. Of course, leased lines and international PDNs can be used to reach IIN's nodes in the United States (or directly to the customer's nodes, in fact).

4.4.4 Electronic Data Interchange

Electronic data interchange (EDI) is one of a number of messaging applications that appears to fit the category of E-mail. According to Caswell, EDI is defined to be the exchange of business trade documents such as purchase orders, invoices, and shipping manifests. Another important attribute of an EDI system is that it uses computers to create and process these highly standardized documents in an automated fashion without human intervention. Unlike administrative computer systems, that a company uses to automate its internal operations, EDI is used to exchange documents among companies. The document is in the form of machine-readable code and is generally not a facsimile or image. Recent innovations in digital processing and optical disk storage support the inclusion of images in EDI messages for particular applications.

The technology of EDI has been around in one form or another for two decades, yet, in 1990, EDI still is not very widespread. This is primarily due to the confusion about standards, which is quite critical because the document format must be very precise to be read by the computers of different companies. If done efficiently, EDI can give a company a strategic advantage over its competitors. For example, an EDI system can reduce the time taken to bring a new product to market, or to alter the product to optimize its fit to an existing market. Costs can be reduced as well because a company can place orders for parts and supplies when they are needed, thus reducing inventories. These functions improve the efficiency of international business activities, where documents are needed to transport products, deal with customs in each country, and handle monetary transactions in different currencies.

Let us consider the EDI standards that have achieved a measure of acceptance among major users and service providers. In 1968, the transportation industry took an early lead in EDI by forming the Transportation Data Coordinating Committee (TDCC), which set rules still in common usage. In 1987, TDCC changed its name to TDCC (the Electronic Data Interchange Association) as a move to broaden its industry base.

A second major industry effort toward a standard EDI structure was initiated under the auspices of the American National Standards Institute (ANSI). Although related to TDCC, the X12 committee (written without a period between X and 12) of ANSI maintains what it considers to be the common set of EDI formats [Canright, 1988]. Literally thousands of companies have adopted the X12 EDI standard. As with interconnected E-mail, the X12 standard requires that the computers of different companies be able to communicate with one another. The X12 standard is the most common in the United States and most vendors of hardware and software for EDI support such applications.

We can divide the potential EDI environment into three groups, the first (and original) typified by the proprietary software systems run on a dedicated centralized computer mainframe system for use by one company or association. A good example is the "just-in-time" ordering system of a major retailer like WalMart and an automobile manufacturer like GM, Ford, or Toyota. To support the application, each has a private "star" network with a central host computer at the hub. Consequently, we would not consider this an E-mail network because there is not a diverse messaging environment. The major player runs the system and has complete control over its users. The second group is represented by a cooperative system used to automate record communication in a real-time information environment. The best example is the electronic funds transfer system used by members of the U.S. Federal Reserve System. There are other such EDT systems in particular markets and financial segments. The operators of EFT networks view EDI as a possible market for diversification, building on their strengths and international connections. The design of the EDI network can be completely proprietary as in the single-hub case. To connect such dissimilar networks without a standard EDI document structure would be very difficult.

The apparel industry developed such an EDI service, based on an agreed-upon protocol suite called Voluntary Interindustry Communications Standard (VICS), which was released in 1987. With this standard, smaller retailers and their suppliers can employ a third-party VICS-EDI network or service bureau, thus avoiding the investment and operating expense of a dedicated system. An important innovation in retailing that has facilitated EDI is the use of the universal product code (UPC). This is the series of bars seen on the label of virtually every product found in stores. These bar codes are put on the product by the manufacturer or packager. Modern cash registers are called "point-of-sale" (POS) terminals, having the ability to read the UPC by using a laser scanning device. The information from all of the POS terminals in a given store is gathered and then transmitted to a central clearinghouse to be used for accounting and reordering purposes. The

Uniform Code Council administers the EDI format, called UCS, which is used with the POS and UPC systems. An EDI message containing the order for more products is automatically sent to the manufacturer at the appropriate time. The manufacturer or wholesaler subsequently identifies each shipment with a delivery code, again using bar codes. This simplifies the routing of the package when it arrives at the loading dock of the retailer. The whole process is nearly a closed loop, reducing time and cost for the retailer and supplier. With the reduced response time of the supply system, some retailers claim an efficiency improvement in the range of 20 to 40% after EDI is introduced.

Major retailers have been compelled to use these automated ordering and accounting systems to remain profitable in very competitive markets, where margins are thin. The largest retailers can build their own proprietary EDI systems, without using industry or international standards for message format and routing. Smaller suppliers are effectively locked into their particular retailers, which is an advantage for the retailers, giving them considerable buying power over suppliers. Standardized EDI, however, would allow suppliers to gain leverage, while permitting them to expand their markets.

The final group is the newest and fastest growing: the public EDI service operating on a distributed computer network. Such a third-party organization solves many of the problems encountered when companies try to build EDI systems through a committee. Public data networks, E-mail services, and computer service bureaus view EDI as a growth market opportunity. To be able to enter this market, however, the standards by which users will access these networks and networks will exchange EDI information need to be carefully developed. IBM announced X12 interface products in 1989, including ExpEDIte for the IBM Personal System/2 and minicomputers in the System 36, 38, and AS/400 series. This software product also can interface with the United Nations standard EDIFACT, discussed below.

U.S. companies have a decided advantage over counterparts in the industrialized countries of Europe and Asia. Recently, the Republic of Korea has begun to adopt X12 because it is a mature standard with software available for the same kinds of industries as exist there and in North America. Furthermore, the United States is Korea's primary trading partner, making it very logical to use the same EDI system to automate the document exchange process.

The situation in Europe is quite different, where governments and companies are seeking to protect various European interests. A different tack is being taken, following the guidelines of the Open Systems Interconnection model. This is attractive from a standards-making perspective. The particular EDI environment is called Electronic Data Interchange for Administration, Commerce and Transport (EDIFACT). EDIFACT is a standard for business messages that has the sanction of the United Nations and the International Organization for Standardization. Quite obviously, X12 and EDIFACT are not compatible with each other, except that the underlying information is basically the same. We thus have the traditional standoff between North America and its trading partners in Asia and South America *versus* Europe and its trading partners elsewhere.

The European Community has seized upon EDI as an element to make Europe more competitive in the post-1992 environment. An organization called EDI/Unite has been established to rationalize the differing EDI systems in different countries to help meet the demand for cross-border electronic trading. As reported in *Communications Week,* they see a common EDI format as a strategic vehicle for harmonizing the community. Two examples of cross-border data interchange have been cited in 1989. One, oddly enough, involves the U.S. company, Electronic Data Systems Corporation (a GM subsidiary), which is branching out from the United Kingdom onto the continent. The other is a chemical industry group called CEFIC. Characteristically, EDI/Unite will push for early adoption and rapid introduction of the EDIFACT standard, both as a way to bring the European Community together and to protect it from the outside. The disadvantage of this strategy is that much more exists in the way of working software products for ANSI X12 than for EDIFACT. Major computer companies and PDNs, however, are bringing interfaces to market to support EDIFACT.

Fortunately, the X.400 MHS protocol suite can operate as the underlying structure for EDI as it does for interpersonal messaging. With proper coding and formatting, EDI data can be transferred on a transparent basis. Perhaps, at some point in the 1990s, the world will gravitate toward one standard. So long as commercial companies and other organizations can conduct their business, however, that there are a number of EDI standards does not matter.

The PDN and E-mail network suppliers of the United States recognize the opportunity that EDI presents, and they are moving rapidly to gain market share. SprintNet is well positioned with its X.25 PDN and SprintMail 400 services to expand in this direction. In 1988, Southern California Edison announced that it had selected SprintNet to provide the backbone for its expansion of EDI. SoCalEd will use EDI to process 20,000 annual purchase orders, linking its purchasing system to 100 suppliers.

Another important vendor of EDI services is IBM Information Network. A review of the range of IIN capabilities is presented earlier in this chapter under the topic of E-mail services. IBM itself uses the network for its EDI Services, connecting to suppliers and dealers throughout the world. Branch marketing offices of IBM are also interconnected. As of 1990, there were plans to include facilities for direct customer ordering and electronic invoicing of IBM products. This network can obviously be interconnected with customers' computer networks, allowing all parties to participate in a growing environment. Access is provided to X.25 networks operated by PTTs in other countries.

As of 1990, the entire EDI marketplace is viewed as less than $100 million of gross revenue. This is not an impressive number when compared to the billions spent on telephone and data communication. EDI is a rapidly growing networking application, however, with the total size of the market doubling in one or two years at a time. Measurable benefits have been quantified by the companies and industries that have adopted EDI. Internationally, the industrialized countries must make EDI services and facilities available so that their indigenous companies can participate in world markets. Japan and the Republic of Korea have been particularly good at recognizing this need and have taken concrete steps to help their industries. The European desire to promote Europe probably needs to be more enlightened because in reality EDI knows no borders. So long as basic data networking principles are followed (such as using the X.25 and X.400 standards), EDI information in whatever format can be routed efficiently over the borders. This will give the advantage to the companies and industries that take the necessary steps to employ proven EDI systems and techniques.

4.5 INFORMATION SERVICES (VIDEOTEX)

Computer-based information services, covering fields ranging from legal research to historical financial data, have been evolving over the years in a variety of forms and various places throughout the world. Typically, a number of information service providers maintain specialized databases on one or more central computers, allowing users to interact with and to select information, using their own display terminals. In the United States, information services mostly use a text interface for which any standard video terminal or personal computer display can be used. The services are provided by literally hundreds of suppliers, which either build their own databases and hosts, or act as a third-party source of data for a service bureau. Videotex is often defined as a specific form of information service that uses a graphic display to add a more pleasing dimension. The quality of the display varies greatly, from rough low-resolution images in black-and-white to high-resolution color graphic images that look like magazine pages. Major information service providers, including PTTs, employ videotex technology to encourage subscribers to obtain and use the service because the majority of the population is typically not familiar with the use of computer terminals.

Our intention here is to consider information services in general and videotex in particular as an application in the international telecommunication environment. Business uses of the more popular consumer forms of videotex are limited because of the consumer orientation of the information and the simplified accessing process. For example, the research department of a corporation needs an information service, such as DIALOG, which can access a great deal of information and has excellent search capabilities. The display approach is secondary, and in fact the user will rather have a clear printout of the results of the search than have a pleasing display in color. The user of a commercial videotex system such as Prodigy, however, wants to be entertained at the same time that information is examined. Prodigy began in 1988 as a user-friendly videotex service, using color graphics with advertising inserts in the display. The promoters of Prodigy are looking for advertising revenue and customer sales of the products offered on the system.

The successful information services in the United States are all textual in nature. For home users, CompuServe and the Source (acquired by CompuServe in 1989) are still the leaders with roughly 300,000 users each. Dow Jones News/ Retrieval is popular among people with a financial interest, and can be accessed directly or through other networks such as Dialcom. By combining a mix of services and capabilities, the surviving players can appeal to as many segments of the home computer market as possible, thus achieving acceptable revenues. A true consumer service not targeted to home computer users has yet to reach a profitable state.

Walter Vignault has identified five different videotex standards in use in different countries and regions [Vignault, 1987]. These standards are as follows: Prestel, developed in the United Kingdom and in use in some other European countries; Teletel, the French system discussed below; CEPT, the European standard, first adopted for the West German videotex system called Bildschirmtext; NAPLPS, the North American Presentation Level Protocol Syntax, first used in Canada; and CAPTAIN, the Japanese videotex standard, meaning Character and Pattern Telephone Access Information Network. Vignault does an excellent job of identifying the various systems throughout the world and evaluating their capabilities.

Strategic units can consider videotex as a means to reach a customer base — the consumer. The French Teletel has achieved such prominence due to the extensive diffusion of Minitel, the consumer version of the national videotex service. Minitel terminals were provided to homes for access to an on-line telephone directory, which for many was the only way to obtain this kind of information. This approach forced people to gain a limited amount of computer literacy. The Minitel terminal can be used to order from a wide range of products offered by a multitude of stores and businesses, including grocery items, on-line news and information, and travel. As of 1988, the number of Minitel terminals in use reached 4.2 million with access provided to more than 10,000 service providers. Teletel, conversely, is very extensively used by businesses to connect to information services offered by third parties.

In 1989, France Telecom took steps to expand the access to Teletel by information providers. Teletel uses the TRANSPAC PDN as the backbone network. The X.29 application layer protocol is used on top of X.25 for the basic link and transport layers. Service providers enter the X.25 network via the higher level protocols, and can thus establish a variety of parameters to optimize service delivery. Among the features introduced in 1989, service providers can reroute connections so that subscribers can reach them under a variety of circumstances. One provider can accept an order for a service, separate the order into components, and subsequently retransmit portions on which other service providers may act. An example is when a prospective vacationer uses an on-line travel agent to book an airline flight, a hotel room, and a rental car. The rerouting feature permits the travel agent to engage even very small hotels by the same network.

As discussed in our previous book [Elbert, 1989], Minitel is approaching critical mass within France and is now being exported to other countries including the United States. A company called Minitel Services Company (MSC) was formed in 1989 as a joint venture of Infonet and France Telecom. The purpose of MSC is to exploit the Minitel technology in the United States and build a very successful videotex business through the recruitment of entrepreneurial information service providers. Infonet (which has a 51% interest in MSC) is responsible for developing the support base and marketing of the information services, which are carried via the Infonet PDN.

Access to information services on an international basis can be problematic because of the wide variety of user interfaces and language differences. From a networking standpoint, most of the videotex and other information services use the PSTN directly or in combination with a PDN. This is important because to make the service profitable the supplier needs to achieve as many users as possible. Politically, videotex is much like broadcasting because individual subscribers can receive information virtually without regard to borders between countries. The European Commission recognized this characteristic in a ruling intended to open borders in 1990 to these services and to allow competition between government and private VAN ventures. The variety of standards presents a barrier to growth of videotex, but at least the more successful services will have a better political environment in which to grow.

Chapter 5
PRIVATE OWNERSHIP OF NETWORKS AND CUSTOMER-PREMISES EQUIPMENT

Private telecommunication networks encompass a wide variety of services provided by third parties as well as equipment and facilities owned and operated by the strategic unit. In the United States, considerable attention has been focused in recent years on intelligent devices that users employ to increase the utility and efficiency of private telecommunication networks. Many offer capabilities that cannot be matched by public networks, whether circuit-switched or packet-switched. In addition, the strategic unit has the option of creating its own circuit-switched or packet-switched environment, and many organizations have made substantial commitments to such facilities.

This chapter focuses on how a strategic unit can add such privately owned capabilities to deliver telecommunication services to a diversity of locations. We cover LANs, *wide area networks* (WANs), PBXs, modems (dedicated and dial-up), digital facsimile (Groups 3 and 4), business video, video teleconferencing, and ISDN terminals. Transmission equipment for global network development, including a variety of multiplexers and packet switching, is covered in Chapter 6.

Implementing a private telecommunication network by acquiring dedicated facilities is a strategic decision, regardless of whether it is to be international in scope. The facilities that are purchased represent a capital investment, which must be operated by the owning organization and "written off" over an extended period of time, typically five to ten years. The purchase usually must be financed by the organization, adding an interest expense. If either the facility does not work as intended or the need changes drastically in the near term, there may be no way to recover the originally invested capital.

Selection of a given type of equipment, such as a packet switch or T1 multiplexer (reviewed in Chapter 6), usually means that products of a particular manufacturer must be used throughout the network. Additional quantities will be acquired in the future when the network is expanded. This is because intermediate nodes and terminating devices must be compatible with each other, which is more easily accomplished if a common vendor is used. Of course, international standards can be rigorously followed in an attempt to create other than vendor-specific networks, although this approach will almost always restrict the number of options. Therefore, the strategic unit must consider issues such as operational flexibility, ease of expansion, vendor reliability, and longevity. For an international network, vendor support outside of the home country is also a major consideration. These factors should make a telecommunication manager pause before embarking on a major capital commitment for installations in several foreign countries. Prior to making any commitment, careful consideration should be given to using a well connected international carrier or PDN for vital communication needs.

In this chapter, we assume that the decision has been made to build a private network using owned facilities for connections outside of the home country. The types of equipment and facilities that are discussed in this chapter and the next are commonplace in North America. Many of the constituents are already available in Japan and Europe. Under certain circumstances and with due permission of the local TA (if necessary), some of the more sophisticated devices can be brought into developing countries. Whether a developed or developing country, we must consider the problems of transporting, implementing, and operating the network facilities under potentially hostile environmental or political conditions. The approach and rationale for doing so are reviewed in Chapter 8.

5.1 DIGITAL INTEGRATION OF VOICE, DATA, AND VIDEO COMMUNICATION

We review briefly the process by which various forms of digital information can be combined, or integrated, for efficient transmission over long-haul media such as transoceanic fiber optic cable and INTELSAT satellite links. This topic is explored in detail in Chapter 4 of [Elbert, 1989]; a brief review is provided in the following section. Our focus is on the international aspects of the techniques available to the telecommunication manager.

The trend from analog to digital communication has taken its full course in the industrialized countries; a similar trend is underway in other nations which are in the process of upgrading their domestic networks. Likewise, the satellite systems and undersea fiber optic cable links between countries are now essentially digital in nature. Strategic units can take advantage of this infrastructure to implement sophisticated domestic and global private networks. Quite obviously, this is more easily done in some countries than in others, because of the stage of development of the domestic public network and the degree of freedom which users enjoy as to what may be connected to the network.

Digital integration is possible because data from a variety of sources can be combined in binary form. The familiar techniques include time division multiplexing (TDM); statistical multiplexing (STAT MUX); packet switching; time division multiple access (TDMA) on satellite links; and the contention-mode and token-passing access methods used on local area networks. Each type of information — voice, low and medium speed data, facsimile, digitally compressed video, *et cetera* — has particularities which the network must accommodate. Due to differences in bandwidth, timing, duration, and routing of information among these applications, digital integration is no simple matter. In fact, no standard architecture has been deployed that does everything for everyone. This should not prevent the more innovative organizations from trying, and efforts such as ISDN and OSI have made significant progress toward a universal structure. The fact remains, however, that users may, due to necessity, take on the additional risk of introducing new technology or approaches. This is particularly the case in international telecommunication management, where the foreign TA may be unfamiliar with what are normal practices in North America, Europe, and Japan.

Figure 5.1 presents an overview representation of the kinds of modern digital telecommunication equipment that can be used within an international private telecommunication network. We assume for simplicity that there are two localities which are connected by a variety of international private lines. Along the vertical edges are different classes of user devices connected to switching or multiplexing equipment used at nodes in the network topology. Data communication devices can access the topology at any type of node along the left or right; each access point (i.e., through the LAN, PSN, T1 MUX, or PBX), however, brings with it a different set of constraints. Voice services enter the international network through a PBX or key system, shown at the top. High-speed data, Group 4 fax, and digital video applications typically enter the network T1 or E1 backbone through the high-level multiplexer.

Here and in the following chapter, we review the capabilities of these devices and identify many of the leading vendors in the international marketplace. We consider voice, fax, and data terminal devices; digital switching systems; and multiplexing systems, all of which convey digital traffic in various forms. Quite obviously, data processing systems are an important ingredient in the international telecommunication picture; those of IBM and DEC are the most common. Rather than repeating the discussion of these environments, we refer readers wishing more background to other works such as [Schwartz, 1987] and [Elbert, 1989]. The information that follows is current as of 1989, so readers should be advised that the particular makeup of the vendor listings will evolve over time. Most of the companies, however, have mature product lines and either are typically part of major corporations or are affiliated with them.

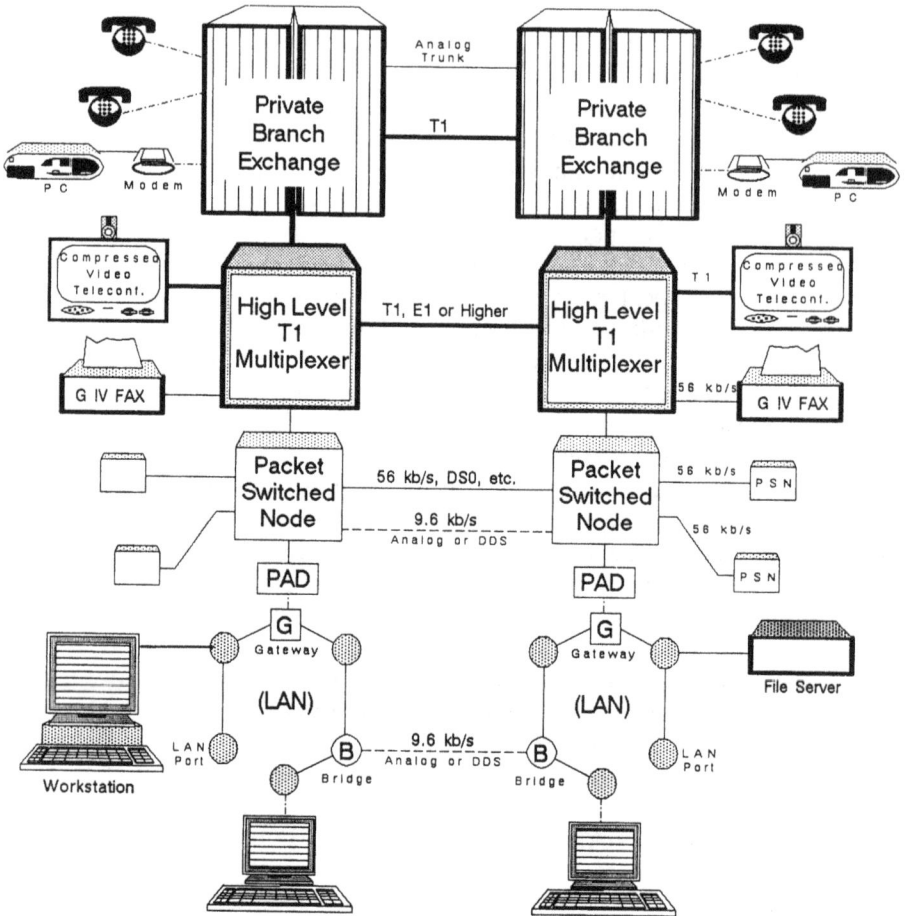

Figure 5.1 Equipment hierarchy in an international private telecommuncation network.

5.1.1 Local Area Networks

Local area networks, depicted at the bottom of Figure 5.1, are growing rapidly in importance as a means to exchange data among users and to employ peripherals such as a laser printer, file server, optical storage system, or page scanner. Throughout North America, Europe, and the developed countries of Asia, LANs have become popular because of the flexibility offered for sharing devices and employing the resources of larger computing systems. Appropriate media for implementing LANs within buildings and industrial parks include coaxial cable; shielded twisted-pair or, under certain circumstances, unshielded twisted-pair cable; and fiber optic cable. *Local area wireless networks* (LAWNs) have also been introduced, employing radio transmission within line-of-sight range. A personal computer or intelligent terminal with appropriate software can be connected inexpensively with a simple LAN tap.

The specific technologies used to transfer packetized data over a LAN are many and varied. As discussed in [Schwartz, 1987], the three most common access mechanisms are *carrier sense multiple access with collision detection* (CSMA/CD); token passing on a cable ring (token ring); and token passing on a bus topology (token bus). Key factors in determining LAN performance include the data rate on the LAN, the maximum distance between access points, and the manner in which contention for the medium is controlled.

The range of possibilities for LAN design and manufacture is extremely wide, owing to the rapidly evolving nature of this burgeoning industry. A brief listing of vendors is provided in Table 5.1, broken down into broad categories of LAN performance. The first LAN to appear on the market was Ethernet, a CSMA/CD scheme developed by Xerox Corporation. Low-speed LANs are geared to relatively small user groups employing conventional personal computers and dumb terminals. Digital Equipment Corporation has achieved a major position through their support of the Ethernet standard. Another large group is within the domain of IBM, which introduced the token ring system for extended IBM SNA environments. Where greater capacity is needed to conduct applications such as computer-aided design and computer-aided manufacturing (CAD/CAM) or very high speed computer simulation, medium speed LANs come into play. This market was still specialized in 1989, with Network Systems Corporation and Proteon having achieved significant user acceptance. The higher speed LANs are still in the developmental stage, relying on fiber optic cable as the transmission medium. One of the more promising standards is called Fiber Distributed Data Interface (FDDI), which is a token ring on doubly redundant fiber optic cable running at 100 Mb/s. Some of the first LAN products employing FDDI are being offered by Apollo Computer and Fibronics.

The continued growth of LANs is ensured by the vitality of the personal computer and high-performance computer workstation, now capable of serving business needs which were only addressed by minicomputers or mainframes in the recent past. Existing low and medium speed LAN technology, however, has significant performance limitations as to the geographic size of the network, the number of users that can be connected, and the amount of data that can reasonably be pumped through it. Some of the basic difficulties are being addressed by increasing the internal data rate of the LAN and by using more effective access protocols. A natural way to grow a LAN so that users in a wider area can utilize a greater set of resources, however, is to install multiple LANs and employ an interconnection of some type to transfer only the appropriate external traffic [Keok, 1989].

There are two classes of devices used to interconnect LANs; these are called *bridges* and *routers* [McQuillan, 1988]. First introduced under the name TransLAN

Table 5.1 Suppliers of Local Area Networks for Data Communication

Low-Speed LANs
(10 Mb/s and under)

Apple Computer, Inc.
AT&T
Banyan
Bull
Digital Equipment Corporation
Hughes LAN Systems (formerly Sytek)
IBM
Micom Interlan
Novell (including Exelan)
Proteon
Standard Microsystems
Tandem Computer, Inc. (including Ungermann Bass)
3Com Corporation
Xerox Corporation

Medium-Speed LANs
(10 to 50 Mb/s)

Digital Equipment Corporation
FiberLAN
Network Systems Corporation
Northern Telecom, Inc.
Proteon

High-Speed LANs
(50 to 150 Mb/s)

Fibronics (FDDI Standard)
Hewlett-Packard (including Apollo Computer)
QPSX, Inc. (IEEE 802.6 Standard)

by Vitalink Corporation and Digital Equipment Corporation, the bridge provides a direct connection, transferring only those packets of data which are not for local devices. Its operation is based on simplified packet routing using the link level of a generalized layered architecture. As shown at the bottom of Figure 5.1, two LANs would be connected by a point-to-point link. The bridge on each side of the link monitors its associated LAN to compile an address database of local devices. This is a continuous process and only requires a few seconds of activity for the database to become reasonably complete. When one of the devices sends a packet to an address not in the database, the bridge recognizes it as foreign and simply transmits it over the point-to-point link to the other LAN. The bridge on the other end applies the message to its associated LAN for carriage to the appropriate device. This operation is quick and transparent to the user. New generations of bridges have the ability to configure the transfer of packets for complex external network topologies automatically, not just point-to-point as in the example.

A router operates at the network layer, using a higher level protocol such as the Internet Protocol of TCP/IP. Due to the rigor of the network layer, the router will only send a properly addressed packet to the distant LAN. The router may send the data through the same type of point-to-point link as used by the bridge or it may access an external network such as an X.25 packet-switched network or a circuit-switched network. Any device on a LAN can reach an associated device on a distant LAN through a bridge without regard to the network layer protocols in use on either LAN. As an example, Network Systems produces a router that can transfer TCP/IP and DECnet traffic via a T1 backbone. Because the ultimate purpose of bridges and routers is basically the same, there is considerable debate about the superiority of one technique over the other.

A gateway, shown at the top of the LAN in Figure 5.1, can be used to build a wide area network capability to interconnect local users with a great variety of distant parties. In this example, the gateway reaches into a privately owned packet-switched network that is operated by the strategic unit. Alternatively, the packet-switching facilities of a PDN could be employed. As discussed in Chapter 4, a packet assembler-disassembler is required to access the packet-switched network if the DTE function is not built into the gateway.

5.1.2 Wide Area Networks

Wide-area data networking capabilities are commonly built for multiple LANs and external computer systems. The user data from the LAN is routed to other addresses in a domestic packet-switched network by the packet-switched node. The PSN in Figure 5.1 is connected to a PSN in the distant network on the right using a dedicated private line, which may be either analog or digital (DDS). This circuit would carry the messages of numerous users and not just those on the LAN shown below. As discussed elsewhere, the protocol on the line between PSNs could either be X.25 or X.75, the latter being the standard for internetworking of public X.25 networks.

The private line circuit can be implemented as a digital time slot assignment provided by the high-level T1 multiplexer shown above the PSN. In a real T1 backbone network, several multiplexers are arranged in a partial or full mesh topology, employing high-speed T1 private line circuits. User data from the LAN and other PSNs are integrated by the multiplexer along with traffic from voice, video, and facsimile users. This develops a greater degree of service integration and can yield excellent economies of scale, provided that sufficient digital traffic is applied. The majority of high-level T1 multiplexers use extensions of the decades-old TDM approach, found extensively in public telephone networks. Another technique, called *fast packet switching,* is also gaining acceptance.

As shown in Figure 5.1, the T1 multiplexer combines a wide variety of digitized communication services with conventional telephone services provided by the PBX. While not strictly a higher level digital communication device, a modern PBX performs an integrating function by combining circuit-switched channels into a T1 digital stream. Individual analog voice-grade trunks may be used when only a few circuits are required. PBX manufacturers are now providing ISDN capabilities, such as the basic rate interface, on digital PBXs for high-speed digital devices such as Group 4 facsimile machines and compressed video. Support of the primary rate interface has also become a popular option.

The hierarchy in Figure 5.1 shows how today's telecommunication applications can be integrated into a diverse network environment. Customer premises applications are reviewed in the following paragraphs; a discussion of network node systems is found in Chapter 6. A given strategic unit will employ some of these elements, relying on the public networks to satisfy the remaining needs. An interesting trend worth paying attention to is the combining of two or more of the directly connected network elements into a single node. This is, in fact, what the high-level T1 multiplexer represents because it integrates narrow-band and wide-band applications and carries them over a thick backbone. Even more sophistication is required to combine the T1 multiplexer with the PSN, an approach being taken by a few equipment manufacturers. The *integrated services switch* (ISS #1) manufactured by Netrix of Herndon, Virginia, provides means to combine time slot switching (akin to circuit switching) with packet switching. Because of the complexity of the problem, such combinations may or may not add value for telecommunication networks. As an example of this difficulty, the Meridian Digital Network System announced by NTI in 1988 was subsequently withdrawn from the market in 1989.

From this overview, we move into a discussion of the devices found on the customer's premises. Strategic units employ such devices for telecommunication applications, like providing telephone service, sending and receiving E-mail, accessing databases, engaging in video teleconferencing, and transmitting high-quality graphic images.

5.2 CUSTOMER PREMISES EQUIPMENT

Customer premises equipment (CPE) can be as trivial as a push button telephone or as complex as a television studio. Basically, CPE consists of any products which a telecommunication user purchases, leases, or rents from the local telco, PTT, or other provider. Users often purchase and install CPE themselves to take advantage of lower cost and technology not available to the general public. Typically included are telephone instruments and handsets, acoustical couplers, telephone answering machines, modems, multiplexers, data terminals, and facsimile machines. We focus on the more sophisticated digital devices that strategic units rely upon for modern business communication. In many countries, the PTT extends its monopoly to the provision of CPE, making innovation through the use of more advanced telecommunication apparatuses difficult for strategic units. Countries that have burgeoning telephone equipment manufacturing industries often resist opening up their markets from fear of loss of sales to imports. This protective attitude is rapidly being broken down, however, particularly in Europe, because of the economic clout of major using organizations and the political power of the European Communities. Nonetheless, PTTs will maintain technical control by requiring pretesting and certification of CPE (a process called type acceptance) before it can be connected to the network.

Any vendor wishing to sell products in France, for example, to be connected to the public network must satisfy the applicable interface specifications. The designated agency tests the unit to ensure that the requirements are, in fact, met. As part of opening up the markets in Europe, a single agency called the European Telecommunication Standards Institute (ETSI) will set standards and perform type acceptance testing.

5.2.1 Telephone Station Equipment

The basic telephone instrument is probably manufactured or assembled in more countries of the world than any other electronic device. Technical specifications of this device are nearly universal through the efforts of the CCITT. Also, most instrument designs are based on original products of major manufacturers such as AT&T, NEC, Alcatel, and Siemens. These companies typically set up local manufacturing operations or joint ventures with indigenous companies in developing countries to satisfy government interest in domestic manufacture. Whether users must rent instruments from the PTT or whether they may purchase them from the PTT or an outside vendor is a matter of local policy. In a number of cases, the PTT requires only that the user must obtain the first telephone on the premises from the PTT, with additional units coming from private sources. For a small office operation in a developing country, these restrictions would not pose much of a problem. If a strategic unit wished to establish a subsidiary headquarters or telemarketing operation, however, PTT restrictions on telephone instruments could be a major problem. Because local policies are rapidly evolving, telecommunication managers should determine the situation in each country where new operations are contemplated. It can be assumed that any user-provided station equipment must be type-accepted by the PTT or other authorized body.

The quality of local and international telephone service is affected by the instruments connected to the local loop. As a technical point regarding basic telephone instrument design, the oldest equipment uses electroacoustic transmitters and receivers [Bell, 1983]. The performance of the internal carbon element transmitter is variable, as many of us remember from days gone by. On an international network, older telephone instruments can cause problems with talking levels and intelligibility. Another aspect of older telephones is the inadequacy of the built-in equalization technique that deals with undefined loop lengths and wire resistances. The telephone must operate with different types of wire and under various moisture conditions, which can effectively short-circuit a conversation in the rainy season. Also, the telephone exchange and its particular switching system is a factor in how the speech will be conveyed. The more modern electronic telephone, now the standard design, is better able to deal with this environment because it contains active electronic circuitry to compensate for variable loop conditions. The electronic telephone improves overall transmission performance of the system, and the instrument is more reliable because it has fewer parts to wear out.

Reviewing again the equipment hierarchy shown in Figure 5.1, we see that the situation with regard to extensions on a PBX (called *stations*) is quite different. Telephone equipment is connected directly to the PBX for access to domestic and international PSTN services. In most digital PBX designs, the station incorporates features not found on standard telephones, such as speakerphones, data communication access ports, push buttons for additional lines, intercoms, character displays, and other signaling keys. Because these features are unique in some cases, stations usually must be obtained from the PBX manufacturer according to various option packages (at additional cost). Taking this one step further toward ISDN, digital telephones wherein the voice is first converted to digital data in PCM or other format would certainly not be found on the local market in developing countries. Assuming that the PTT allows the user to purchase such a PBX, the station equipment would also have to be permitted or else many of the benefits of modern PBX technology would be lost.

A class of telephone important to retail business is the transaction telephone, which can automate credit verification, authorization on credit-card purchases, inventory control, and electronic funds transfer. Similarly, autodialers are useful telephone instruments that are activated with a pushbutton. Some autodialers are themselves operated automatically as part of a PBX, to convert from abbreviated dialing to the appropriate domestic or international dialing system. This is not unlike building your own software defined network. Since the functions performed by the transaction telephone and autodialer are definitely business-related, getting PTT approval can be crucial to a particular foreign end of an international private telecommunication network. More sophisticated business telephone CPE is employed by traders in foreign exchange, commodity, and securities markets.

There is an interim step between the use of individual telephones on the PSTN and the application of a fully functional PBX. This is the key system, which is a simplified switching unit to allow several telephone stations to share a set of public access lines. A complete discussion of modern key systems is provided in [Elbert, 1989]. The original purpose of the key system was to allow one telephone to handle multiple incoming lines. Also, a key system permits users within an office complex to talk to each other over an intercom without placing calls through the telco's switch. The telephone instruments have many of the advanced features found with PBXs. The central element, called a *key system unit* (KSU), contains line switching equipment and provides power to the instruments. Telephone access lines terminate on the KSU.

Recently introduced "hybrid" key systems include microprocessors in the KSU, giving the system computing power for more advanced services like least-cost routing and automatic call diversion. These are features formerly only found in more expensive PBXs. Telcos and PTTs should prefer key systems over PBXs for the simple reason that key systems rely on the switching capabilities of the local public telephone exchange. Principal among today's applications for the key system is its use as a front-end for private switching capability provided by the TA (referred to as CENTREX in North America). Key system technology is relatively easy to reproduce since the complexity tends to be lower than that of the PBX or other intelligent device. Hence, key systems are manufactured in and exported from the Republic of Korea, Hong Kong, and Taiwan.

5.2.2 Voice-Band and Wideband Modems

The purpose of a modem is to transfer digital data from a terminal or computing device through the bandwidth of an analog transmission medium such as a voice circuit or radio link. The word "modem" is simply a contraction of modulator-demodulator, indicating that it is functionally two pieces of equipment. The modulator section converts an input digital data stream to an analog signal using a bandwidth-efficient modulation technique such as *phase-shift keying* (PSK); the demodulator section recovers the data again from the analog signal. The two complementary functions are typically combined into one package for ease of connection between the source and destination of the digital data and the access to the analog transmission medium. This also facilitates other functions such as timing, synchronization, error control, data compression, and encryption. A list of leading modem manufacturers is presented in Table 5.2.

One of the main impediments to full use of voice-band modems on the worldwide PSTN is the policy of most PTTs to supply or own CPE. This is particularly the case with modems, perhaps because the rental of modems is a growing source of revenue to PTTs in developing and developed countries alike. Aside from the inflated cost of such rental, users can get by as long as the more conventional CCITT Recommendations are followed. The problem is accentuated

Table 5.2 Suppliers of Modems for Data Transmission Over Analog Links

Voice-Band Modems
Alcatel
AT&T
Codex Corporation (Division of Motorola)
Fujitsu
Hayes Microsystems
IBM
Microcom
NEC Corporation
Paradyne
Racal Vadic, Inc.
Siemens

Wideband Modems
Alcatel
Codex Corporation
Comstream
Fairchild Data Corporation
Fujitsu
Harris Corporation
Hughes Network Systems
IBM
NEC Corporation
Racal Vadic, Inc.
Siemens

when a user needs to push the performance limit by introducing specialized voice band modems like the V.42 standard capable of 19,200 b/s.

Modems play a vital role in international data communication because they work through the analog telephone networks of the world. Voice-band modems typically have data transfer rates in the range of 1200 to 9600 b/s. Digital compression can increase the throughput by a factor of two to four, depending on the composition of the data (i.e., the amount of redundancy in the data) and the particular compression algorithm. A pair of voice-band modems, one on each end of the circuit, provides a link-level connection for any variety of asynchronous and synchronous applications. In a subsequent section, we will review Group 3 fax, which is one of the principal applications of voice-band modems. The telecommunication manager has the same options on international and domestic telephone networks for setting up link: either through dial-up service over the PSTN or through an analog private line leased from carriers. Availability of these services on a global basis is treated in Chapter 3.

The following discussion of modem characteristics is fairly detailed and will be of greatest interest to technically oriented readers. Table 5.3 presents a brief

Table 5.3 Summary of Voice-band Modem Characteristics

Standard	Speed (kb/s)	Modulation	4-Wire or Dial-up 2-Wire	Modes
Bell 103, CCITT V.21	300	FSK	Dial-up	Duplex
Bell 202, CCITT V.23	1200	FSK	4-wire, Dial-up	Full-duplex, Half-duplex
Bell 212A	1200	DPSK	Dial-up	Duplex, self-equalizing
Bell 208, CCITT V.27	4800	DPSK	4-wire, Dial-up	Duplex, Half-duplex
CCITT V.22 bis	4800	QAM	Dial-up	Duplex, self-equalizing
CCITT V.29	9600	QAM	4-wire	Duplex
CCITT V.32	9600	QAM, Echo cancellation	Dial-up	Full-duplex, self-equalizing
	19,200*	QAM, TCM	4-wire	Half-duplex
CCITT V.42*	19,200*	QAM, TCM, plus Error correction and Data compression	4-wire, Dial-up	Full-duplex Half-duplex

* Not adopted as standards as of January 1990.

summary of key characteristics of voice-band modem standards for speeds in the range of 300 to 19,200 b/s; all are in common use as of 1989. Indicated in the left column is the CCITT Recommendation number for each particular modem standard, with Bell System designations also shown for earlier low-speed designs. Speed is quoted in bits per second, which is different from the baud rate designation more common in the past. The distinction is that baud refers to the transfer rate of the modem in terms of the number of symbols per second. A symbol is a constant condition on the analog transmission line, where the amplitude and phase of the modulated signal are held constant. The maximum symbol rate is limited by the bandwidth of the telephone channel, which is fixed at 3100 Hz. To implement higher rates of transfer, each symbol contains multiple bits so that the channel bandwidth is not exceeded. In modem designs operating at 2400 b/s and less, the baud rate and the data rate (in b/s) are the same. Data rates higher than 2400 b/s require the use of complex modulation formats which combine amplitude variations along with the more reliable frequency and phase variations found in older designs. A 9600 b/s modem with a baud rate of 2400 encodes four bits per symbol, increasing the data rate over the telephone channel. Generally speaking, the higher the data rate, the more complicated the voice-band modem design. The third column of the table indicates the type of modulation specified in each recommendation. For more information on modem modulation formats see [Humphrey, 1988].

When using a high-speed modem, the technically easiest approach is to employ a four-wire private line circuit because the transmitting and receiving directions are carried on separate paths from end to end. The circuit is established ahead of time and would not change in its characteristics. Because four-wire circuits must be leased from the TA on long-term basis, their cost to the user may not be justified if only short data communication sessions are required. Another difficulty is that they are subject to outages of significant duration. A two-wire circuit, which results from a dial-up connection, brings with it the complexity of dealing with variable transmission parameters and echo on any telephone connection. (Actually, two wires are used on the local loop, as opposed to four wires in the long-haul connection between exchanges.) Echo impairment, which is unacceptable for duplex data, is a consequence of imperfect conversion from two-wire to four-wire that occurs at the access point to the long-haul portion of the circuit. Perfect circuit balance is impractical in the dial-up network, so the echo path must be blocked with an electronic device. In the 1970s, dial-up modems had to be operated in the half-duplex mode on international circuits so that echo suppressors (which block the echo by opening the circuit, preventing bidirectional communication) could be disabled on both ends. This is still the case with one of the most popular standards, V.29, which is used to transfer Group 3 fax messages at 9600 b/s. The V.32 standard is suited for interactive data communication at 9600 b/s on four-wire leased lines and two-wire dial-up connections. Incorporated in the modem is a digital echo canceller, the device now prevalent on long-distance telephone circuits. The echo canceller has the ability nearly to eliminate echo and at the same time to allow the data to be conveyed in both directions without disruption. Older echo-suppressor technology cannot allow information to flow in both directions at the same time, and hence is not compatible with duplex data transmission.

The cost of a V.32 modem has dropped below $1000 as of 1989, making it a viable candidate for the majority of business applications. There is now the prospect that home computer users will be able to have inexpensive V.32 modems for use with E-mail and videotex, which will significantly improve the usefulness and friendliness of these services. Users are already enjoying V.32 performance at 19,200 b/s over four-wire leased lines, through the technique of *trellis-coded modulation* (TCM). In TCM, an extra bit is added to each symbol set of bits. The demodulator at the receiving end uses this bit to resolve ambiguity caused by line impairment. The concept is similar to that of an error correcting code, except in this case the correcting process is done within the modem and does not involve the network.

The most advanced standard for voice-band modem design is defined in Recommendation V.42. As shown at the bottom of Table 5.3, current V.42 modems are capable of transferring 19,200 b/s on a full-duplex basis using a four-wire leased line. Half-duplex operation on a two-wire circuit at the same data rate is also provided. The throughput performance of V.42 modems is enhanced by the application of error detection and correction by the modems on both ends of the circuit. Using a synchronous protocol, the error correction is both transparent to the user and highly effective in reducing the error rate performance of the link. The two competing error correction protocols are: Link Access Protocol D (LAP D), which is related to HDLC; and Microcom Network Protocol (MNP), which is a proprietary protocol developed by the modem vendor of the same name. Both protocols are carried as alternates in Recommendation V.42; selection of one approach is hampered by the large existing base of modems using MNP *versus* the preference of some vendors to pursue a more standard course (i.e., OSI) with LAP D.

Dial-up connections are particularly convenient in the international environment because the PSTN is available on a continuous basis in literally every country. The speeds of transmission, of course, are restricted due to available bandwidth and limitations of the two-wire local loop on both ends. The technical parameters that affect data transfer are amplitude distortion, envelope delay distortion, and *signal-to-noise ratio* (SNR). These are usually within typical bounds on the international PSTN. The higher speed dial-up modems in Table 5.3 all have the ability to equalize the circuit automatically prior to the initiation of data transfer. This corrects for envelope delay distortion so there is no need for a preequalized class of leased line, such as that specified in CCITT Recommendation M.1020. Dial-up service has built-in alternate routing because the user can easily reinitiate the call if circuit quality is inadequate to pass data. Unattended operation is possible with an autocall modem that automatically places a call to a predefined distant location to establish the data communication session. Likewise, the distant end can interrogate a terminal by dialing into the modem which has a direct connection to the PSTN. A significant measure of security against unauthorized access is provided by a "call-back" feature which causes the device to call back a previously registered number after it first receives an inquiry from a computing device. This security fails, however, if the telephone at the registered location has been secretly placed in the call-forwarding mode.

Generally speaking, modems used for dial-up service use asynchronous transmission, and therefore their throughput capability is less than that of a synchronous duplex modem on a dedicated private line. The delay associated with setting up the call for each session must also be considered. Synchronous data transfer is growing in importance on dial-up modems, particularly for use as a backup to leased line service. In many data communication applications, the time delay involved in setting up a call on the PSTN cannot be tolerated. Also, telephone service in many developing countries is still poor. Their telephone exchanges are so overloaded that a user may have to wait minutes or even hours just to get a dial tone. These are applications for which private lines and dedicated modems become a necessity.

The modem approach is also followed when analog bandwidths of greater than 3000 Hz are used to carry high-speed data. Usually, the demands on the modulation method are lessened because more bandwidth is available to support the link. Terrestrial microwave systems and satellite communication networks employ modems with data rates in the range of hundreds of kilobits per second up to several megabits per second. Such modems are very specialized and are supplied as part of an overall network architecture. For example, the modem within a very small aperture terminal is integrated into the indoor unit of the terminal. Telecommunication managers should be aware of the type of modem being employed and be sensitive to issues such as redundancy and sparing. The supplier of the network facility, however, will generally take pains to design or select modems with the right performance.

A somewhat different situation exists if a user supplies high-speed modems for a wideband leased line. PTTs in Europe offer private line services covering a baseband frequency range of 60 to 108 kHz, which can support a modem operating at 56 or 64 kb/s. CCITT Recommendation V.35 defines the characteristics of modems which can employ this bandwidth. Obviously, a telecommunication manager must arrange to use compatible V.35 modems on both ends of such an international circuit. This concern can be alleviated if the PTTs supply the V.35 modems; the user then interfaces with the baseband V.35 standard.

Modems used on leased lines are usually provided with some form of network management facility; we refer to this as *monitor and control* (M&C). In [Elbert, 1989], the alternatives for creating an on-line M&C system are reviewed in detail. There are basically two alternatives when it comes to providing remote M&C facilities for modems. In the first case, modems specifically designed for use on leased lines often come with built-in M&C capability. This is achieved with a separate data communication path which rides beside the actual user traffic. Measured transmission performance is sampled and sent back to a central M&C controller, while control information is returned to the modem over the same type of path. The M&C data can be carried either in a narrow slice of bandwidth below 400 Hz, or it is multiplexed in with the user data traffic. Usually, the modems are supplied by one vendor because of the uniqueness of these M&C systems. Four of the largest vendors of integrated M&C modems are IBM, Codex, Infotron, and General Datacom. In the second case, the M&C channel can be added to a modem network with the use of "wrap-around" units, which insert and remove the necessary separate channel.

Monitor and control of modems used in dial-up service is not as critical because the circuit can be changed at a moment's notice. The leased line, however, is much more difficult to deal with, particularly if the user has no alternative routes to take. An international leased line for data communication is a much more significant problem when one considers M&C. This is where an on-line M&C link is particularly important. Even if the M&C alarm and test facilities are working properly, however, it may be difficult to get someone to do something about the problem. Automatic rerouting through dial backup is an accepted way to respond to the problem of leased line interruption. The modem has connection to the leased line and an access line to the PSTN, and can automatically place a call to a prescribed number to reestablish the data circuit.

Some users believe that the motivation to push modem performance is deflated by progress in the area of switched 56 kb/s services and ISDN. Users need high-speed connectivity today, however, and cannot wait for worldwide deployment of ISDN. Pushing modem performance to the theoretical limit is the aim of technologists and entrepreneurs who understand how to meet needs with practical and inexpensive hardware.

5.2.3 Group 3 and Group 4 Facsimile

The facsimile is probably the fastest growing segment of the worldwide CPE market. Currently, fax machines employing the Group 3 standard are everywhere and the technique has effectively reached critical mass. Impending is the Group 4 standard which many believe will be an important application of ISDN.

Availability of at least one compatible fax unit on a business's premises has become a necessity. Two factors have propelled this telecommunication application to such proportions: first, the acceptance of the Group 3 digital fax standard, and second, the plethora of low-priced units, some selling for as little as $400 in 1989. Although these low-end units are not suitable for heavy business usage, more typical full-featured fax machines sell for prices in the range of $1000 to $3000. Importantly, the Group 3 fax unit transmits and receives business documents and graphics with acceptable quality in 20 seconds per page (or less, depending on the graphic detail) over virtually any segment of the international PSTN. In addition to guaranteed speed of delivery, document transmission is decidedly cheaper by fax because of the relatively low cost of using the PSTN as opposed to higher cost express mail and commercial overnight delivery services. It has been estimated that there are a total of at least 10,000,000 Group 3 fax units in the world as of 1989, which is almost five times the number of telex lines that ever existed at one time.

Evolution of fax technology has taken more years than people realize, even though its recent rise has been so rapid. A thorough discussion of fax development and technology can be found in [McConnell, 1989]. The first true fax machines were sold to the U.S. Army just after World War II, but these devices were cumbersome to use and had poor reproduction quality. Transmission time was measured in multiples of minutes. The business fax machine appeared in the 1960s when Xerox offered their first telecopiers for an application they called Long Distance Xerography. Quality was compromised for speed, but it still took minutes per page. The CCITT adopted the Group 1 and Group 2 facsimile standards in 1976, defining analog formats which used frequency modulation and amplitude modulation, respectively [Voros, 1989]. Group 1 machines have all but disappeared, although a few Group 2 machines are still being used by businesses. The policy of the CCITT is to make any standard "backward-compatible," so current Group 3 machines can exchange documents with the Group 2 machines still in use. The time of transmission in Group 2 is again measured in minutes and reproduction is noticeably inferior to Group 3.

5.2.3.1 The Group 3 Standard

The standards-making efforts of the CCITT were a decided factor in the success of Group 3 facsimile, which were completed in December 1988 and are part of Recommendations T.4 and T.30. The following discussion of the characteristics of Group 3 facsimile can be skipped by readers not requiring technical information. Group 3 fax encodes the graphic information into a totally digital format and uses a medium speed modem to convey the data over the PSTN. Under the standard, a variety of transmission modes and speeds are permitted, and vendors can even build in proprietary features which only their machines can employ. All Group 3 fax units, however, must be compatible with each other at the basic level of performance. The most common modem design follows Recommendation V.29 and transmits at 9600 b/s in half-duplex. This design is an option under the Group 3 standard but it is essential to rapid document transmission, giving a speed of roughly 20 seconds per page.

The compatible parameters are determined during an initial set-up routine which two paired machines automatically go through before document transfer. A prime parameter is the speed of transmission, which starts at 9600 b/s and steps down through 7200, 4800, *et cetera,* until a workable rate is established. The modem's "step down" feature ensures that a workable path can be established even if the circuit has a high noise level or impaired bandwidth. Other aspects of this initial "handshake" establish if any proprietary error-correction or compression modes are available. An important feature of Group 3 is that the called machine will transmit back an identification which the calling machine can print locally. Such a confirmation would be useful for business purposes, operating in a fashion similar to the telex "answer back" feature.

A thorough review of the structure of the Group 3 standard and its variations is found in [McConnell, 1989]. The quality of reproduction is established by the resolution of the system, specified in terms of the number of black or white picture elements (pel) per inch. This is also the way in which resolution is specified for dot matrix and laser printers. The standard resolution for Group 3 is 203 pels horizontally and 98 pels vertically. As most readers already know, this produces text copy that is easily readable. Resolution can be enhanced with an optional high-resolution mode where the vertical resolution is increased to 196 pels, which is essentially the same as the horizontal resolution. Whether standard resolution or high resolution, this performance is substantially better than that which results with Groups 1 and 2. With roughly 200 by 200 capability, the Group 3 fax machine compares favorably with the 150 by 150 resolution of the HP laser jet, one of the most popular laser printers on the market.

The Group 3 machine scans one page of the document at a time, converting black-and-white images into a basic bit stream. The key to success in terms of speed of transmission is the system used to compress the data. Compression is necessary so that the relatively narrow bandwidth of the PSTN can be used for practical fax communication. Compression is done both in the horizontal and vertical directions, and the unit has the sophistication to adapt the compression to the particular situation. The techniques that are used are called *modified Huffman run-length coding* and *modified read coding* (MH/MR). Complete explanations of the techniques are provided in [McConnell, 1989]. Briefly, a horizontal scan generates 1728 bits but usually contains mostly white pels. Run-length coding means that instead of sending a long sequence for a constant white line segment, the coder sends a code word that simply indicates the length of the run of white bits (the "run length"). The compression can be by as much as 100 to one, depending on the composition of the particular line. Compression in the vertical direction is accomplished with the MR coding scheme, which checks for correlation between adjacent horizontal lines. Group 3 fax is conservative in its implementation to account for disruption on the circuit; MR coding is only done on every other line. In Group 4 fax, the entire page is coded in the vertical dimension using modified read coding, but a communication link with lower error rate and higher speed is preferred.

Group 3 fax has some simplified means to recover from errors in data transmission. If a given line is totally corrupted, the previous line is repeated. Alternatively, depending on the particular manufacturer, a white line may be substituted. The encoding algorithms are very effective in this regard and have been carried over to the Group 4 standard with slight modification. The availability of the V.29 modem standard at 9600 b/s has also helped enhance Group 3 fax performance because lower speeds mean longer transmission times per page.

To send a document, an operator places one or more pages in a loading bin, enters the telephone number of the called party, and pushes a send button. The called fax unit automatically answers and applies a 2100 Hz tone back to the calling unit for three seconds. This causes the sending unit to engage in the handshake protocol to establish the common parameters for the transfer of the document. The handshake lasts approximately another three seconds, then the document is fed into the scanner and the data transmitted. The called fax unit is able to check for errors; if reception is acceptable, the called unit so indicates. Additional pages are sent and at the conclusion, both units can hang up.

The automatic operation of Group 3 has a number of important benefits for international service. As in telex, the receiving unit answers automatically and no person need be present. This is the familiar mode of international telex service, allowing the recipient to pick up a message in the morning and act upon it while the sending party is asleep. The more full-featured machines provide automatic sheet feeders on both ends, which improves office efficiency. Basically, the machine acts like a modern copier. More sophisticated fax machines can automatically send documents at night to take advantage of lower telephone calling rates. Capabilities such as polling and broadcast transmission (to multiple recipients) are also available. This is facilitated by extensive memory capacity built into the unit to store images of entire pages.

For the telecommunication manager and business person, the establishment of this excellent standard and the reaching of critical mass are blessings when one engages in international telecommunication. Interestingly, as much as 20% of all business telephone calls in Japan are in actuality fax calls; this is perhaps the highest percent in the world, one theory being that fax is more useful there because of the Japanese written language. Fax communication in business was really proven in Japan, however, and that nation is perhaps just ahead of the rest of the world in its deployment. Comparable percentages in the United States and Europe are in the range of 5 to 10%, leaving further room for growth. The fax boom may have only just begun.

Approximately 90% of all Group 3 machines are manufactured in Japan, with most of the remaining 10% coming from the Republic of Korea, Singapore, and Hong Kong. Major business equipment manufacturers in the United States and Europe obtain Asian machines for renaming under other equipment manufacturer (OEM) agreements. Because the market is still expanding rapidly around the world, the number of Asian manufacturers is too numerous to mention. The most basic feature on all machines allows documents to be sent and received (but not simultaneously). Another form of Group 3 facsimile is the fax board which can be introduced into a PC. These are most useful for transmitting screen images from a PC to a distant fax machine. Also, the fax board is useful for transmitting the same message to many users because the PC can be preprogrammed with addresses and telephone numbers.

For the international traveler, the portable fax machine should prove to be particularly attractive. These machines were just appearing in 1989, their introduction slowed by a lack of standards for reduced page size. The basic idea is that a traveler can carry a fax machine that easily fits within an attaché case. Upon arrival in a foreign city, the person can use the fax to receive documents by way of the telephone in the hotel room. The potential applications include: receiving proposals for hand delivery, sending orders on standard forms, and even getting copies of current newspaper articles. Those who travel in developing countries know that it is not uncommon for local censorship to prevail, and the fax machine is basically uncontrollable (assuming that the traveler can enter the country at all).

The networking aspects of Group 3 fax machines are self-evident because of their design for use over the PSTN. This is the strongest attribute of Group 3 fax for the international telecommunication manager. In Chapter 3, we mentioned dedicated fax services which are provided by common carriers such as MCI and Infonet. There are advantages to these networks in that they provide extra services such as message store and forward and remote printing for users without fax machines. In many countries, the PTT provides a telegram-like service using Group 3 machines called Bureaufax.

5.2.3.2 *The Group 4 Standard*

The Group 4 standard was primarily developed by the CCITT as a graphical communication medium under the seven-layer OSI model. Networking of facsimile communication would be by way of a packet-switched architecture with the capability to transmit information at up to 64 kb/s. This speed is necessary to be able to deliver a page of high resolution copy in a matter of ten seconds or less. As mentioned previously, the black-and-white encoding scheme of Group 4 is essentially the same as that of Group 3, relying heavily on compression through the removal of redundancy in the image. Much of the interest in Group 4 relates to its integration with the 64 kb/s B-channel that could terminate in virtually every business in a country where ISDN is fully deployed. At this speed in conjunction with ISDN's other service features, Group 4 can deliver nearly perfect copy quality over a very flexible network. The basic fax machine design of Group 3, although geared toward the analog PSTN, is, however, adaptable to a digital 64 kb/s transport medium. In the high resolution mode and at this speed, a Group 3 machine can perform exactly the same basic function as a more sophisticated Group 4 machine.

The Group 4 standard covers how the machine could be used to function as a Teletex receiving terminal, accepting ASCII input and printing the associated document consisting of mixed text and graphics. The only problem here is that Teletex is still very much on the drawing boards and may be another example of a solution searching for a problem. The ability of a standard Group 4 machine to print mixed text and graphics could represent an important benefit to future users.

The positive attributes of Group 4 are many, giving the system a decided appeal from a technical standpoint. For compatibility with Group 3, the basic resolution of Group 4 is 200 by 200. Two other high resolution modes are available: 300 by 300 and 400 by 400. The 300 by 300 case provides quality that is noticeably better than 200 by 200, and to a casual observer, a document could pass for an original. The 400 by 400 resolution, however, reproduces fine detail and delivers documents nearly indistinguishable from originals. Because of the effectiveness of the encoding scheme, 400 by 400 requires an increase in data throughput of only one-third as compared to 300 by 300.

In summary, there are three classes of Group 4 units [McConnell, 1989]:

Class 1 units can send and receive facsimiles which have been scanned and coded with the approved system. The minimum resolution required is 200 by 200; 300 by 300 and 400 by 400 are, however, available as options.

Class 2 units include the features of Class 1 units, adding 300 by 300 resolution as mandatory. In addition, Class 2 units must be capable of receiving Teletex information and mixed mode documents.

Class 3 capabilities include those of Classes 1 and 2; in addition, they must be able to send teletext and mixed-mode documents.

While specifications for all three classes are contained in the CCITT Recommendation for Group 4, the only units being manufactured are those which comply with Class 1 (i.e., for strictly facsimile services).

Machines with Group 4 resolution capabilities were used by Federal Express in their now-defunct ZapMail service. The quality of the documents delivered over the system was very close to that of the originals. Federal Express experienced major problems in bringing up a private network operating at 56 kb/s at a time when the long-haul networks in the United States were just being converted to digital microwave and fiber optic cables. The service worked as promised but Federal Express was just not able to meet their objectives for revenue at the same time that transmission costs were building almost exponentially. Originally, all machines were located in Federal Express offices and documents were distributed locally with their existing fleet of vans. A strategy of providing Group 4 machines on customers' premises was even tried. By virtue of ZapMail, Group 4 machines were introduced in North America and many users were given a taste of what could become commonplace. An important factor in the demise of ZapMail was the rapid growth in popularity of Group 3 facsimile, which did not have the same networking problems that Federal Express experienced. It also became clear that Group 3 fax was cheaper and more convenient to use.

With its close ties to ISDN and teletext, Group 4 is experiencing user-acceptance problems even in the industrialized countries. This is not difficult to understand, particularly because Group 3 facsimile has already reached the critical mass stage. Group 4 can be used within a private telecommunication network environment, where 56 and 64 kb/s circuit-switched or private line services are available. One of the attractive features of Group 4 machines is that they can act as a gateway for Group 3 messages. According to Gary Voros of Fujitsu, a strategic unit may have hub locations equipped with Group 4 fax machines linked with high-speed lines over a private backbone [Voros, 1989]. Then, the hubs could collect Group 3 traffic from branch offices and route it over the backbone by way of the Group 4 hub machines. A given Group 4 machine can distribute documents to any number of designated branches which only have analog PSTN service available. The cost of using the backbone to send the fax traffic could be significantly lower than using the PSTN, particularly if the information rides piggyback with a larger cross section of private network traffic. This model can easily be extended to the international environment, where the backbone uses international private lines or IBS service and the hubs are located in major cities around the world. Extensions over the PSTN would work nicely to distribute fax messages using the Group 3 standard.

5.2.4 Video Teleconferencing Equipment

Video teleconferencing has gained a decided foothold in North America, with many corporations, government agencies, and university systems relying on the medium on a day-to-day basis. Perhaps the most popular is distance education, where closed-circuit television broadcasts allow an institution greatly to expand class size and to reduce travel time and distance for students, particularly those who must work for a living. Another established application is two-way interactive video for long-distance meetings. For those who accept interactive television in lieu of face-to-face meetings, the teleconferencing route essentially eliminates travel time and cost, and allows more people to participate in the discussion. The considerable financial and technical barriers have been reduced through the availability of inexpensive *television receive-only* (TVRO) satellite dishes, low-cost digital transmission, and the introduction of more efficient video processing equipment. Video teleconferencing is still costly to use and is not expected to reach critical mass even by the year 2000.

Carrying these developments over to the international environment is a difficult task because the previously mentioned barriers exist at even higher levels. The benefits of video teleconferencing, however, could be greater in international telecommunication than they are domestically. Travel distances are greater, as are expense and time wasted on airplanes. This must be traded off against incompatibility of time zones, as discussed later in this chapter and again in Chapter 8. Branch office personal at remote locations are in more need of interaction with the home office for current information on new products, procedures, marketing campaigns, and the like. One of the first concrete examples of international video teleconferencing is the extension by Hewlett-Packard of its U.S. domestic video network to Europe. Through the French Telecom 1 satellite, HP broadcasts its private analog video telecasts to European sites. An INTELSAT satellite acts as the bridge between Europe and North America.

Facsimile service, while excellent for quick document delivery, cannot satisfy the need for color graphic image transfer and personal interaction which the teleconferencing medium offers. There are large time zone differences, however, which make it difficult to coordinate teleconferences. A typical teleconference requires one to two hours of on-air time, which is difficult to arrange for offices in different hemispheres. This is one of the reasons why unattended fax communication is so popular in the international business arena.

5.2.4.1 Teleconferencing with TVROs

Most video teleconferencing activity in the U.S. is with conventional analog FM transmission on domestic C- and Ku-band satellites. Satellites in Western Europe, Australia, and Japan operate at Ku-band, and those in Latin America, Southeast Asia, and the Pacific operate at C-band. TVRO service is a one-way broadcast for use in private programs for employees scattered around the country and for other corporate communication purposes like press conferences. A return audio channel is often implemented through the PSTN. Literally hundreds of transponders are available on these satellites and can be rented on an occasional basis. This eliminates a major hurdle to building a private video network, that of the investment in one or more transponders.

The cost of an hour of transponder time is typically in the range of $250 to $500 in the United States, depending on the satellite used, time of day, and the frequency band (Ku-band transponders are more expensive to rent because they offer higher radiated power). Service in Europe or Japan may be higher in cost, but the same principles apply. To this is added the cost of using a transmitting earth station (the *uplink*), which can be rented on an hourly basis at a rate comparable to that of the transponder. Some users, like General Motors and WalMart, implemented their own uplinks so that access would be guaranteed. Before rushing out to buy an uplink, telecommunication managers should be prepared for an expense as high as $1 million and the complexity of obtaining the necessary license and permits. The studio for the broadcast can also be rented, installed at great expense on the customer's premises, or created in a make-shift fashion in a conference room or classroom.

The receiving side of a conventional satellite television link is with an inexpensive TVRO terminal of the same design as the home satellite dishes which consumers can buy for $1500 or less. TVROs do not have to be licensed with the FCC but may have to be licensed in other countries, depending on local regulations. Building permits and landlord permission to install the dish are needed in virtually all cases. A nice feature of analog video reception is that it can be added to an existing small earth station such as a VSAT used for data communication. Dedicated TVROs are used extensively by university systems in Virginia, West Virginia, California, and Oklahoma, and other states are quickly pursuing the same route. Although behind in terms of full development, educational networks are forming in Japan and Western Europe.

The analog TVRO approach certainly has taken off in North America, and is on its way toward a similar success in Japan and Western Europe where domestic satellites provide the needed capacity. International video transmission is an entirely different matter, explored in detail in Chapter 6. Suffice it to say here that the cost of obtaining the uplink, satellite, and downlink capacity is many times greater than for domestic service. Also, much more complex is the matter of initially setting up the teleconference. Therefore, international teleconference transmission is best done by using compressed digital video so that digital private lines can be used.

5.2.4.2 Freeze-Frame Television

Another technology which we mention only in passing is *freeze-frame* television, also called *slow-scan*. The concept is similar to facsimile in that a fixed image is transmitted to the distant end. The difference is that the picture can be of almost anything because it is captured with a video camera. Both color and black-and-white freeze-frame systems have been in use for a number of years. Telecommunication managers have provided freeze-frame systems for use by marketing departments wishing to inform branch offices of new products. For mass merchandisers, the medium is used to transfer full color "stills" of new clothing lines and accessories. What makes freeze-frame so attractive is that it can employ a voice-band modem to transmit over the PSTN. As in fax, the modem speed determines how quickly the image will "paint" on the remote receiver. Picture telephones are already on the market, employing freeze-frame television to add the picture element to voice communication. They represent a stripped down version of AT&T's Picturephone®, with a black-and-white capability but costing as little as $500 each.

Peripheral (add-on) boards for personal computers are on the market for an application called PC conferencing. Lacking a precise definition, PC conferencing includes the use of the graphic capability to transmit pictures over the PSTN. This differs from fax and normal data transfer because the image is typically obtained from a video camera or image scanner. The add-on board stores the image and allows it to be transmitted over a telephone circuit using a voice-band modem. The conferencing element comes in because the user can add text which is entered from the keyboard. Voice communication can be included over a second telephone circuit. An innovative image-data-voice environment called Freestyle was introduced in 1989 by Wang Laboratories. PC conferencing is still in the germinating stage as of 1989, and no standards have been produced. In time, PC conferencing could evolve into an important application on its own, or it could simply end up as an element of ISDN.

5.2.4.3 Interactive Video Teleconferencing Using Digital Compression

Interactive two-way video is usually done with bandwidth-efficient digital processing equipment. In satellite communication, this allows several video channels to share the bandwidth of one transponder. Alternatively, terrestrial private lines at the T1 rate (or fractions thereof) become viable. Prices of digital video encoding and decoding equipment (*codecs*, a contraction of coding and decoding) have been halved over a five-year period. A codec operating at T1 speed has dropped from approximately $80,000 in 1982 to approximately $40,000 in 1989. This represents the high end of the market. Less expensive devices which produce reasonably good motion reproduction on a fractional T1 circuit can be purchased for approximately $30,000. Fractional T1 services, which are available in multiples of 64 kb/s, can be obtained at significant cost savings, as discussed in Chapter 4. The fractional T1 codec did not even exist in 1982; the marketplace is clearly expanding. The cost of putting in a teleconferencing room has also been drastically reduced. Readers may be familiar with studio installations, which innovators such as ARCO and IBM installed in the early 1980s at costs of up to $1 million each. Users, however, find that the systems can be installed in large cabinets on wheels which can be used in a properly prepared conference room. These "roll-around" systems are much cheaper and can be moved within a building without major construction expense.

The video codec is the critical technical element and is also one of the most costly in a teleconferencing system. Codecs can be divided into five classes: broadcast quality (45 Mb/s and up); full-motion T1 (760 kb/s to 1.544 Mb/s); fractional T1 (112 to 568 kb/s); ISDN (56 to 64 kb/s); and slow-scan or freeze-frame (4800 to 9600 b/s, voice band). A summary of the vendors offering codecs in these ranges is presented in Table 5.4. All of these systems can reproduce a color image with the kind of resolution that normal television sets are capable of displaying. Where they differ, however, is in the ability to reproduce motion and changing imagery.

Table 5.4 Suppliers of Digital Codecs for Video Teleconferencing

T1 and Fractional T1
Compression Laboratories, Inc.
GPT (U.K.)
NEC Corporation
NV Philips
PictureTel

56 and 64 kb/s (ISDN-Compatible)
Compression Laboratories, Inc.
NEC Corporation
PictureTel

Slow-Scan and Freeze-Frame
Concept Communications
Mitsubishi
NEC Corporation
Widcom

The telecommunication manager must realize that the only way to understand how good the motion performance is of a given codec is actually to see it used in an appropriate setting. Color photographs of images are not adequate for evaluating codec quality.

The codecs used in all these systems take a conventional television picture, convert it into digital data, and compress the data rate down to the specified level. The compression function is typically what degrades the motion performance. The shortfall of highly compressed images does not show up in a still picture. When something moves in the picture, however, the image can become blurred or distorted. Developers of compressed codecs try to reduce motion distortion through a variety of compensation techniques. In some applications, some amount of motion distortion is tolerable. Near perfection is required in commercial broadcasting where viewers expect to be able to see all of the action in a basketball game or motion picture.

An often overlooked aspect of teleconferencing is audio operation and quality. The codec is typically the element which is responsible for encoding the audio and carrying it along with the video, typically with time-division multiplexing. Assuming that adequate bandwidth is provided, audio performance should be as good as or better than a good telephone circuit. Problems arise, however, when using microphones and speakers in a conference room to permit many people to take part. One of the principal problems is echo, which will come about even with short propagation delay. This is because the codec introduces a significant processing delay, amounting to as much as a half second or more. Fortunately, echo cancellation is available within the teleconferencing system.

The 45 Mb/s codecs currently in production provide excellent motion performance, essentially indistinguishable from broadcast television. The required transmission capacity (an entire DS3), however, is probably not justifiable for private teleconferencing applications; we defer further discussion of this system to Chapter 6. T1 codecs were first installed in the early 1980s on private teleconferencing networks of major corporations in the aerospace, energy, and manufacturing industries. Their use was somewhat limited, due to the high expense of the equipment and the transmission links. By fine tuning the performance of the codec, a single T1 can now support two video signals. This is mainly useful in distance education where a company wants to add a second channel for alternative media (e.g., a picture of a person talking simultaneously with a slide presentation) or to carry two classes emanating from one location. Alternatively, a single T1 could support a compressed video channel plus integrated voice and data, as indicated in Figure 5.1.

As of 1990, the leading suppliers of T1 codecs in the world are NEC of Japan and a small American company called Compression Laboratories, Inc. (CLI). The two companies have about equal market share. CLI has a major presence in its home country, the United States, whereas NEC is more known as the international supplier of teleconferencing systems. The Japanese domestic market is dominated by NEC and Fujitsu. Unfortunately, there were no standards at the time that these T1 codecs were developed and incompatibility pervades. In fact, different generations of the products of a given company are incompatible. GPT of the United Kingdom produces a T1 codec for the European market and had at one time an early lead in the United States.

The proprietary nature of T1 codecs has made it difficult for telecommunication managers to select a vendor. Almost all applications are for private networks, permitting all locations to employ codecs from the same manufacturer. In time (and as technology improves), the buyer is faced with the problem of upgrading and expanding the system. For the same vendor to cease production of one model, offering instead an improved but incompatible replacement is not uncommon. Once you make a codec decision, be prepared to adhere to it for a considerable period of time. In terms of background, the CLI codecs have gained a strong foothold in U.S. government and aerospace industry applications, yielding a *de facto* standard. NEC, conversely, has sold codecs to IBM and others in the United States, and NEC offers the same products in Japan and elsewhere. Technology aside, the choice of CLI *versus* NEC should depend on connectivity and vendor support. International teleconferencing networks which are non-U.S. governmental may best employ the NEC codecs because the company has an excellent presence around the world.

Real economies come with the fractional T1 codecs, which provide reasonably good motion performance within one-tenth to one-quarter the bandwidth of the original T1 systems. There is visible motion distortion when a person or object is moved rapidly on the screen. Also, the motion of a person's lips while talking will not be reproduced very accurately. This is a compromise, but it is one that users are learning to accept in exchange for greatly reduced transmission cost.

The leading supplier of fractional T1 codecs is PictureTel. The model C3000 codec provides reasonably good motion performance at 112 kb/s, using the *hierarchical vector quantization* (HVQ) algorithm. Their earlier model, the C2000, effectively established the market and employed the *motion compensation transform* (MCT) algorithm. This particular algorithm is the basis for the CCITT $P \times 64$ standard, discussed in the next section. At 112 kb/s, the C3000 would employ the capacity of two 56 kb/s circuits. This is attractive because the user can set up a short teleconference using switched 56 kb/s services. PictureTel's codecs are gaining acceptance in the United States and for teleconferencing applications from there to Europe because of resale agreements with AT&T and US Sprint.

The 56–64 kb/s market is the most notable; it is also the one of greatest promise in international telecommunication management. We refer to this as the "ISDN market segment" because it would be compatible with the B-channel of the BRI and PRI. Many envision the time when interactive video teleconferencing will be commonplace via ISDN, raising the possibility of critical mass being reached. Because this represents an attractive future for codec manufacturers, most of the development effort on 56 kb/s has been shifted to 64 kb/s. This rate is also important because the higher rate codecs operate at $N \times 64$ kb/s, giving the user the option of tuning performance to the available bandwidth. The price of an "inexpensive" 64 kb/s codec is still not appreciably lower than that of one operating at 768 kb/s. This is probably because the signal-processing and compression circuitry are comparable. Only with the development of custom VLSI of the needed complexity, manufactured in very large production volumes, will truly low-cost codecs reach the market. If we had a standard like Group 3 fax, codec prices probably would soon drop below $10,000.

Realizing this, occasionally a new company appears claiming to have a technology and even a product for low-cost, full-motion digital video teleconferencing. One of the popular ideas is to employ a personal computer as the platform for teleconferencing. The user need only purchase a circuit board that can be plugged into the PC's expansion slots. An external camera and high-resolution monitor are also needed. From a technical standpoint, this concept is feasible. Of course, the PC may need to be a high-speed machine using the Intel 80386 microprocessor, for example; perhaps even an engineering workstation or desktop computer may be needed. These machines could be used to enhance the service, allowing graphics to be inserted into the teleconference or allowing parties to exchange files. These products would meet a real need and should provide large revenues for a successful supplier, but none had actually appeared on the market as of 1990.

5.2.4.4 Digital Teleconferencing Standards

Study groups within the CCITT have developed a framework for several standards in the *audio-visual* (AV) aspects of telecommunication on digital networks such as ISDN [Matsumoto, 1989]. Two standards that define key services were adopted at the 1988 Plenary Assembly of the CCITT: F.710 for general teleconference service, and F.721 for videophone service. Recommendation H.261 contains an encoding algorithm based on MCT of PictureTel, which is being considered as standard for $P \times 64$ codecs. The first step, however, is believed to be agreement on standards for 384 kb/s, which happens to occupy six 64 kb/s channels. A practical codec device would adhere to the standard and operate much like a Group 3 fax machine, going through an initial handshake to determine if the basic CCITT algorithm would be used or if the two machines were made by the same manufacturer and could use a more advanced proprietary algorithm to improve motion performance.

Eventually, a standard for the ISDN basic rate interface (2 B + D) is needed so users can have full-motion video teleconferencing at one or 2×64 kb/s. Ideally, the video and its associated audio would use one B-channel, leaving the other B-channel for other services such as Group 4 fax and data transfer. This reduces the capacity available for the video itself, however, which degrades motion performance. Another approach is to leave one entire B-channel for video and use the other for audio and graphics. The most conservative approach would be to use the full capacity of the two B-channels (e.g., 128 kb/s), robbing only as much as is needed for the audio portion. Interestingly, participants in the proceedings referred to this as the "to be or not to be" problem (or, more correctly, *2B or not 2B* for the codec's use).

In terms of real teleconferencing systems and equipment, the market is established for specialized applications, as cited above. Leaders in the market are NEC, CLI, GPT, and PictureTel. Other companies are coming into the market now that applications are taking hold. Fujitsu, Mitsubishi, Matsushita, and Philips will likely become strong players, particularly as ISDN connectivity at 64 kb/s is expanded. Quite possibly, freeze-frame video will evolve into videophone, allowing the PSTN to serve yet another function for much of the population. This could become the next fax-like market, which would be a welcome addition to the international telecommunication environment.

5.2.5 Integrated Terminals for ISDN

The technology of ISDN offers the opportunity to deliver a wider range of services than possible with conventional switching and transmission over the PSTN. PBX and local telco exchanges already offer ISDN service in terms of the basic rate interface. We have discussed how the B-channel can be used for Group 4 fax and eventually for audio visual services like digital video teleconferencing. These require special equipment that usually must be purchased from a vendor. In this section, we discuss a number of versions of the more general *integrated voice-data terminal* (IVDT).

An IVDT is either a single piece of station equipment or a closely coupled set of devices which can be connected to the BRI of ISDN. The loop from the BRI would connect either to an ISDN-compatible PBX or telco exchange. Both possibilities already exist in scattered areas throughout the world. The purpose of the IVDT is to place on the user's desk a single package that simultaneously delivers telephone, data, and possibly video communication services. Conceptually, an IVDT would look a lot like a modern personal computer with a telephone set built in. Because the capability of ISDN in general and user access services in particular are very much subject to change, the precise configuration of IVDTs is still fluid.

One of the more important attributes of ISDN is the commonality between PBXs and public exchanges that exists under ISDN. With D-channel and SS-7 signaling, user access is facilitated from the PBX side or from the public network side. A user can use an IVDT at home in one city to connect to a compatible ISDN device on an ISDN PBX in a different city. This effectively makes ISDN into an integrated data communication environment provided through the public network. Today, this kind of environment would otherwise have to be provided by a strategic unit acting on its own. There are many possible applications, including database access or E-mail. With the growth of international ISDN service within Europe and between Europe, North America, and Japan, these possibilities could force a rapid delivery of IVDT devices to the field.

Another important feature of ISDN is that it can be integrated into a corporate digital backbone network which could already be in existence. Modern digital facilities provide T1, E1, and DS3 transmission on a private line basis, as discussed in Chapter 3. Strategic units purchase the types of intelligent multiplexer discussed in Chapter 6 to use the long-haul media efficiently. When ISDN services are delivered by the PBXs, the existing backbone can be used to provide the requisite trunking. This would most likely be in the form of the PRI, which corresponds to the capacity of a T1. Telecommunication managers should realize that some existing T1 multiplexers will need to be upgraded to support 23 B + D, even though the bulk data rate is the same.

Early models of IVDTs were literally "thrown together" by major switch manufacturers simply to demonstrate ISDN capabilities. Considerable time and experience are needed for IVDT configurations to stabilize for users to obtain units that fit real-world needs. For the time being, pioneering users are still experimenting with current IVDT configurations. Two configurations are prevalent: the new IVDT terminal made especially for this purpose; and a means to adapt a personal computer and telephone into an IVDT. Some of these will be discussed below. More detailed information on IVDT arrangements and ISDN systems can be found in [NEC, 1987].

5.2.5.1 ISDN Terminal Adapter

The purpose of the terminal adapter is to connect existing non-ISDN devices to the BRI operating at 144 kb/s. In addition to the basic multiplexing function of combining two 64 kb/s channels and one 16 kb/s packet-switched channel, the terminal adapter interfaces the local devices with the protocols for signaling and data transfer particular to the B- and D-channels.

The terminal adapter is how digital devices employ the 64 kb/s capability of the circuit-switched ISDN network. Group 4 fax units may require the terminal adapter to provide the interface, as would personal computers to be used in data transfer. We have already mentioned that standards are being developed which define how digital video teleconferencing will employ either one or two B-channels. Presumably, a terminal adapter could be required. There is still controversy in the

United States as to whether the terminal adapter is a piece of telephone network gear and hence the responsibility of the RBHC or other telco, or is customer premises equipment to be provided by the user. Pacific Telesis had developed terminal adapter technology under its Project Victoria; the U.S. Federal Court, however, barred the company from proceeding further with development and manufacture. In almost any other country, PTTs are likely to maintain the terminal adapter as part of the network.

The voice telephone function can be added to the terminal because the terminal adapter contains the requisite PCM codec. A "dumb" computer terminal can also be interfaced to the D-channel with a PAD contained within the terminal adapter. Adaptation to different data rates is provided as are alternate physical interfaces such as RS-232C, V.35, *et cetera*. This flexibility is built into the unit by virtue of microprocessor control. Quite obviously, there is not one design for the terminal adapter because of the variety of possible subscriber configurations. All terminal adapters, however, appear the same on the ISDN network side of the connection.

5.2.5.2 *True ISDN IVDT Terminals*

An integrated voice-data terminal for ISDN has within it the terminal adapter function previously described. Therefore, the IVDT can be connected directly to the basic rate interface, typically on twisted pair cable. Integration of voice and data means that the user can employ the telephone function and conduct a simultaneous data communication session with a remote computing device. Due to the wide variety of instrument feature combinations, the range of cost could be as high as ten to one. In its simplest form, the IVDT consists of what looks like a typical PBX station unit with key pad, handset, function keys, and perhaps a speakerphone. This would be supplemented with a *liquid crystal display* (LCD) for network control and data monitoring purposes. In the Fujitsu SRS-2000 ISDN telephone set, the LCD works with internal memory to display the numbers (and possibly names) of missed calls. More elaborate versions add a video display or computer screen which allows pages of information to be received. A keyboard is necessary if the user is to input data as well as receive it. Internal computing power and mass storage could be useful in heavy data processing applications. One IVDT can employ the full capability of the 2 B + D access line; or one of the two B-channels can be left open for another device such as a Group 4 fax machine or freeze-frame teleconferencing system.

Readers should be sensitive to the early stage of IVDT evolution that exists in 1990. There is no doubt that any capability can be built into the terminal at some price. As stated previously, production volumes are still low, making manufacturing costs difficult to reduce. There also is confusion in the user community about what ISDN is capable of doing for strategic units. To make matters worse, early adopters of ISDN in the United States are not satisfied with the reliability and capability of the first IVDTs to be provided along with their ISDN PBXs and CENTREX service. They have had to discover how the terminals operate in the real world by setting up laboratories where prospective users can experiment with the capabilities of the equipment and the network. Credibility has been weakened by a few vendors who have marketed IVDTs that are not even ISDN-compatible.

This confusion is only temporary, as evident by the large commitment to ISDN standards and switching systems by the RBHCs and long-distance carriers in the United States, the international carriers, and the major TAs of Europe and Asia. ISDN terminals, particularly IVDTs, will be designed to meet the needs of real-world users who have found value in the intelligent combination of voice and data communication. The foundation for this may be the plug-in boards that have appeared on the personal computer market, as discussed in the next subsection.

5.2.5.3 Personal Computer Plug-In Boards

Not surprisingly, plug-in circuit boards are already on the market in 1990 to give personal computers direct access to the ISDN basic rate interface. In essence, the plug-in board acts like a modem, although transmission is completely digital and does not employ modulation. An early unit by NEC converted asynchronous data at 9600 b/s into the synchronous data protocol at 64 kb/s for the ISDN B-channel. In addition, the device provides the D-channel signaling necessary to establish either a circuit-switched connection or to send packet-switched information. Newer units on the U.S. market are much more sophisticated, allowing the PC to employ the full capacity of the 64 kb/s channel in the synchronous mode without being throttled down to 9600 b/s.

Another example of an ISDN plug-in board is manufactured by Teleos. The product comes with communication software that is loaded into the computer's memory before the initiation of a session. Applications in telemarketing and data-file transfer can be written around the capability of the board, thus tailoring the terminal to the needs of the particular user. This is precisely the approach that is needed at this early stage of the development of ISDN applications.

Chapter 6
TELECOMMUNICATION NODE EQUIPMENT
IN PRIVATE NETWORKS

Strategic units can purchase the same types of node equipment that telecommunication service providers use to aggregate and route traffic. Other more specialized devices are produced for use specifically by strategic units wishing to improve their utilization of public networks or to gain a business advantage. As part of a private telecommunication network, node equipment delivers advanced services for internal and external users. In most countries, node equipment is treated as just another form of CPE, subject to type acceptance and other forms of regulation. In some, node equipment is considered to be outside the realm of customer-owned devices and may not be legally connected to the PSTN. Many of the devices, particularly PBXs and multiplexers, are available from the TA on a rental or service basis. Therefore, the telecommunication manager can anticipate doing a good deal of research before committing to an expensive private international infrastructure based on owned nodal systems.

The general purpose of nodes is to combine traffic for efficiently loading the long-haul network, achieving economy of scale. The interconnecting links are usually derived from leased lines, but flexible digital services such as ISDN and switched 56 kb/s, can be incorporated as well. This ability to integrate the metropolitan area and wide area transmission capabilities of public and private networks is a key attribute of today's modern nodal devices. Included in this chapter are VSAT satellite communication networks because users can derive benefits from employing small earth stations as nodes for interactive data communication and business video.

The topology of a private telecommunication network that is international in scope is typically different from one that is domestic. As illustrated in Figure 6.1, T1 and DDS lines in domestic private telecommunication networks are usually

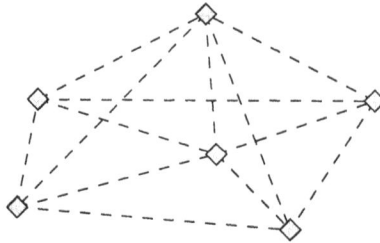

Figure 6.1 Mesh topology used in domestic private telecommunication networks, with links provided by point-to-point private links (T1, 56 kb/s, *et cetera*).

arranged in a mesh structure. Depending on the requirements and the available budget for varied point-to-point links, the mesh offers diverse routing for reliability and bypassing traffic overload. Trends in costs of digital bandwidth are definitely down within the United States, Japan, and parts of Europe. Therefore, mesh structures are certain to become more prevalent. This is in contrast to an earlier preference among data communication users for multipoint-multidrop lines to support host-to-remote applications. Data communication networks are less hierarchical and more distributed in nature, giving further support to the mesh approach [Lippis, 1989].

International private networks are burdened with much higher transmission costs and restricted availability of long-haul facilities. For example, a strategic unit may wish to have diverse routing over transatlantic fiber optic cable between the United States and West Germany; as of 1990, however, only the TAT-8 cable is operational. IBS circuits can be used effectively as an alternative route to TAT-8 for strategic units that demand full-time availability. France Telecom is approaching this potential demand with a unique offering of diverse routing on IBS and TAT-8, which they call TAT-SAT, but some applications are not compatible with the time delay of a satellite link.

In general, international private networks use star architectures, with a major hub located in a central country where good telecommunication facilities are available at reasonable prices. This is illustrated in Figure 6.2 where the spokes from the hub (shown as dashed lines) are connected to several remote locations within a country or region where services are required. The larger the network geographically, the more stars that are installed. The hubs are then connected together using wideband digital links (shown as solid thick lines). A partial or full mesh structure for diverse routing, while desirable for reliability and speed of data transfer, may not be justified on a cost basis. The determination of optimum hub location and backbone topology often requires a thorough analysis of comparative tariffs, such as that suggested in Chapter 3.

Building a network from node equipment is reasonably straightforward in North America and Japan. Doing so on an international scale or even just within

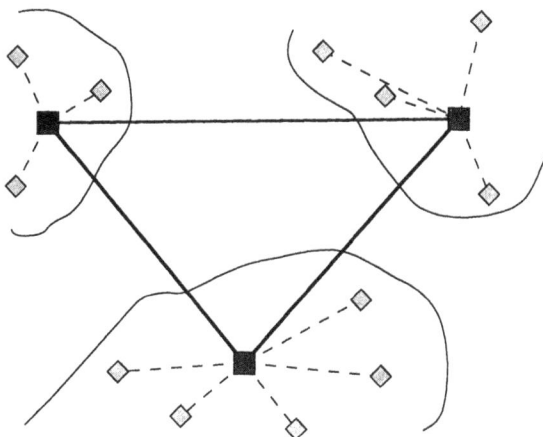

Figure 6.2 Star topology used in international private telecommunication networks, with interconnected hubs.

Western Europe is entirely another matter. Nodal devices have generally been the domain of the service provider, particularly the PTT. The devices may not be available in the country of interest, and maintenance can be a real concern. Some devices are more common around the world than others. For example, packet-switching systems have been deployed in more than 100 countries for use in PDNs which the PTTs operate. Some vendors are more aggressive than others to homologate their products for the particular country and subsequently to arrange for the type acceptance and local distribution. Telecommunication managers may have to take a more open view when extending a private network overseas, engaging in cooperative ventures with the PTT, not unlike the tack taken by Infonet (see Chapters 4 and 8).

In the following paragraphs, we review the most common nodal devices that strategic units employ in private networking. The private branch exchange is key among them because of its flexibility as a switching system for voice and data communication services. Once traffic is aggregated for transmission, multiplexers are used to integrate channels for efficiently loading long-haul media. We review conventional time division multiplexers as well as the intelligent "resource managers" used in T1 networks. These resource managers represent a strategic asset because of the commitment to a particular architecture and possibly even a single manufacturer. The *digital access and cross-connect system* (DACS) is becoming more standardized because of heavy use by telcos and long-haul carriers; hence interoperability between vendor designs is the rule more than the exception. The DACS provides the functionality of the old patch panel for time-slot channel assignments within the T1, E1, or DS3 time frame.

Packet-switching nodes are a good way to load data from a variety of low- and medium-speed data communication devices and applications onto a common digital backbone. Because packet switches are used extensively on international private networks, we review the capabilities and sources in some detail. A related technology called fast-packet switching integrates wider bandwidth services like voice and digital video with data, and is a hybrid of T1 multiplexing and packet switching. *Asynchronous transfer mode* (ATM) has been coined as a technical term from an even higher speed version of fast-packet switching which could some day revolutionize long-haul links and even the local loop. ATM potentially comes into play on international links where the aggregate rate of 140 Mb/s is available. This capacity can be employed in *broadband ISDN* (B-ISDN) to integrate broadcast quality video, voice services, and any data communication application. Much of the work in this area is still at the experimental and definition stage, although some pioneering products are on the market. A further discussion of ATM can be found in Chapter 10.

Bandwidth on international long-haul facilities is still relatively expensive, so telecommunication managers are always searching for ways to squeeze more services into a smaller "pipe" and avoid the expense of a full T1 or E1 circuit. The basic approach is to use a new class of bandwidth-efficient multiplexer which subdivides a 56 or 64 kb/s leased line into separate voice and data channels. Then, data at rates of 9600 or 19,200 b/s can be combined with voice channels for efficient utilization of the available bandwidth of the international leased line. Speech compression technology exists which can squeeze an intelligible voice channel from 64,000 down to 8,000 b/s. This type of circuit would be acceptable for normal voice communication, particularly where users are accustomed to the noise and clipping associated with old analog circuits on coaxial cable and satellite links. A digital 8,000 b/s compressed voice circuit sounds good to these listeners because impairments are nearly absent. This narrow bandwidth, however, cannot be used to transfer data using voice-band modems.

There is an important concern about using digital multiplexing systems on international circuits. Timing of the bit stream is critical to the operation of the nodes on the ends of the circuits because references used by various countries in various regions can be appreciably different. It is fairly common for data transmission to become impaired when nodes are connected for the first time due to inconsistent timing or synchronization. This problem will eventually be diminished as the networks of the world are synchronized to the same time standard. Today, the problem is critical and telecommunication managers must consider how the equipment is to take care of the slight timing differences that exist. Equipment can be obtained which includes buffering and elastic clocks that account for timing differences, which will reduce the frequency of synchronization loss. The actual applications on the network would still have to respond to such loss by retransmitting the data or, at the worst, reestablishing the communication sessions.

6.1 PRIVATE BRANCH EXCHANGE

Private branch exchange (PBX) and private automatic branch exchange (PABX) are two terms that refer to the same system. The principal purpose of the PBX is to provide switching of telephone calls within an organization and to permit calling to and from the outside world [Elbert, 1989]. Many more features and capabilities are included, allowing the PBX to play an important strategic role for the organization. An example of this is telemarketing, where the PBX works as a call distributor to manage a workforce. Employees can be supplied with digital workstations which integrate voice and data so that customers can be assisted with speed and consistency.

Today's PBX is a versatile electronic time-division switching system which operates under the control of redundant computer processors. Many are downsized telephone exchanges, sharing technology and even hardware elements with systems used by telcos. Telecommunication managers now look to the PBX to handle local switching services for data users who previously might have employed a separate data PBX or even LANs. One of the properties of a PBX is its ability to act as a tandem node, i.e., allowing calls from other PBXs to be routed through it as an intermediary switching point. It can also interface directly with analog voice lines, converting the information into a standard PCM format. These attributes give the telecommunication manager the option of relying on the PBX for virtually all nodal functions as opposed to T1 multiplexers and data-switching devices.

PBX manufacturers are now expected to deliver systems which can interconnect and interoperate with the PSTN in all its variations. For the most part, PBXs can satisfy signaling and interface standards contained in appropriate CCITT recommendations. We emphasize that connection to public ISDN services is offered as an option by the leading suppliers of PBXs, which are companies that have major stakes in the carrier switching market. Interface flexibility, however, has its price in financial terms because PBXs tend to be expensive on a per-line basis. Part of the cost is in the line adapter and port cards, and part is in the complex software that must be loaded into the processor. Additional computer power and memory may have to be included, particularly for voice-mail, network-management, and call-distribution services. The more voice-data functions that are performed, the higher the per-line investment which the strategic unit can be expected to make. A standard system may well cost approximately one-half of the price of a fully configured ISDN system of the same capacity [Sulkin, 1989]. On top of this is the higher cost of digital telephones and IVDTs, as reviewed in Chapter 5.

Leading manufacturers in North America, Japan, and Western Europe are listed in Table 6.1; they produce PBXs for domestic and export markets, making the PBX a particularly hot product. Importantly, PBXs give the leading telecommunication manufacturers of the world another outlet for their technology and production capacity. Current units are directed toward a digital infrastructure

Table 6.1 Manufacturers of PBX Equipment

North American
AT&T
Harris Corporation, Lanier Voice Products
InteCom, Inc.
Mitel, Inc.
Northern Telecom, Inc.
ROLM (IBM, Siemens)

Japanese
Fujitsu
Hitachi
NEC Corporation
Toshiba

European
Alcatel
GPT Plessey
Italtel
L.M. Ericsson
Siemens

where the PBX can be connected directly to the TA through T1 and E1 access lines. ISDN is available as an option to give strategic units an advantage in evaluating how to employ such capabilities. As PBXs installed in different countries may be supplied by different manufacturers, ISDN compatibility is attractive because of the potential for uniform voice-data services, regardless of location.

Because this marketplace is occupied by major telecommunication equipment manufacturers and because PBX systems are "big ticket" items, vendors are continually upgrading their product lines and software features. Upgrades should be made by inexpensive changes, such as loading a new software release, adding memory, or exchanging particular circuit boards. One would hope that new capabilities can be added without replacement of major portions of the physical hardware, a process which is facetiously called a "fork-lift upgrade." As an example, Northern Telecom's SL-1 PBX can be upgraded and modernized by any one of the three following options:

1. Adding memory and loading a new software release;
2. Replacing the processor with a new system in the form of a shelf;
3. Replacing the entire main cabinet.

Each option carries with it substantially more capability and potential for future expansions.

6.1.1 PBX Architecture

Figure 6.3 provides a simplified block diagram of the basic PBX design. The terminology and exact configuration differ from manufacturer to manufacturer and even between product lines of the same manufacturer. A typical PBX consists of the following principal components: a time division switching matrix; a control processor with memory; interface modules for lines and trunks; an operator terminal; data communication facilities for interoffice signaling and remote access; and supporting multiplexing and encoding equipment as required. The heart of the system is the time division switching matrix, which does the actual interconnection of lines with lines (internal communication within an office complex), lines with trunks (calling to external locations both on the network and off net via the PSTN), and trunks with trunks (e.g., tandem switching). The control processor

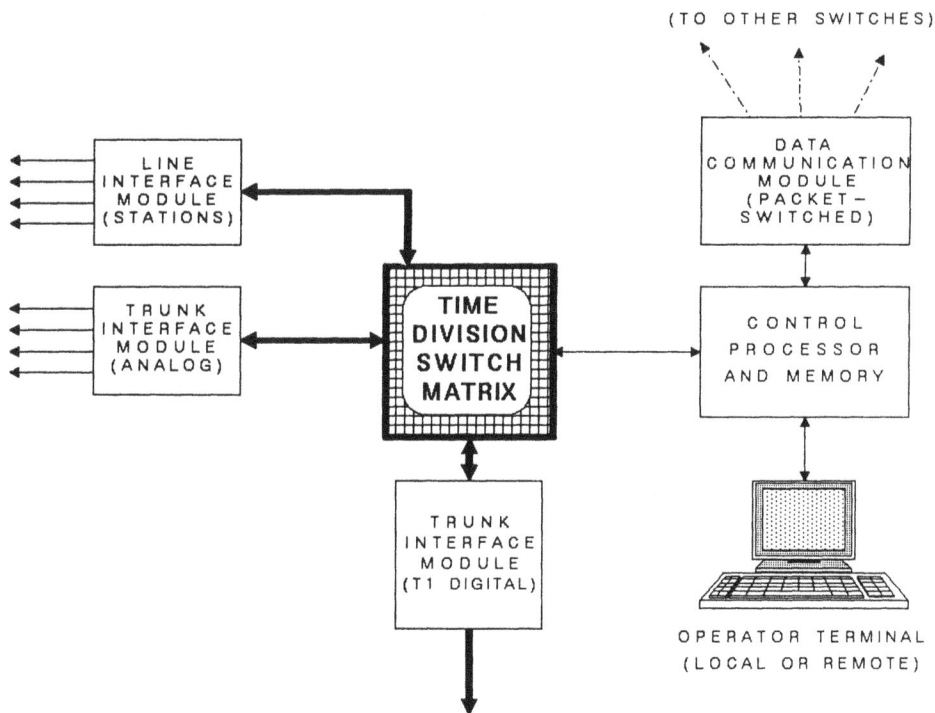

Figure 6.3 Basic architecture of a private branch exchange (PBX).

may be dual redundant where either computer can run the switch in the event of failure of the operating unit. The database containing number identification, least cost routing, and administrative records can also be maintained by the control processor, although it is possible to carry these functions over to a separate mini-computer or even a PC. Northern Telecom is working closely with Digital Equipment Corporation to integrate DEC's VAX computers for this type of purpose. A similar tack is being taken by Rolm and IBM, where the associated computer is the A/S-400 series. Integration of computing and switching is being pursued for such new service concepts as *automatic number identification* (ANI), potentially valuable in telemarketing applications.

The programming of the PBX's computer is very complex because the control functions must respond to the dynamics of telephone calling, answering, signaling, and a host of other functions which are continually expanding in scope. With central processor control, the dynamics of organizational growth are easily accommodated. The feature called "adds, moves, and changes" is how a PBX can be conveniently reprogrammed when individuals are moved within a building or between buildings. Accounting for calls, including the management of telephone service almost like a business, is also available as an option.

The leading PBX manufacturers are very active in Europe where the tele-communication equipment market is being deregulated. Northern Telecom, AT&T, and Fujitsu, among others, have taken the necessary steps to win technical approval of their products. To do this, the vendor must tailor the interface between the standard product line and the requirements of the public network, a process called *homologation*. Fortunately, the PBX is a versatile device, and homologation usually only amounts to programming changes. Signaling is an important consideration, but the European PTTs have rationalized their respective systems. As discussed in Chapter 3, SS-7 is quickly becoming the international standard for interexchange signaling on the continent. Major vendors have demonstrated the interworking of SS-7 for their highest priced PBXs.

The telecommunication manager has many options to consider when making a move in the PBX marketplace. As a basic alternative, it is possible to obtain the capabilities of the PBX without making a direct commitment to the investment. The service called CENTREX is offered by U.S. telcos to give the strategic unit most or all of the capabilities of the PBX without the operational headaches. Financial risk may or may not be diminished, because a long-term commitment to CENTREX service may still be obliged. CENTREX capabilities are offered in European countries, particularly the United Kingdom, as well as in Japan and leading Asian cities.

We will review the suppliers of PBX equipment in the three key regions of the world. Many of these companies export their products to other countries. A comparison of the market shares of the leaders is presented in Figure 6.4 for the year 1988. In this discussion, we focus on those suppliers with a decided

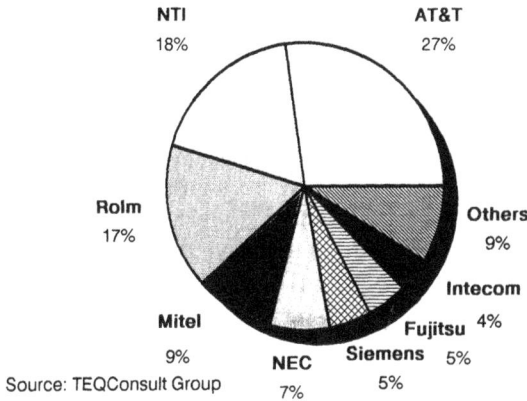

NTI 18% AT&T 27%

Rolm 17% Others 9%

Intecom 4%

Mitel 9% Fujitsu 5%

Source: TEQConsult Group NEC 7% Siemens 5%

Figure 6.4 Market share of major PBX manufacturers.

international presence. PBXs are rated in terms of the maximum number of lines for stations and trunks which connect to the telco or other PBXs. Ideally, the switching matrix should be of the nonblocking type, meaning that incoming calls will be guaranteed access to the switch (but not necessarily the external line). Blocking designs introduce an additional source of contention for facilities, manifested by dialing delay experienced by the local subscriber or a busy signal for an incoming call over a trunk.

6.1.2 North American Manufacturers

North American PBX manufacturers really created the PBX market in the first place. The first PBXs were small electromechanical telephone exchanges manufactured by Western Electric, but the true digital PBX was introduced by much smaller companies wanting to develop a market. One of the first was produced by a startup company with the obvious name of Digital Telephone (now Lanier Voice Products Division of Harris Corporation). This company defined the small- and medium-sized PBX as a versatile digital machine with features not found in public exchanges or PBXs of date. This was created with digital coding and the programming sophistication and flexibility of the minicomputer. The market has grown tremendously to the point where approximately $2 billion is spent on PBX hardware and software every year. Because PBX sales are not likely to grow in North America, local manufacturers are targeting overseas markets where deregulation is taking hold.

Principal among North American companies is Northern Telecom, a Canadian company which broadened its business base through sales of PBXs and long-distance switches to the U.S. market. Their products have become ubiquitous in North America and subsequently in Europe. Northern Telecom's foremost PBX product is the SL-1, having a maximum capacity of 7000 lines. Stepping up from this basic system, the SL-100 can provide essentially all of the capabilities of a central office exchange and is comparable to the Number 5ESS of AT&T, discussed below. The SL-100, with a capacity of 30,000 lines, is closely related to NTI's DMS-250, a central office switch popular among long-distance carriers and international carriers as well. The NTI PBXs as a group have generally been more compatible with digital standards including SS-7 and ISDN. For example, it was possible to obtain a true T1 interface for the SL-1 and SL-100 before several of the other leading manufacturers made it available.

AT&T has not been left behind after having created the PBX market in the first place. In 1989, they combined their System 75 and System 85 product lines to create the Definity Generic series. Maximum line capacities of the Generic 1 and Generic 2 models are 1,600 and 32,000, respectively. ISDN access includes both the basic rate interface and the primary rate interface. The BRI connects to ISDN station equipment and terminal adapters so that users can employ 2 B + D services. One prime example is ANI for incoming customer calls. AT&T can configure the PBX to perform the functions of an automatic call director. On the PRI side, the switch interfaces directly with another PBX or central office exchange to employ SS-7. For example, to use ANI, the local telco must provide a packet-switched routing capability over SS-6 or SS-7 between the PBX and AT&T's centralized database.

AT&T has successfully gained market share through an aggressive pricing policy. Prices of stations and line cards have been reduced ten to fifteen percent since 1987 and that of the common control complex about 70% since 1984 [Sulkin, 1989]. Prices of the combined Divinity series have been reduced even further. As in the mainframe computer business, the PBX market is very competitive and manufacturers understand the importance of market share and production volume.

Larger users may be able to justify the purchase of the *Number 5ESS* central office exchange, which AT&T has sold to companies such as American Express and General Motors. Number 5ESS is the fifth stage of evolution of AT&T digital switches and is one of the most purchased central office exchanges in the United States. Number 5ESS represents AT&T's "calling card" for ISDN because of the ease with which customer-owned and telco-owned systems can be connected together. The 5ESS switch can be considered for PBX installations where 10,000 or more lines are required. One of the obvious benefits of going to this switch is its large growth capability. In addition, AT&T is committed to this market segment which supports telcos and their own domestic and international long-distance business.

After unsuccessful attempts at internal development, IBM joined the established PBX marketplace with their purchase of leading PBX manufacturer Rolm.

Originally seen as a major strategic move into business telecommunication, IBM subsequently divested itself of the Rolm manufacturing arm through a complex transaction with Siemens. Siemens and IBM continue as 50-50 partners in the service _ ortion of the business although Siemens handles all equipment sales. IBM completed an extensive redesign of the Rolm *computer branch exchange* (CBX) product line, giving it a definite "Big Blue" tint. On the acquiring side, Siemens (discussed below) moves strongly into the U.S. PBX market and worldwide to a position in close proximity to leaders AT&T and NTI.

Designed for the U.S. corporate telecommunication market, the 9751 PBX product line was developed by the Rolm division under IBM's watchful eye. The 9751 was hoped to bring IBM up to the level of the two leaders in the industry. Quite plausibly, IBM users would view the 9751 with the same favorable eyes as they do IBM mainframes and PCs. After all, "no one ever got fired for choosing IBM," as the slogan says. The system has a high capacity capability, being expandable to 20,000 lines. Users of earlier CBX switches with the Rolm label were given an opportunity to change over to the 9751. For voice services, users have been very satisfied with the performance, reliability, and cost of the newer series. The 9751 can operate as a data switch where computer terminal users do not need a modem or telephone to establish a circuit-switched connection. Commands to the switch are issued through the keyboard itself in the same manner as with a data PBX. This is a proprietary system and would not be directly compatible with ISDN. IBM, however, has announced an ISDN interface available in 1990.

Prior to the Siemens acquisition, IBM had announced that it would introduce a related PBX product line for the European market; these development efforts, however, were abandoned as of 1989. It is likely that Siemens will use the 9751 as its North American entry and rely on other PBX product lines developed for application in Europe. Siemens products, originally developed for the European market, are discussed subsequently.

The other principal North American PBX manufacturers are Mitel of Canada (an affiliate of British Telecom PLC), InteCom, and Lanier Voice Products (a Harris Corporation subsidiary). Interestingly, all three of these companies were entrepreneurial ventures (as was Rolm). Other important suppliers to the U.S. market include Fujitsu, NEC, and Siemens. These companies are discussed in subsequent paragraphs. PBX products for the United States have to be more sophisticated than those designed for other countries (with the possible exceptions of Canada and Japan) simply because there are so many more options already available in terms of telecommunication services and competing PBX manufacturers. Here in the United States, we have three main long-distance carriers with several others vying for the business on a regional basis. Each carrier has several different bulk calling packages, as discussed in Chapter 3. Local telcos and RBHCs are also introducing new service packages to expand their business and protect their market in the face of bypass and possibly the day when competition will be allowed on the local loop. Signaling System No. 7 and ISDN are two capabilities which local and long-distance companies are offering to strategic units that have the appropriate PBX and CENTREX installations.

To take advantage of the options, PBXs contain software and memory to evaluate the lowest cost long-distance service for each individual call. Referred to as *least cost routing* (LCR), the owner-operator loads the memory of the PBX with routing tables containing the order of preference for selecting a carrier for outgoing calls. The same control processor can support administrative operations and switch maintenance. A flexible offering in this regard is the GTE Network Management Control Center, which is discussed in [Elbert, 1989].

6.1.3 European Manufacturers

The European manufacturers listed in Table 6.1 are developing their home market as telecommunication equipment ownership is deregulated. All have embraced the U.S. market, however, because of its sheer size and diversity. Any product that meets the American challenge is bound to be viable in Europe. To this end, Northern Telecom and AT&T have made strong showings on the other side of the Atlantic. This has given the European manufacturers a lot of competition within their home territory at a time when local markets are opening up in preparation for 1992.

The leading indigenous supplier in Europe is Alcatel, the French-based electronics conglomerate that absorbed ITT's European operations. Alcatel One is the U.S. version of Opus 300, a system that has had great success in Europe [Knight, 1989]. The product has a maximum capacity of 895 lines; it can be configured for normal voice service or for data service at up to 4800 b/s.

Siemens intends to become a world leader in the PBX market by combining their product line with that of Rolm. The Saturn models II E and III are Siemens' current PBX products, offering a modest capacity of 864 lines. This pales in comparison to the much higher bound of 20,000 lines of their newly acquired Rolm 9751 PBX.

The main domestic switch manufacturer in the United Kingdom is GPT, a large corporation formed by the merger of Plessey and General Electric Company of the United Kingdom. The Stromberg Carlson subsidiary in the United States manufactures switches primarily for independent local telcos.

The Scandinavian telecommunication power, LM Ericcson, supplies PBXs to the European and North American markets. This company competes with Alcatel on a worldwide basis for the public telephone exchange business of PTTs. Therefore, they understand the telephone switching problem in all its dimensions. The MD 110 has a maximum capability of 20,000 lines, putting it up there with the largest PBXs available from AT&T, NTI, Rolm, and the Japanese. Ericcson claims to have installed two million lines in more than 40 countries and emphasizes that the MD 110 has a distributed architecture to simplify geographic expansion.

6.1.4 Japanese Manufacturers

Japanese electronics companies have a way with digital communication systems, particularly in voice, video, and facsimile. They also have invested heavily in ISDN technology and can offer this capability with their equipment. The Japanese government has recognized that the export of telecommunication equipment in general and switching systems in particular represents an avenue of growth. Joining NEC and Fujitsu, the leaders in switching systems, are Hitachi, Toshiba, and Oki making it difficult for any one company in Japan to dominate.

The largest supplier of PBXs and central office switches in Japan is NEC, and they have a reasonably good position in the U.S. PBX market as well. NEC is a long-standing advocate of the thoughtful coalescing of communication technology with computers and this campaign appears to have paid off. The NEAX 2400 is gaining respectability as a large PBX with up to 12,000 lines. NEC uses this system as a platform for integrated voice and data services and ISDN. The 2400 is a relatively mature switching system because it benefits from years of Information Network System field trials by NTT in Japan. We have already discussed NEC's development efforts in the area of IVDTs and terminal adapters. Several U.S. companies and universities have selected the NEAX 2400 in recognition of NEC's reputation in ISDN and digital communication in general.

Sales in the United States are through the NEC American subsidiary, which assembles or manufactures the PBXs and other telecommunication equipment. NEC has for more than twenty years been developing the U.S. market and is now a well recognized brand name, particularly in telecommunication. A European subsidiary is set up in France and in the United Kingdom.

NEC has emphasized the capabilities of the NEAX series in the area of integrated network management. As discussed in [Elbert, 1989], NEC offers an on-line computer system to manage both the technical aspects of the network and the administration of services which the network provides. The second area is usually overlooked by equipment vendors who only provide remote monitoring and control facilities, leaving the telecommunication manager to his or her own devices when it comes to call accounting, cost allocation, and inventory control.

A second substantial supplier of switching and computer systems, Fujitsu, has several established product offerings in the international PBX market. The U.S. market had been approached through a joint venture with GTE; Fujitsu, however, has now taken over responsibility for its own U.S. marketing and manufacturing. The highest priced F9600 is an ISDN compatible PBX with a capacity of up to 10,000 lines. Smaller models such as the F960 PBX and Focus hybrid key system are particularly popular because of the high ratio of performance to cost. Fujitsu products are sold throughout the world, with manufacturing and distribution centers in Europe.

Two other Japanese manufacturers, Hitachi and Toshiba, concentrate mainly on the smaller systems and hybrid key systems. Their small PBXs are popular with cost conscious telecommunication managers who are not looking for ISDN capabilities and expansion to a large system configuration. Hitachi introduced the HCX5000 series of PBXs in 1989, offering an architecture which is compatible with the ISDN specifications of AT&T. Up to 3000 lines can be supported with the product, placing it in the middle range of the U.S. market. Both analog and digital stations can be connected, including a line of proprietary IVDTs. Another player is Oki Electric, one of the oldest telecommunication manufacturers in Japan. With a primary emphasis on the Japanese domestic market, they offer the IOX-1000 PBX as the core of the Oki Digital Information Network System.

6.2 DIGITAL ACCESS CROSS-CONNECT SYSTEMS (DACS)

With all of the T1 and E1 channels crisscrossing the continents and oceans, it becomes essential that the telecommunication service providers and strategic units be able to connect and route basic DS0 (64 kb/s) time-slot assignments efficiently. This can be accomplished by a digital "patch panel" with functionality not unlike an old-fashioned manual switchboard. Such a device does not operate in real time to set up and take down calls or route packets based on the dynamics of user demand. Instead, the basic routing and connecting function is fairly static, changing perhaps a few times a day or even much less frequently.

This type of capability is provided by the *digital access and cross-connect system* (DACS), a term coined by AT&T and used generally throughout the industry. Another acceptable term is *time-slot interchange* (TSI), referring to the way that this multiport device operates. The time slots are individual channels within the T1 frame, each containing one or more bytes or words of data for a particular channel. A DACS is rated by the number of ports that can be cross-connected: a typical small unit may have eight ports for private network applications; or as many as 128 ports occupying several equipment racks for use by TAs. The schematic diagram for an eight-port DACS is shown in Figure 6.5; the time frames have been abbreviated for simplicity, but the figure shows how individual DS0 channels on any T1 channel can be connected through within the DACS to any other T1 channel. In effect, the channelization can be completely rearranged, under software control with an internal processor. A complete discussion of DACS technology can be found in [Minoli, 1988].

DACS technology can benefit a strategic unit by reducing the number of T1 circuits needed to access a major network. Without the DACS, a separate T1

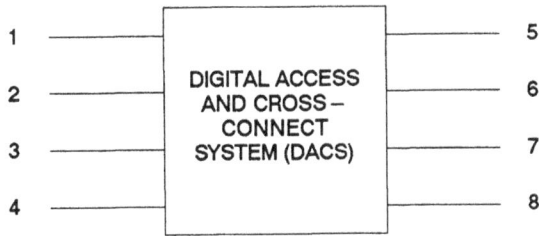

(A) SCHEMATIC DIAGRAM OF DACS

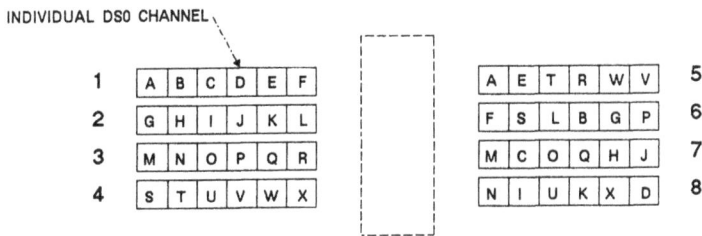

(B) RELATIONSHIP AMONG TIME SLOT ASSIGNMENTS BETWEEN PORTS
WHICH CAN BE CHANGED BY PROGRAMMING

Figure 6.5 Application of digital access and cross-connect system (DACS) for use in T1 time-slot interchange. (a) Schematic diagram of DACS. (b) Relationship among time-slot assignments between ports which can be changed by programming.

access line would be needed for every destination. This multiplies the cost of the network particularly because T1 access lines tend to be much more costly per unit distance than long-haul circuits. As illustrated in Figure 6.6, a T1 line coming in from a user's location can contain channels for a variety of destinations. The carrier uses the DACS to route the DS0 channels to the particular outbound T1 channels. Changing the division of channels among destinations and even adding destinations

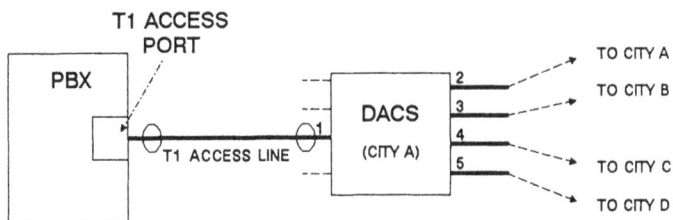

Figure 6.6 Application of telco-owned DACS for efficient access to a digital long-distance network.

are also simplified because the local physical circuit can remain the same (unless the aggregate capacity exceeds the total available). In this age of backhoe fades on terrestrial fiber links, carriers make good use of DACS to provide redundant paths since entire T1 channels can be reconnected without making physical changes. Being a programmable device, a given switching pattern can be preloaded into the DACS and then activated only when the need arises. Smaller DACS configurations (i.e., four ports) could be purchased by the strategic unit for use at a major international node to achieve the same aims of TAs.

The leading vendors of DACS are listed in Table 6.2. DACS technology was pioneered by AT&T and they continue to dominate the market, particularly with regard to central office installations. Their systems as well as those of DSC, NTI, and Rockwell tend to be large. More compact DACS are offered by Telco Systems, Tellabs, and Coastcom. International participants in the market include NEC and Siemens, making this product fairly common in the industrialized countries.

Because DACS need to be interconnected with public networks, interface standards and network interoperability are vital to consider during specification and acceptance testing. Units made for the U.S. market almost always support standard T1 framing, commonly referred to as the D4 frame. Within the frame are 24 eight-bit bytes plus a framing bit (i.e., the 193rd bit). In the *extended super frame* (ESF) specification of AT&T, the 193rd bit is used as a distinct data channel operating at 8000 b/s to improve transmission performance and maintain timing reference. Compatibility with ESF is becoming more important in the United States because the specification permits the full eight bits of data to pass through the DS0 channels in a transparent mode without regard to the 1s density restriction mentioned in Chapter 3.

There are instances where the T1 channel is arranged differently on the private backbone so that the DACS cannot change time-slot assignments as transparently as before. For example, many of the high level multiplexers discussed later in the chapter use proprietary frame arrangements, incorporating a unique common signaling channel. A DACS could break the continuity of such a frame, completely disrupting the operation of the private T1 network. Before choosing a high level multiplexer, telecommunication managers need to consider if DACS are employed in the network and, importantly, if the multiplexer is compatible with their operation. Compliance testing is a must.

Table 6.2 Digital Access and Cross-Connect System

AT&T
Coastcom
DSC
Northern Telecom, Inc.
Tellabs
Timeplex (division of UniSys)

6.3 HIGH-LEVEL (T1) MULTIPLEXERS

The high-level multiplexer, introduced at the beginning of Chapter 5, is a sophisticated manager of digital traffic on an integrated voice-data-video network. Other names that are applied include T1 multiplexer, intelligent multiplexer, and T1 resource manager. In this section, we use T1 in reference to both the North American–Japanese T1 and the European (CEPT) E1 because nodes can serve either standard. The high-level multiplexer can sense when traffic is present (as with a statistical multiplexer) and assign bandwidth as needed. In comparison with the DACS, which is mainly a device for telcos and other service providers, the high-level multiplexer is clearly targeted to the strategic unit because of the flexibility and versatility that are provided.

What high-level multiplexers of different manufacturers have in common is the ability to combine digitized voice, data, and video signals from CPE and direct the information over a T1 backbone. Each node is equipped with one of these multiplexers because the system has a built-in signaling system to monitor performance and control the network configuration. Owing to the proprietary nature of the individual vendor's hardware and software, the T1 network is completely homogeneous. The good news is that the network can be expected to operate smoothly, regardless of location and service provision; the bad news, however, is that all nodes must be supplied by the same vendor and future additions must be with compatible nodes.

What the high-level multiplexer replaces in the telecommunication manager's inventory is the conventional fixed-assignment TDM multiplexer and PCM channel bank. These forerunners are installed for a specific network requirement and cannot be altered without reengineering of the network followed by a major physical changeout process. The time-slot assignments must be fixed at time of installation either with jumper wires or mechanical switches. While digital transmission quality is excellent, traditional TDM equipment is only slightly more convenient to employ than antiquated FDM equipment so common prior to 1980. Telcos and long-distance carriers continue to employ conventional TDM equipment because of heavy investment and because most of the flexibility in their networks is built into the telephone switches (which are digital) and DACS. For the telecommunication manager working in the domestic or international environment, the only multiplexers that are of value for investment purposes are the intelligent devices that are reviewed in these pages.

A high-level multiplexer is basically a high-speed computing device with data storage and an internal processor with software to control the many facets of operation. The data storage acts as the digital switch to take synchronous bytes of data from the local devices and route them to the appropriate T1 trunk line that connects to a distant node. This particular function is essentially the same as the time-slot interchange used in the DACS or the switching matrix found in a PBX (see Figure 6.1). The key aspect in this case is that the high-level multiplexer is extremely flexible in its operation and programmability. Memory is used to store the network configuration so that the user can input changes in circuit routing and priority. The typical high-level multiplexer design has a data bus architecture where all data are applied to an internal common high-speed channel (not unlike a LAN). This permits additional devices to be attached as required to support the network features and growth. Computation and data storage are critical to the operation and functionality of the multiplexer, hence software is a major constituent of the development cost. Being software driven, it is relatively easy to modify the capability of the multiplexer to provide new services or to interface with a different network. As a cautionary note, modification of software for a new interface specification, such as that posed by a PTT, could require extensive software development by the manufacturer.

Figure 6.7 illustrates the configuration of a high-level multiplexer and its application in a partial mesh network. An important attribute of the multiplexer, shown in Figure 6.7(a), is the modularity provided by port and trunk interface cards, which are printed circuit boards, for various user devices and subnetworks that are to be integrated into the T1 backbone. This modular approach is generally taken in PBX and conventional TDM multiplexer configurations because of the flexibility it affords the user. Port cards provide the connection for local lines to CPE on the loop side; illustrated are connections to a PBX using a T1 line, high-speed data at 256 kb/s, low-speed data (9.6 kb/s typically), and voice tie lines. The trunk cards provide connection to the T1 circuits between nodes on the line or trunk side of the multiplexer. Because of the programmability of the interfaces to the system, port and trunk cards can be fairly complex little digital machines, benefiting from the use of microprocessors and firmware. Quite logically, cards are not interchangeable between manufacturers (a shortcoming carried over from the nodes themselves).

The established manufacturers or suppliers of high-level multiplexers are listed in Table 6.3. In 1990, the only non-U.S. source was Fujitsu of Japan, which responded to the rapid growth of private T1 and fractional T1 private networks in that country. These companies have taken steps to homologate their products for the European market. Manufacturers build their systems to be able to accommodate eight T1 channels as a minimum, although capacities of sixteen or thirty-two T1s are not uncommon. The primary functional characteristics common to high-level multiplexers include the following:

Integration and combined transmission of digitized voice, medium- and high-speed data, and digitized full-motion video;

Dynamic bandwidth allocation to increase the ratio of combined input data capacity to aggregate capacity on the T1 trunks (similar to, but not the same as, statistical multiplexing);

PORTS   **T1 TRUNKS**

PBX (T1 ACCESS) ——

HIGH SPEED DATA
(256 kB/s and up) ——

LOW SPEED DATA ——

VOICE TIE LINES ——

HIGH
LEVEL
MULTI–
PLEXER
(T1)

—— TO B

—— TO C

—— TO D

—— TO E

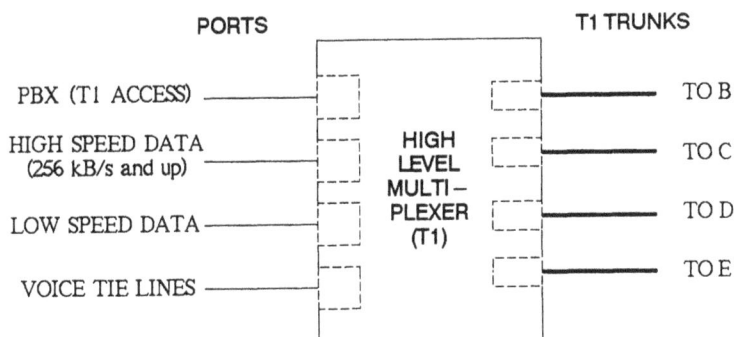

(A) FUNCTION ARRANGEMENT OF NODAL EQUIPMENT

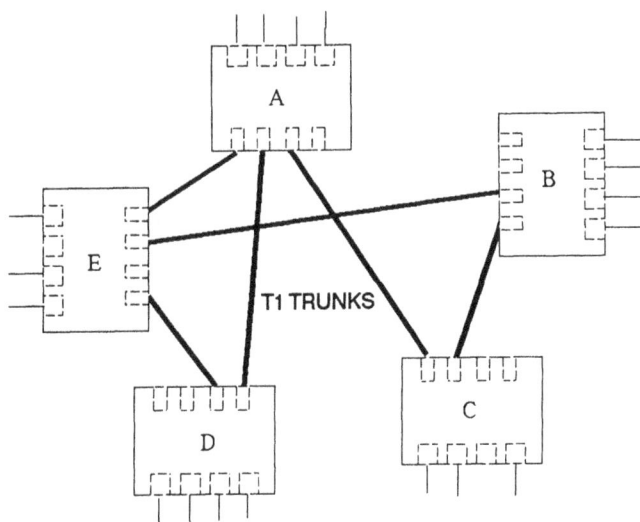

A

B

E

T1 TRUNKS

D

C

(B) EXAMPLE OF A MESH NETWORK

Figure 6.7 The high-level multiplexer used in backbone T1 networks. (a) Function arrangement of nodal equipment. (b) Example of a mesh network.

Automatic alternate routing on T1 trunks to maintain service when individual trunks are congested with traffic or are inoperative;

Synchronization of clocks to external or internal sources, based on user option and availability;

Ability to create a network channel assignment matrix and routing (e.g., the network map) through a reasonably simplified human-machine interface such as a personal computer;

Table 6.3 High-Level Multiplexers (T1 Resource Managers)

North American
Avanti
DCA, Cohisive
General Datacom, Inc.
Network Equipment Technologies
Stratacom
Tellabs
Timeplex (division of UniSys)

Japanese
Fujitsu

Integrated network management facilities to simplify monitoring and control as well as administrative management of the T1 network;

Voice channel multiplication through compression (such as ADPCM) and *digital speech interpolation* (DSI);

Modular design with redundancy and a wide variety of access port options for line speeds, interface standards, and loop signaling conditions.

Equipment from different manufacturers provide these features but often in differing ways, which is why dissimilar units are incompatible. The area where incompatibility is intolerable is in regard to the interfaces with user devices and with the public networks that provide the T1 trunks.

The pioneer of this technology is Network Equipment Technologies (NET), the creator of the original products in the form of a closed architecture. The market is beginning to mature to the point where telecommunication managers can now consider using other equipment in concert. For example, it might be desirable to use a DACS on the trunk side of a high-level multiplexer in order better to load the T1 channels. At least one vendor, General Datacom, allows users to connect their high-level multiplexer to less expensive TDM multiplexers at nodes which do not require sophisticated network control features.

Dynamic bandwidth allocation and voice channel multiplication are two features which must be considered carefully when using these multiplexers with PBXs and other intelligent communication devices. The most efficient interface between PBX and multiplexer is the T1 or E1, where the multiplexer actually processes time slots directly from the PBX. The dynamic bandwidth allocation function of the multiplexer assigns DS0 capacity to active voice-data channels at the port access lines; this demand-related process, however, might defeat the operation of the PBX or external subnetwork. A case in point is how the multiplexer can be programmed by the user to assign input time slots and outbound trunks, much like the function of the DACS. To obtain circuit multiplication, however, the number of input channels must exceed the number of available output channels on the trunk side of the interface. Believing that a particular port channel is assigned to a particular destination, a PBX would apply a telephone call from a local user to a particular channel. Suppose that no time slots are available on a trunk to that destination, even though the input channel is configured in the multiplexer's database. This should cause an indication of a blockage to be fed back to the PBX. The key point is that the PBX must understand this message and be able to handle the call appropriately, possibly by feeding a "fast busy" tone to the caller.

Issues such as these are many and varied and the telecommunication manager will have to deal with them when considering intelligent nodal devices such as the high-level multiplexer. For international T1 networks in particular, the possibilities are multiplied by the number of countries involved. It would be important to have the ability to use both T1 and E1 channels on the same network. Also, the system for timing and synchronization must be flexible enough to deal with variations between countries and regions. The problem is not insoluble because several manufacturers and users have broached it successfully, judging by the size of the market and its continued growth. Manufacturers of the equipment are having to deal with real world interface problems and so the "bag of tricks" is continually getting larger. The key point, however, is that these issues must be faced early.

Vendors are taking steps to remain ahead of the competition and respond to new desires of telecommunication managers. For heavy domestic backbones, a DS3 interface has become a keen area of interest for applications such as CAD/CAM and broadcast video switching. The leading vendors have either embarked on internal development programs or have aligned themselves with companies with DS3 technology through mergers and marketing agreements. For the DS3 to become important on international backbone networks is doubtful, but one ought not rule out the possibility that such a service may be needed in particular cases.

6.4 PRIVATE X.25 PACKET SWITCHING

Many of the motivations that cause strategic units to purchase PBXs for private switching and T1 multiplexers for private wideband networks are at play in diverse data communication networks as well. Corporations and government agencies are implementing private packet-switched networks along lines parallel to the PSNs discussed in detail in Chapter 4. Telecommunication managers see that such a network is very flexible in routing data traffic on domestic and international private lines, allowing diverse terminal and computer equipment to communicate almost without regard to protocol constraints. The primary reason for this is that both the private packet-switched network and the PDN adhere to CCITT Recommendation X.25, which specifies the user interface. (See Table 6.4.)

Table 6.4 Packet-Switching Equipment

North American
AT&T
BBN Communications
Hughes Network Systems
Micom Systems
Netrix
Plantronix
Northern Telecom, Inc.
Telematics
Telenet
Tymnet

European
OST
Philips
Siemens

Japanese
Fujitsu
NEC Corporation

The significance of X.25 is profound. Every major computer manufacturer has recognized the importance of X.25 and provides both hardware and software to employ its flexible and reliable network routing facility. As presented in [Stemmler, 1988], telecommunication managers recognize the following strengths of X.25 as they relate to a private network implementation:

Corporate interoperability — a variety of unrelated data communication applications and subnetworks can be integrated onto the same backbone network. This can be superior to the STAT MUX, which is more of a point-to-point technology (switching STAT MUXs notwithstanding).

International connectivity — X.25 is an established international standard for use in public and private networking. Private networks in different countries can theoretically interface with each other and with domestic and international PDNs as well.

X.25 applications — the network, link, and physical layer protocols of X.25 fit below higher layers of the OSI seven-layer stack. This almost guarantees that as applications like X.400 E-mail and *file transfer access method* (FTAM) mature, the X.25 network will increase in value to the strategic unit.

IBM support of X.25 — with the largest worldwide customer base of any computer company, the acceptance of X.25 by IBM guarantees that a private network using compatible nodes can achieve economy of scale relatively quickly for companies which are IBM customers.

Maturity of X.25 as a standard — after some ten years of development and application, X.25 is thoroughly tried and tested. Telecommunication managers can select from nodal equipment and terminal devices with several years of experience.

For the largest organizations with the need to conduct interactive data communication for a large user population, private packet-switched networks can reflect the PDN in size and scope. This kind of investment is clearly beyond the need and means of many strategic units; packet-switching equipment is produced which can, however, match more modest needs. In this section, we treat this range of possibilities, focusing on what is available to the telecommunication manager in North America and in the international market.

6.4.1 Overview of Private X.25 Networking

The basic structure of a packet-switched network adhering to X.25 standards was presented in Chapter 4 and is highlighted in Figure 4.2. Access points to the network which actually implement the X.25 protocol are called data circuit terminating equipment (DCE). A user device acting as data terminal equipment is connected to a DCE port to transfer data packets each addressed for a particular destination. Service is rendered by way of a virtual circuit which is established for the duration of the active connection to the network. On the network side of the DCE, packets are handed off from node to node according to an optimum routing algorithm. Upon completion of message transfer, the virtual circuit is terminated.

An important point for telecommunication managers to consider is that the protocol in use within the network itself need not adhere to X.25; this gives rise to proprietary routing approaches which are geared to maximize speed of transfer, reliability, or total throughput. Manufacturers who wish to sell packet-switched nodes compete on the basis of price and performance. They are able to optimize the design of their equipment because they have control over virtually all the details except the DCE interface. As a consequence, packet-switched nodes of different vendors are basically incompatible on the network side of the interface (i.e., within the cloud shown in Figure 4.2). The point in the network where all devices interface regardless of manufacture (or at least where they are supposed to interface) is on the user side at the DCE port. For this purpose, CCITT Recommendation X.75 was devised and is in use among PDNs.

So why do manufacturers not use X.25 with their networks and thereby make everything standard and fully compatible? The main reason is that, while X.25 is a good interface standard, it is not particularly efficient in the way data are processed for transmission. Speed of transfer is enhanced in some proprietary approaches by stripping off the address information and sending it through a common signaling channel. Reliability can be increased by using the datagram mode (as in TCP/IP) instead of the virtual circuit mode specified by X.25. And cost can be minimized by using existing technology like statistical multiplexing within the network. These are only examples of the options that are open to the manufacturers of packet-switching equipment for the private telecommunication market. A listing of many of the active packet-switch manufacturers is presented in Table 6.4.

A private packet-switched network is a very flexible platform for diverse data communication applications. Major international banks, airlines, oil companies, and information service providers have well-established networks of this type. The approach is not the answer to every data communication need, however, and even if it is at one point, times change and the packet network may outlive its usefulness. A careful examination of current needs and the future direction of user requirements should be accomplished before even the first few nodes are installed. One important consideration is the upper limit to the amount of data that can be pumped through the nodes and leased lines. At some point, the network may not be adequate for day-to-day operations, and, as user bandwidth requirements expand for services like Group 4 fax and video teleconferencing, the private X.25 network will no longer support the demand. As long as the requirements are in terms of low- to medium-speed data rates, however, the X.25 approach is very viable. Greatly expanded bandwidth would be better served with the high-level multiplexing and fast-packet technologies discussed elsewhere in this chapter.

6.4.2 Common Equipment Elements

A strategic unit, after completing a requirements definition and network design exercise, would purchase packet-switched nodes from an appropriate vendor. The node has the built-in hardware and software to support data traffic for a quantity of local DTE ports. In addition, the node can perform intermediary routing of packets which enter the network through other nodes. User devices would adhere to the DTE specifications of X.25, ensuring that packets are assembled and disassembled properly.

In cases where the user device does not support X.25, a separate PAD would be needed. Alternatively, specialized software can be obtained from the terminal equipment or packet-switch vendor. Protocol and speed conversion would be performed, if necessary, by the PAD. As an important example, the IBM SNA environment can provide the necessary conversion from SNA/SDLC to the DCE requirement. It is the responsibility of IBM to ensure that the addressing and packetization are properly converted. The same comments can be made for Digital Equipment Corporation's DECnet.

Another type of conversion is needed for users who wish dial-up access to the packet-switched network. There are modems available with built-in PADs; their use, however, cannot be assumed in all cases. Packet-switch vendors can supply PADs that convert from asynchronous transmission and a variety of modem protocols. Telecommunication managers should appreciate that anything connected between the user and the packet-switched node will increase both the cost of service and the time associated with the data transfer. Reliability is also a consideration; the fewer external pieces of equipment there are, the fewer things that may go wrong.

6.4.3 Performance Measures

Packet-switched nodes are rated in terms of the capability of the device to transfer data packets between one input and any desired output. Internal to the machine is a high-speed processor which reads packet addresses and performs a search through a database to determine routing. Routing tables are maintained within the database, permitting the processor to assess the optimum outbound route to reach the final destination. A given processor would be able to service a given maximum number of packets per unit time for a given number of incoming access lines and network links. Since the processor is basically a sequential device, every input port has a storage buffer to hold packets while they wait for service. The time that the packet must wait for service is called the queuing delay, which is a significant contributor to the end-to-end delay of packets through the network. The other significant delay is due to the time of transmission over the trunks themselves.

Delay can be reduced in two basic ways. First, the processor can act quicker by increasing its speed of computation and database hookup. A way to do this is to perform parallel processing with multiple processors which are attached to a common data bus within the node. The second basic approach to reduce overall delay is to increase the speed of data transmission on the links between nodes. Ideally, node processing delay and transmission delay are balanced, where neither is determining overall performance. Users can pay for bandwidth on the links and thereby reduce transmission delay; this may be unnecessarily expensive, however, or the bandwidth may simply not be available. Packet-switched node processing delay is usually inherent in the delay of the particular unit, which is why processor speed is an important factor in vendor selection. The overall network must still be engineered properly, a process which is not particularly simple because it is so dependent on the particularities of the applications running over the network [Elbert, 1989].

The attributes of packet-switching nodes which tend to be the most important for private network applications include the following [Stammler, 1988]:

Maximum number of trunks and access lines that can be configured, including the port speeds and whether they can operate simultaneously;

Maximum throughput in terms of the number of packets of a particular size (such as 128 bytes) per second;

Maximum window size (which is the number of packets that the final node can maintain as still outstanding before requesting retransmission);

Maximum packet size permitted, which according to X.25 could be as large as 4096 bytes;

Maximum number of call setup packets per second without a busy signal being returned to the originating node DTE;

Maximum number of virtual circuits per node;

Whether an internal PAD is available;

Packet throughput delay, which is related to the processing speed.

Comparing node equipment on these objective criteria is a good way to get started with the procurement decision. As is common with network architecture, a selection of a particular vendor's hardware will have long-term consequences. Besides these facts and numbers, telecommunication managers must consider the management aspects of the procurement. One company which proceeded rather thoughtfully in the process is Schlumberger Ltd, the noted oil exploration services company [O'Reilly, 1988]. Schlumberger employs nearly 50,000 people worldwide and relies heavily on the gathering, processing, and transportation of information. The company uses a wide variety of computers, including PCs, IBM minicomputers and mainframes, DEC VAX, and even Cray supercomputers. Their telecommunication management concluded that an international private packet-switched network was the most effective way to integrate multiple data communication applications and to provide a common E-mail system throughout their worldwide operations.

6.4.4 Suppliers in the International Marketplace

The principal suppliers of packet-switching equipment to the world market along with their relative shares of this market are given in Figure 6.8. There are three distinct classes of packet-switch manufacturers that can supply equipment to strategic units wishing to create a private packet-switched network environment. First, there are the PDN companies who design, manufacture, and integrate these systems primarily for their own use. Both BT Tymnet and Sprint International (manufacturer of Telenet® packet switches) are leading suppliers of packet switches mainly to TAs that are implementing domestic PDNs. Not surprisingly, their equipment is geared toward the high end of the market where speed and port capacity are

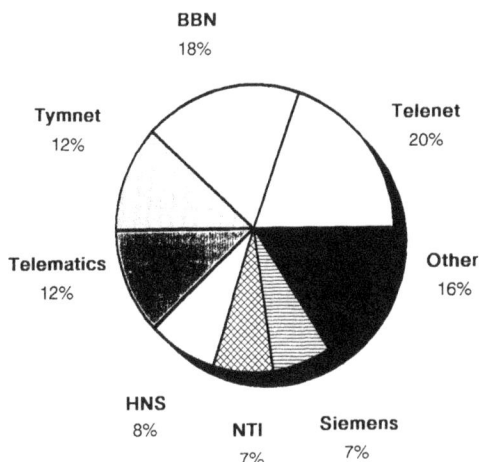

Figure 6.8 Market share of leading packet-switch manufacturers.

particularly vital. The sheer cost of the equipment along with the quantities involved give these suppliers a significant market share even though the number of customers in a given year is low.

A second important group of suppliers includes major telecommunication vendors who deal primarily in the common carrier marketplace along lines similar to the PDN manufacturers cited above. Included in this group are BBN, Telematics, HNS, NTI, and Siemens. The customers include telcos, particularly the RBHCs, PDNs other than the manufacturers, international carriers, and TAs. Private networks have become an important target for these suppliers, although only the largest and most diverse networks can justify these switches. The switches are attractive for international applications because of the strong commitment to standards and because the product lines are used extensively around the world. As a case in point, the Schlumberger network reviewed in the previous section employs packet switches made by BBN, which pioneered packet-switching technology for the ARPANET.

An important trend has appeared in the private packet-switching marketplace, namely that toward smaller networks using relatively inexpensive nodes. Among the second group of switch vendors, HNS has developed low-end products to extend private networks relatively inexpensively to remote locations. Telematics has also done a relatively good job of diversifying its product line [Korostoff, 1988].

We identify a third group of companies which focus almost entirely on the low-cost, low-capacity sector of the private packet-switching market. Aside from HNS and Telematics, most of the key players are below the threshold of market share shown in Figure 6.8; in aggregate, however, they are doing a brisk business.

Companies in this group include OST, Netrix, Plantronics, Micom, and Amnet. Although the market for the PDN class of switch is flat or waning, that of the relatively inexpensive private packet switch is rapidly expanding. OST is an interesting company, being the U.S. subsidiary of the leading supplier of packet switches in France. Amnet sells a PC-based X.25 packet switch that was chosen by Foxboro, an international company that manufactures instrumentation and test equipment, for a network comprising fifteen offices in the United States and abroad.

There are literally dozens of manufacturers in the packet-switch manufacturing business. As we stated before, the internal workings of the packet network itself are proprietary to the particular supplier. What is standard is the X.25 interface where users connect DTE to the nodes. Therefore, selecting a supplier is a major decision where the telecommunication manager basically makes a bet on the future.

6.5 FAST-PACKET SWITCHING AND BROADBAND ISDN

Unlike X.25 packet-switching networks which are designed strictly to support low- and medium-speed data communications, fast-packet switching systems operate at T1 rates and higher. Thus, wide bandwidth is assigned quickly enough to pass voice, data, and even digital video traffic in real time. The packet approach is used so that all types of traffic can be intermingled without regard to speed, protocol, and frequency of activity. Products on the market today emulate the functionality of the high-level multiplexer. Also, CCITT Recommendation I.122 is moving out of the conference room and into the field for an application called frame relay to transfer digitized traffic between telephone exchanges. In the future, fast-packet technology will operate at 140 Mb/s and higher, bringing with it the basis for broadband ISDN over the public networks and even into the home. In the following paragraphs, we briefly review these possibilities.

6.5.1 T1 Fast-Packet Switching

Fast-packet technology has gained a foothold in the area of private T1 backbone networks, primarily in the U.S. market. The leading supplier of fast-packet switches is Stratacom, Inc., offering the *integrated packet exchange* (IPX). A simplified block diagram of the basic node (called a *system unit*) is presented in Figure 6.9. Nodes connect packets to different T1 trunks by means of the address included with each packet. Because of the packetization, bandwidth need not be assigned ahead of time to particular channels. The T1 transmission speed, framing, and synchronization are retained to be consistent with the public networks [Elbert, 1989].

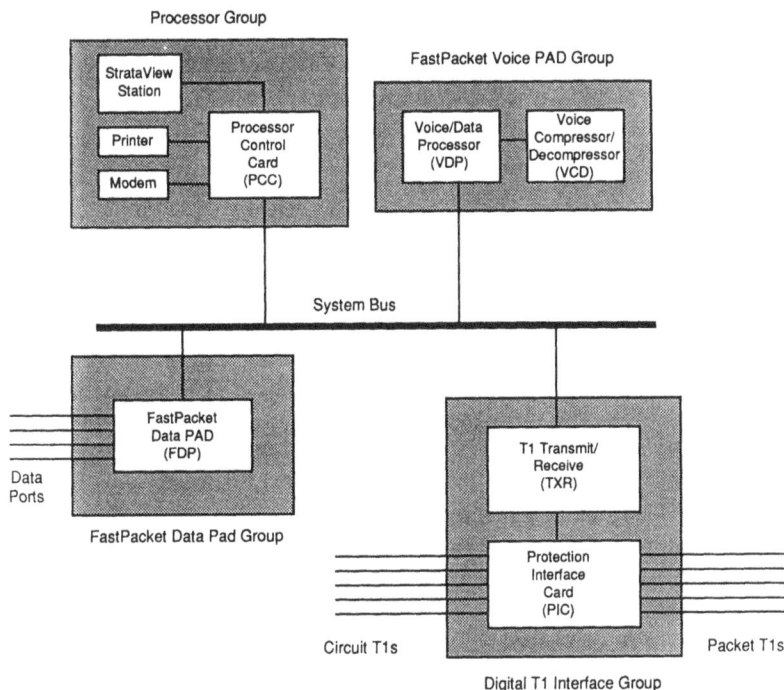

Figure 6.9 The system unit of the integrated packet exchange, providing T1 high-level multiplexing using fast-packet switching (courtesy of StrataCom, Inc.).

Port access cards are provided which convert voice, data, or digital video information into packet form. The packets are addressed and then applied to the T1 trunks, which are used to reach other nodes in the backbone network. Voice packetization is particularly interesting because it automatically produces the same effect as digital speech interpolation. If a talker is not speaking, then no packets are sent over the network. The network only transports live traffic even though users believe that they have a fixed connection. The speed of processing and transmission are sufficiently high so that time delay is not noticeable to subscribers. Another benefit of the fast-packet approach is that it is more tolerant of inconsistent timing on public networks which connect between different nodes. Basically, a frame loss caused by timing buffer overflow will only affect one channel of data, as opposed to affecting all channels in a typical T1 frame using TDM.

International application of the fast-packet switch is certainly possible, particularly in situations where high-level multiplexing can be justified. The technology is still evolving, so telecommunication managers should assume that they would be more or less on their own. In a case where fast packet is being used domestically, it would be reasonable to extend the network to a limited number of foreign locations. In an interesting combining of efforts, SprintNet and Stratacom have teamed to offer an integration of standard X.25 packet switching and T1 fast-packet switching. Having wider bandwidths available through fast-packet switching will potentially overcome the bugaboo of low-speed data rates associated with X.25 packet switching.

6.5.2 Frame Relay

The concept of integration of digital services through high-speed packet switching for private networks is being offered telcos and long-haul carriers by AT&T and Northern Telecom [Bush, 1989]. Frame relay is a spin-off of ISDN that uses the *link access protocol for the D-channel* (LAPD). This link layer protocol has very little overhead in terms of excess packet routing and control bits and hence provides high throughput. The links must be point-to-point, however, and should have an error rate generally below 10^{-8}. This requirement is generally met on fiber optic cables and ISDN-compatible satellite links. The confluence of high transmission speed, low error rate, and good throughput gives the foundation for the integration of voice, data, and video services by the telecommunication service companies.

AT&T's offering is called *integrated access and cross-connect system* (IACS) and is targeted toward the RBHC market. The local telcos would purchase IACS switching equipment to make their services more attractive to strategic units planning private networks. As with the high-level multiplexer and the IPX, the AT&T technology provides dynamic bandwidth allocation so that users can vary port speeds and assign packet sizes depending on bit rates. For example, a user who requires service at 9600 b/s for a particular call would be assigned a 20 kb/s packet-oriented connection rather than the 64 kb/s connection of a standard TDM link. At this stage of development, the links between IACS switches would operate at the standard T1 rate using either the D4 frame or the extended super frame.

Northern Telecom has been in the frame relay business for a number of years, employing the concept in their SuperNode digital switching product line. As mentioned earlier in the chapter, the DMS 250 is NTI's large digital telephone exchange offered to long-haul carriers and telcos. The DMS SuperNode, illustrated in Figure 6.10, uses the frame relay principle on the wideband digital trunks between compatible DMS 250 switches. This is in effect a closed architecture not unlike the approach taken by vendors of X.25 switches and high-level multiplexers. As long as the carriers buy NTI exchanges, they can use frame relay to integrate the long-haul traffic. Digital Equipment Corporation is teamed with NTI, providing DECnet/OSI routers which feed the heavy links with lower rate data communication traffic. The first major user is NyNex, the RBHC that includes New York Telephone Company and New England Telephone and Telegraph Company.

Providers of frame-relay technology and services hope to exploit the private network marketplace by selling equipment to telcos. The switch and computer

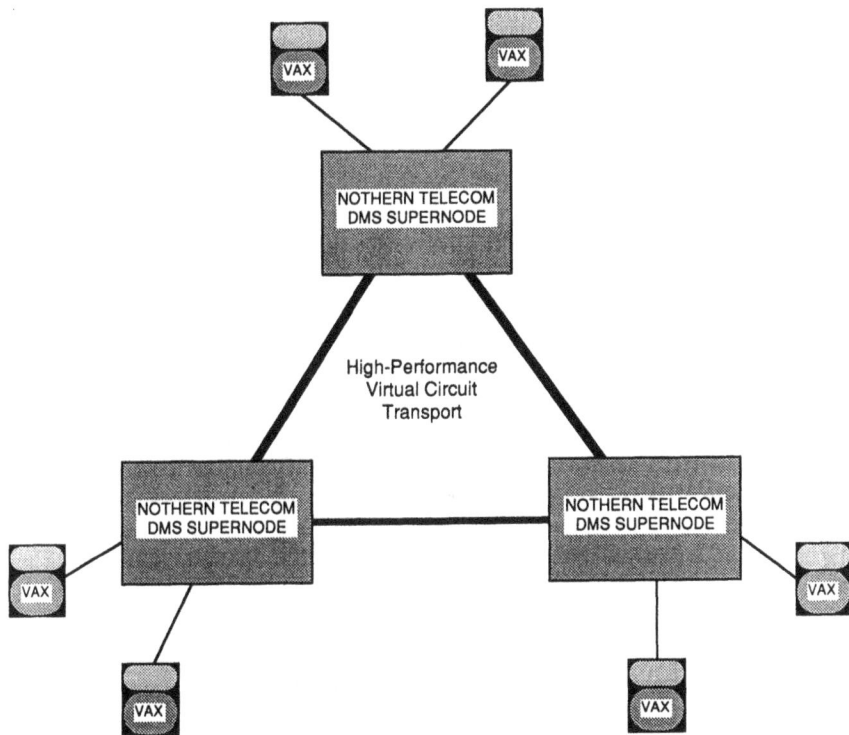

Figure 6.10 Application of the Northern Telecom DMS Supernode on a high-performance virtual circuit transport network. Computational support is provided by DEC VAX minicomputers and the Supernodes exchange traffic using the frame relay principle.

manufacturers stand to gain additional sales of equipment, whereas the telcos can possibly prevent defections by major corporations needing greater telecommunication flexibility. From an international perspective, the CCITT has already drafted Recommendation I.122, called a "Framework for Providing Additional Packet Mode Bearer Services." This defines the basic frame-relay structure that telcos, long-distance carriers, PTTs, and international carriers could employ as a cohesive whole.

6.5.3 Asynchronous Transfer Mode (ATM)

Looking farther into the future, telecommunication service providers hope to implement B-ISDN capabilities. Research on the necessary high-speed processing and transmission systems is underway by the major manufacturers in the United

States, Japan, and Europe. The method for integrating wideband services like broadcast-quality television, telephone, and high-speed data has been dubbed *asynchronous transfer mode* (ATM, not to be confused with automated teller machine). The links would operate at the 140 Mb/s speed already adopted by the CCITT for international ISDN links between domestic networks. To cope with a wide range of services, public networks require built-in control and flexibility in the way resources are allocated [Carter, 1989]. This sounds a lot like fast-packet switching and frame relay, but at much higher speeds and for a greater range of services.

Developers of the ATM methodology around the world have identified three structural layers, not unlike those used in X.25 packet switching [Toyoshima, 1989]. A conceptual view of the ATM architecture is given in Figure 6.11. The *virtual circuit layer* provides a logical connection between two switching nodes. It is set up and then released on a call-by-call basis, as one would expect from the name of the service. Moving down the architecture, the *virtual path layer* controls transmission between pairs of switching nodes, and therefore is akin to the link layer of the OSI stack. This layer also does multiplexing and cross-connecting functions familiar in common carrier networks. The lowest layer is the *section layer* that defines the physical connection between nodes. Distances can be extended indefinitely through the use of digital line repeaters, so that international services can employ the ATM architecture.

Owing to extensive research by switch manufacturers and leading telecommunication service companies, the ATM methodology could become a facet of the ISDN environment of the late 1990s. As these organizations implement broadband services, strategic units will find it attractive to move some services from private facilities to the public networks. Ultimately, B-ISDN service could become available on the local loop. Telecommunication managers should keep a watchful eye on the developments in this area; at the same time, however, the opportunities to use private networking approaches in the present should not be overlooked.

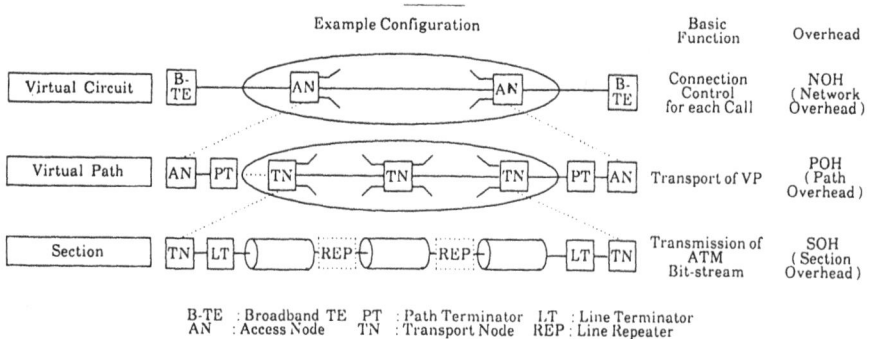

Figure 6.11 Layered structure for ATM transport architecture.

6.6 BANDWIDTH-EFFICIENT MULTIPLEXERS

Private line circuits with relatively narrow analog or digital bandwidths can support multiplexed traffic using bandwidth-efficient encoding and transmission schemes. One of the most successful devices is the statistical multiplexer, which was devised to allow several medium- and low-speed data communication devices to share a common modem and leased line. The STAT MUX created an industry and formed the foundation for other resource-sharing approaches, particularly high-level multiplexing. With the cost of wider digital bandwidths such as T1 and DS3 coming down, statistical multiplexing and other bandwidth-efficient technologies should be falling from use. It was the popularity of IBS services at multiples of 56 kb/s and now the introduction of similar bandwidths on fiber optic cables, however, that gave bandwidth-efficient multiplexing a "new lease on life," so to speak. As discussed below, strategic units can use multiplexers which stuff four voice channels along with 19,200 b/s of data into a single 56 kb/s leased line.

6.6.1 Statistical Multiplexers

First appearing on the U.S. market in the early 1970s, the statistical multiplexer provides a means to share a leased voice line or 56 kb/s circuit by several devices and independent services. Timeplex gained an early lead in the market, to be joined by Codex, General Datacom, and Infotron. The market grew so quickly that it supported a large number of manufacturers, each with its own unique way of implementing the basic concept.

The statistical aspect of the device is its ability to compress the data in such a way that the sum of the individual input rates (called the *aggregate rate*) is greater than the rate over the modem and link (called the *combined rate*). Data are transferred to the circuit in packets at more or less random times. Hence, the delay through the link is a random variable. In contrast, a TDM multiplexer has a constant and predictable time delay, but the composite rate cannot exceed the aggregate rate.

The loading of data on the circuit is increased in statistical multiplexing in a number of ways, some of which can be combined for multiplicative benefit. First, only the data from active inputs are applied to the link. Second, start-stop bits contained in asynchronous characters are removed at the input side and then replaced at the distant end. Parity check bits can be handled likewise. Third, the data can be compressed using an algorithm which removes blank spaces; this approach usually only applies for synchronous input data.

Depending on the nature of the data, the ratio of aggregate-data rate to combined-data rate is at most a factor of four (and at the least a factor of one). A typical STAT MUX might have four or eight asynchronous inputs at 9600 b/s for connection to the synchronous modem and link at 19,200 b/s. Higher speed units are available, which operate at combined rates of 128 kb/s. The individual inputs may or may not have to be of the same rate, depending on the design of the particular STAT MUX.

At times, enough inputs could be active to require more data to be sent than the modem and circuit are capable of handling. Flow control is the process by which the input devices are commanded to restrict transmission so that data are not lost. Technically speaking, flow control of asynchronous inputs is possible using either the "data set ready" or "clear to send" lines of the RS-232C interface. Care in selection of interface conditions is appropriate as some terminal devices may be incompatible with the method of flow control being provided by the STAT MUX. Without flow control, data will literally overflow at the STAT MUX input. For additional background on STAT MUX technology, see [Elbert, 1989] and [Curran, 1989].

The multiplexing process between STAT MUXs is done in one of two ways. A conventional TDM frame is employed on the circuit but data assignments to time slots vary from frame to frame, depending on the activity at the input. The assignments are communicated to the distant STAT MUX with a separate common signaling channel which occupies a small amount of bandwidth within the TDM frame. Alternatively, a packet-switching scheme is employed wherein input-to-output assignments are given in the packet headers. The data link between STAT MUXs is typically bit-synchronous using a reliable protocol such as HDLC. Unfortunately, there are no accepted standards for the STAT MUX compression algorithms and virtually all are proprietary. This means that the STAT MUXs on the ends of the circuit must be from the same manufacturer, an important consideration for the international telecommunication manager. Having the same device on each end is also necessary because of the signaling that is conducted between them to control transmission back and forth. In [Curran, 1989], the author strongly recommends that throughput and time delay be measured experimentally under conditions of real or simulated traffic loading before committing to a particular STAT MUX design or configuration.

The principle of the STAT MUX has been applied in many other types of device and the concept is hardly static. For example, a variant called the switching STAT MUX appeared on the market in the mid-1980s and was deployed in many data communication networks in the United States. The leading supplier, Codex, provided systems which have features found in packet-switched networks, allowing several locations to be interconnected in a mesh network. One of the advantages of this approach is that the compression and multiplexing processes are particularly efficient and can yield throughput which is greater than that of X.25 and other strictly packet-switched approaches. The main issue, however, is that a network of STAT MUXs is closed to the outside. Codex has devised an X.25 interface to their proprietary architecture, so that users can at least have a gateway to a PDN or private packet-switched network.

Regardless of the lack of standards, the STAT MUX will still be a valuable device on international private lines. For a relatively modest price, a strategic unit can subdivide an analog leased line. Multiple data communication applications can be serviced, each of which can run on entirely separate computer systems and terminals. The STAT MUX integrates dissimilar and unrelated data for transmission. Banking institutions in particular benefit from this because every branch is typically served with three or more independent computer applications, including automated teller machines, savings and checking account maintenance, credit applications, *et cetera*. The marketplace is dominated by U.S. companies because the technology was created in the United States. To find a U.S. manufactured STAT MUX in European countries or in developing countries in Asia is not unusual because the STAT MUX is designed specifically to build point-to-point data communication networks using analog voice lines. (See Table 6.5.)

6.6.2 Narrow-Band Voice Multiplexers

The bandwidth required to transmit the human voice with acceptable quality and intelligibility has been reduced several times over. Modern compression algorithms based on a technique called linear predictive coding can deliver near toll-quality voice with as little as 8000 b/s as contrasted to 64,000 b/s normally allocated to PCM in a T1 frame. Today, voice compression is no longer only the domain of the government where secure voice transmission is needed at any cost. Conceptually, the encoder removes redundant speech patterns which arise because the human voice does not occupy every frequency in the 300 to 3400 Hz range. The complexity and cost of implementing the technology have also been reduced due to the availability of *digital signal processing* (DSP) chips produced by Texas Instruments and others.

A voice coder which operates at, say, 9600 b/s could be connected to an analog leased line with a standard voice-band modem. This may not seem like

Table 6.5 Multiplexers for Data Communication

Time Division Multiplexers
Gandalf Data
General Datacom, Inc.
Infotron
Micom
Racal Milgo, Inc.

Statistical Multiplexers
Codex (division of Motorola)
Infotron
Timeplex (division of UniSys)

such a practical idea because the coder could be bypassed and the voice connected directly. The benefit comes from combining several voice conversations by using the statistical multiplexing principle. As with the fast-packet switch discussed previously, circuit multiplication by a factor of approximately four is achieved because of the bursty nature of human speech.

The voice coder combined with a STAT MUX has rapidly gained favor in international telecommunication using leased lines at 56 and 64 kb/s. Such circuits are available on IBS links and international fiber optic cables, as discussed in Chapter 3. It has been estimated that approximately half of the IBS circuits operating in 1989 were employing this class of multiplexer. Users are therefore becoming accustomed to the quality of 8000 b/s coded speech. In comparison with the analog voice lines previously used internationally, the digital voice circuits sound very good. Due to the high degree of processing within the coder, however, medium- and high-speed modems cannot be used to pass data directly through the channel. This is where the STAT MUX element comes into play. Rather than using voice band modems, the data lines are connected to the same multiplexer used by the voice coders. All of these sources appear like data to the STAT MUX, which simply operates on the inputs in the manner discussed in the previous section.

The leading supplier of voice-data STAT MUX for international circuits is Republic Telcom Systems of Boulder, Colorado. Its RLX-8 Private Line Extender connects to a 56 kb/s circuit on each end, providing up to eight voice channels. Figure 6.12 illustrates how the RLX-8 can be used to integrate telephone and data communication services on a 56 kb/s private line circuit. Built into the coder is a digital echo canceller for each channel. Combined voice and data configurations include six voice plus 9600 b/s of data capacity, or four voice plus 19,200 b/s of

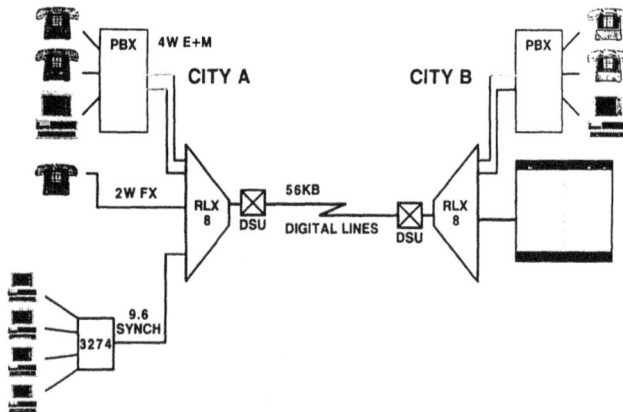

Figure 6.12 Application of narrowband voice multiplexer to combine multiple voice and data channels on a 56 kb/s international private line (courtesy of RepublicTelcom).

data capacity. The data connection itself could come from a submultiplexer or a communication controller (such as the IBM 3274). The 56 kb/s line can be a leased DDS circuit or a DS0 channel of a T1. Republic Telcom also produces the model RLX-40 which allows up to forty simultaneous voice conversations on four 56 kb/s lines. A higher ratio of aggregate capacity to combined capacity is a consequence of the larger bundle of voice traffic.

Coded speech at 8000 b/s along with separate data multiplexing have become acceptable in international telecommunication. Already, variants are appearing which offer greater utility. One concept is to provide switching of the narrow-band voice circuits within a hub. Recall that international private networks often use a star topology, where long-haul circuits connect from the hub of the star and the spokes extend to neighboring countries. A hub arrangement gives a great deal of flexibility and can significantly reduce bandwidth requirements. This is very attractive in Europe, where leased line prices can be high in some countries.

6.7 VERY SMALL APERTURE TERMINAL (VSAT) SATELLITE NETWORKS

Since the writing of [Elbert, 1987], applications of VSAT satellite communication networks have increased dramatically in North America. Receive-only VSATs are applied throughout the world in substantial numbers to receive data broadcasts from domestic satellites and even INTELSAT. We restrict this discussion to the two-way interactive VSAT because this technology is more challenging to implement on an international basis. As indicated in Figure 6.13, there are more than 15,000 two-way VSATs in operation in the United States, with nearly 40,000 having been ordered in total. These figures, provided by London-based consultancy Communications Systems Ltd (COMSYS) demonstrate the popularity of the technology on one continent. VSAT networks grew rapidly in the United States because of three factors: the technology was developed primarily there; there is considerable regulatory freedom, particularly with regard to satellite communication; and North American telecommunication managers have a willingness to innovate. Regulatory problems and a lack of familiarity with the technology have hampered the development of these networks in other parts of the world, although these shortcomings should be overcome in the 1990s.

Information on the technical characteristics of VSATs and their application in domestic private telecommunication networks for data, video, and voice services is contained in [Elbert, 1989]. In terms of review, a VSAT is a small earth station designed for unattended operation at a remote location. The antenna is installed on the roof or on the ground adjacent to the building with a specialized cable link running indoors to a controller unit. As a combined system, the VSAT operates

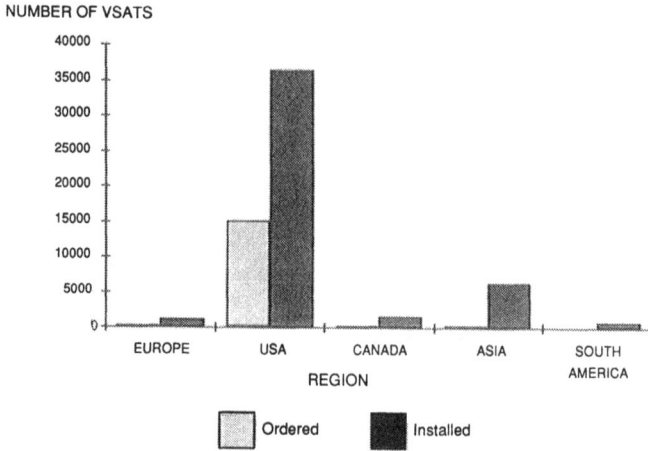

Figure 6.13 The interactive (two-way) VSAT market, indicated by the number of earth terminals and region of the world (courtesy of COMSYS).

as a data communication node or multiplexer; video reception (and possibly transmission as well) can be added by attaching either a standard TVRO type of receiver or a digital video codec.

Due to its small antenna size and low transmitting power, the VSAT must communicate with a larger "hub" earth station, which is maintained either by the user or by a third-party service provider. The network configuration is therefore a star with the VSAT located on the end of a spoke. A given network could be composed of a host computer connected by a terrestrial data line to a dedicated or shared hub earth station, and as many as 400 to 4000 remote VSATs. Applications are in host-remote data communication often using IBM's SNA/SDLC protocols. To find the X.25 packet-switching protocols in use over the VSAT network is increasingly common. The hub to remote link is over a single satellite hop, whereas connections from remote to remote (which are not generally required in typical applications) would employ a double hop.

Most individual VSAT networks are homogeneous, meaning that a single computer environment and set of applications are supported. One of the most successful uses is in retail sales for inventory control and credit checking. In some cases, retail stores are located in outlying areas where good data lines are not available. Also, business video is included to enhance the merchandizing and training efforts of the user. International service via the VSAT network is usually limited to bringing into the network a few more stores which are located across the border in a neighboring country. This is the case with Chrysler Corporation's network that services their automobile dealers in the United States and Canada. As discussed in Chapters 1 and 8, transborder service is usually subject to regulatory and political difficulties. The technology, however, is conducive to transborder service because the satellite coverage pattern can be expected to provide sufficient spillover.

We focus here on the prospects for development of VSAT networks in regions of the world where medium powered Ku-band satellites are in operation. C-band satellites support a limited number of quasi-VSAT networks (i.e., with relatively large diameter antennas) in countries without Ku-band satellite coverage. The reason for limiting our investigation is that VSAT technology is suited to installations with small Ku-band dishes, typically 1.2 to 1.8 m in diameter. (As a technical point, antenna size is smaller at Ku-band primarily because downlink transmit power substantially higher than at C-band is permitted by international regulation.) In addition, we are concerned with the transmit-receive variety, as these systems are useful for interactive data communication. Such VSAT networks can substitute for data lines in areas where terrestrial service is more costly or unreliable. Another significant aspect of VSAT systems is that they can be upgraded at modest cost to receive full-motion color video for business television applications.

6.7.1 VSAT Manufacturers and Service Suppliers

The international VSAT market is composed of two classes of suppliers: the manufacturer and the service supplier. Manufacturers are typically experts in digital communication and have experience in satellite communication as well. Some are affiliated with satellite hardware manufacturers while others have a telecommunication manufacturing parent. The number of VSATs installed or on order is still relatively small, so high volume electronics manufacturers are remaining on the sidelines for the time being. The second class of supplier approaches VSATs as a service, offering the facilities on a rental basis. This is advantageous to a user who does not want and cannot afford to make a major commitment to such satellite communication networks. Service is needed in three areas: the space segment (which must be obtained on a lease basis regardless of the size of the network), the use of a hub earth station, and the VSAT installation itself. Most common is that the user makes a commitment of from three to five years to the remote VSAT to be economically satisfactory.

Leading manufacturers of VSATs in 1989 are listed in Table 6.6 (information supplied by COMSYS). The bulk of the VSATs installed to date and on order were supplied by the combination of Hughes Network Systems, Tridom, Contel/ASC, and Scientific Atlanta. A large fraction of the outdoor units, composed of the antenna, radio frequency (RF) amplifier, and low-noise converter, were manufactured by NEC, the company which is also the leader in the Japanese domestic market.

European companies are engaged in joint ventures and marketing agreements with manufacturers in the United States and Japan. Alcatel, in particular, used

Table 6.6 Manufacturers of Very Small Aperture Terminals for Interactive (Two-Way) Data Communication

United States Manufacturers
ComStream
Contel/ASC
Hughes Network Systems
Scientific-Atlanta
STM
Tridom-AT&T
European Manufacturers
Alcatel
Multipoint
Japanese Manufacturers
NEC Corporation

foreign technology as a platform to evolve a uniquely European VSAT product set and is able to manufacture the hubs and remote terminals. Dornier of West Germany (part of Deutsche Aerospace) has concluded a marketing agreement with Hughes Network Systems, a leading U.S. supplier. Also, ANT, a subsidiary of Robert Bosch, has an agreement with a small Southern California VSAT manufacturer, Satellite Technology Management (STM). The advantage that these European companies have is that they are intimately familiar with non-U.S. telecommunication environments and are powerhouses when it comes to TA marketing. Traditionally, leading European manufacturers have had protected markets in their home countries, although this advantage is supposed to diminish due to efforts of the European Community (see Chapter 10 for more information on this subject).

6.7.2 Prospects for VSAT Development Worldwide

Operators of domestic and regional satellite systems throughout the world recognize the attractiveness to users of VSAT business networks. Consequently, they are usually supporters of the technology, because the bandwidth required will help fill (and pay for) satellites in orbit. TAs find it more difficult to embrace VSATs, however, because they tend to view it as competition for traditional services like private lines. A related concern is that the cost of the VSAT network is artificially inflated due to space segment charges that have excessive markups. As VSAT networks are usually justified through prospective telecommunication cost reductions for the strategic unit, these hindrances can make the technology economically unattractive. There are also instances where the domestic satellite operator is also the TA, making things even more difficult for the potential VSAT user.

While there is ample satellite capacity available, the TAs in Europe have been reluctant to allow private companies to exploit the technology. The pressure of the European political will is causing change in some countries. As of 1990, it is possible to employ VSATs for domestic applications in West Germany, Italy, and France. In addition, the TAs in Spain, France, and the United Kingdom are considering allowing the use of VSATs for international services (i.e., permitting the installation of a "foreign" VSAT which uses a hub in another country). The most progressive may be the Federal Republic of Germany, where private companies are being allowed to go into the VSAT service business. This entails installing one or more hub earth stations and offering access to the hub and the terrestrial network as a service. Companies such as Dornier and ANT acted upon this opportunity as it also helps them sell their VSATs. In the United States, Hughes Communications similarly supports the marketing efforts of sister company Hughes Network Systems by providing both shared hub services and Ku-band satellite bandwidth. Both of these companies are wholly owned subsidiaries of Hughes Aircraft Company, a leading manufacturer of communication satellites. Moving back to the West German market, space segment services are still the exclusive domain of DB Telekom, the domestic TA and international carrier.

Except for one or two markets in Europe, users must obtain the equipment and supporting service from the TA. This allows the TA to maintain some control over what actually is a competitor to its national infrastructure. EUTELSAT recognizes the advantages of VSATs for filling satellite transponders with data carriers and video traffic. This organization, however, is owned by the TAs and hence cannot market directly to end users. As discussed in Chapter 1, users would like EUTELSAT to act and market available capacity more aggressively; this position is also supported by the European Commission.

The Japanese domestic VSAT service market is definitely open to competition. Two satellite systems are in operation and the capacity is very suitable to VSAT network development, particularly because the downlink power level is significantly higher than in the United States. (This results from the more highly focused antenna beam aboard the satellite, taking advantage of Japan's smaller geographic size.) The appearance of several Type I and Type II carriers, however, has stirred the telecommunication markets so much that users are still confused by the alternatives. Japan also has an excellent digital infrastructure as a result of previous investment by NTT. At some point, an attractive VSAT niche will develop in Japan, although where these applications will appear is not yet clear.

Aussat, the Australian Ku-band satellite operator, offers VSAT networks and services to their domestic market. Users have a choice of hardware designs and can optimize the performance for a particular application. The second generation of Aussat satellites, planned for launch in 1992, will enhance the capacity and reliability of these services due to increased downlink power.

Chapter 7
INTERNATIONAL TRANSMISSION
OF TELEVISION AND AUDIO

Perhaps the fastest growing segment of the broadcasting industry is that which deals with the development and distribution of international television and radio programming. According to COMSAT, the number of full-time transatlantic television channels has doubled on average for the period 1985 to 1988. In the overall context, international television and audio is a small part of the industry, but important players are making moves to grow this market and sources of programming supply.

For years, the major U.S. networks have employed international television to carry sporting events and news from around the world back to the homeland. A classic example was the CBS coverage of events in Beijing on Tiananmen Square in June 1989, as was the coverage of Mikhail Gorbachev's visit to Cuba, which was not a uniquely Western event, but was carried live on Western television. This type of coverage is the predominant application of international television, facilitated by satellite communication which is the principal mode of transmission. Through hundreds of earth stations around the world, broadcasters and private network operators can provide service from or to any nation on the globe. Television coverage can be limited to a single event under occasional video service, or it can be for recurring or full-time transmission.

Many countries make do with news gleaned from the wire services and what they obtain from the leading U.S. and Western networks. Within Europe, terrestrial systems were first employed on a cooperative basis, giving rise to the European Broadcast Union (EBU) as an important force. Traditional approaches are fast giving way, however, as satellite systems are expanded around the world. This allows even a local broadcaster to create its own international delivery medium.

U.S. media companies have long dominated television and film, and still provide more than fifty percent of the movies and television dramas viewed on the world's television screens. The opening up of European markets in the 1990s, however, is providing a dynamic force which could slow what detractors say is another form of American imperialism. The European Commission is adopting rules for open television markets which, in addition to simplifying the broadcast of programs across borders, are hoped to encourage the growth of domestic European media. This "television without frontiers," with as many as 200 full-time services employing as many satellite transponders, has been projected [Owen, 1989].

There are instances where international video transmission will be required in the teleconferencing or business video area. In Chapter 5, we discussed business applications of TVROs; these are inexpensive receive-only earth stations commonly used for one-way private broadcasts. Carrying such networks on an international scale will require many of the techniques and services described in this chapter. Furthermore, the proprietary content of these transmissions demands some form of privacy to prevent others from listening in on the satellite link. Techniques such as scrambling and audio encryption are used routinely in the United States, Japan, and Europe, where business video is developing rapidly.

Audio services are also growing even though radio is a very mature medium. An innovator in international radio networking is National Public Radio, a non-profit corporation which produces and distributes what has been called "news for people who think." Included in the daily schedule are the BBC and West German news (in English). In the future, NPR anticipates carrying programs from Latin America over their transponder on a U.S. domestic satellite. ABC radio network typically sends crews to the Olympics and to summit conferences. Unlike telephone service, the radio networks require wider audio bandwidth to be compatible with normal over-the-air quality for the FM stereo format.

The prospects of continued expansion of international television and audio are excellent. To take part, an organization must understand how signals originate and are subsequently carried from country to country. Applications range from event coverage in sports or news to full-time transmission of services like CNN International and NHK, to teleconferences for strategic units wishing to apply the same techniques proven successful in the United States. In the following paragraphs, we review relevant technology and facilities which can be applied to these problems. Concrete examples of services which are currently in use today are also provided. Regional broadcasting unions, which are cooperative associations geared to the need of domestic and international broadcasters, are covered at the end of the chapter.

7.1 VIDEO COMMUNICATION TECHNOLOGY

We begin with a review of the the three color television systems in use throughout the world: NTSC, PAL, and SECAM. This section is fairly technical and may be difficult for some readers; due to the importance of the differences between systems, however, the material should at least be scanned for key points. The basic three standards relate to the way the color image is encoded and modulated onto a radio frequency carrier for broadcasting to the home either over the air or via a cable television distribution network. With different systems in existence, the broadcast channel and the TV set must be configured for the particular format. There are two camps in the world, which evolved from the power line frequency in use, namely 50 and 60 Hz. (For a detailed review of television technology, see [Noll, 1989].) In North and South America, Japan, and parts of Asia, 60 Hz power is used, and the television system employs a frame repetition rate of half that number, or 30 times per second. A total of 525 scan lines are presented in a frame; only 483 are available to the picture, however, as the rest allow for internal operation of the television set. Resolution, which is measured with a standard test pattern, is effectively reduced further to 350 lines due to bandwidth limitations of the entire system (including the home receiver).

The rest of the world derives ac power at 50 Hz and hence this frequency is carried over to the television systems. A frame repetition rate of 25 times per second, or one-half the power frequency, is employed. Scan lines total 625 for the complete frame, which increases the overall resolution by 20% as compared to the 525 lines per frame of the 60 Hz system.

A frame rate of 25 or 30 Hz is so slow that the human eye would easily detect the "flicker" of light on a television screen. To reduce the disturbance to viewers of flicker, the frame is split into two *fields* each of which contains half of the scan lines. This produces an effective image change rate of 50 or 60 Hz, as appropriate, which is much less of a bother. Each field contains every other scan line and the two fields are said to be *interlaced.* The electron beam which is doing the scanning within the picture tube is modulated in its intensity by the brightness of the original image (this is called the *luminance component* of the picture). Picture brightness on the screen is directly proportional to the beam intensity.

The three standards provide a high level of enjoyment and can transfer a wide variety of visual information. At a frame rate of 60/s, the North American system contains little visible flicker in the picture; flicker from the 50 Hz rate of the European standard is somewhat more noticeable. To compensate for flicker impairment, the developers of the European standard chose 625 lines of resolution. In some ways the 50 Hz picture is superior to the 60 Hz. Graphic images are clearer simply because there are more lines of resolution. We provide a brief overview in the chapter on a new technology called *high definition television* (HDTV), which will approximately double the number of lines of resolution.

In addition to the basic image resolution and brightness (the *luminance*), we must consider the manner in which color is conveyed (the *chrominance*). As stated previously, there are three color systems in use around the world, two of which employ the 50 Hz scan rate. The system for 60 Hz is called the National Television

Standards Committee (NTSC) standard, which was developed by RCA Corporation. In the Federal Republic of Germany, an improved system was developed under the designation Phase Alternate Line (PAL); and the third system, also for 50 Hz, is the French Sequential Couleur avec Mémoire (SECAM). SECAM and PAL both benefited from NTSC having gone first and offer some improvement in color accuracy. We review the three systems in the following paragraphs. From an international transmission standpoint, any of the three standards can be converted into any other by a device called (appropriately) a standards converter. Multiple standard television sets can be purchased by consumers for use in areas where two or three can be received, such as in areas of Western Europe.

7.1.1 Color Television Formats

A summary of the color television formats in use in different countries is presented in Table 7.1. This is an important consideration when an organization considers engaging in the transmission of television into or from a country, unless a complete set of television production equipment is brought in. U.S. companies manufacture test equipment for the PAL and NTSC systems, whereas almost all SECAM gear is manufactured in France. Most European electronics companies focus on PAL. Japanese manufacturers, however, produce production equipment for all three systems. The systems are reviewed in brief in the following sections.

7.1.1.1 NTSC (525 line, 60 Hz)

The NTSC color television standard was approved in December 1953 by the FCC, confirming the preeminence of the RCA-developed technique. One of the most important attributes of NTSC is its compatibility with the monochrome (black-and-white) system already in use in North America. This means that all existing B&W sets could receive the color telecast unimpaired. (In fact, this author personally wrote to WNBC to inform them that the quality of the black-and-white picture actually looked better!) NTSC is both backward and forward compatible, meaning that NTSC is viewable on a black-and-white set, and that a black-and-white picture is viewable on an NTSC color set. In addition to receiver compatibility, the transmitted signal occupies the same bandwidth, which means that the designers of NTSC had to be clever in how the color information was to be added. Compatibility is a very important issue, particularly in the case where new formats for HDTV are being evaluated.

The NTSC system was developed in the age of analog communication electronics. All circuits were based on electron tubes and discrete components like resistors, capacitors, coils, and transformers. One of the principal drawbacks of

Table 7.1 Color Television Systems in Use in Selected Countries Around the World
(*Source:* International Telecommunication Union)

PAL B AND G	PAL I	SECAM B AND G	NTSC M
Australia	Angola	East Germany	Bermuda
Austria	Hong Kong	Egypt	Bolivia
Belgium	Ireland	Greece	Canada
Denmark	South Africa	Iran	Colombia
West Germany		Iraq	Costa Rica
Finland	PAL N	Saudi Arabia	Cuba
Ghana	Argentina		Ecuador
India	Paraguay	SECAM D AND K	Haiti
Indonesia	Uruguay	Afghanistan	Honduras
Israel		Bulgaria	Japan
Italy	PAL D	Czechoslovakia	Mexico
Jordan	China	Hungary	Netherlands Antilles
Kenya		Poland	Nicaragua
Luxembourg	PAL M	USSR	Panama
Malaysia	Brazil	Vietnam	Peru
Monaco			Philippines
Netherlands	PAL D	SECAM K1	South Korea
New Zealand	North Korea	Benin	Surinam
Nigeria	Romania	Burundi	United States
Norway		Congo	Venezuela
Pakistan		Ivory Coast	
Portugal		Senegal	
Singapore		Zaire	
Spain			
Sweden		SECAM L	
Switzerland		France	
Syria		Luxembourg	
Thailand		Monaco	
Turkey			
Yugoslavia			
Zimbabwe			

NTSC was the variability of the color of images on the screen. Faces commonly appeared in yellow instead of flesh tones and items supposed to be green (like grass and trees) appeared bluish in color. The hue information is encoded into the absolute phase angle of the color subcarrier. Color shift was caused by lack of an absolute phase reference in the transmitted signal, so that receivers had to be adjusted manually. Even if the correct colors were obtained, the older sets would "drift" away in a matter of minutes or hours.

Instead of repealing NTSC and starting over, the color drift problem has largely been cured by greater sophistication in the set. The stability of receivers has improved as solid state devices and complex integrated circuits have replaced older discrete components. A color reference signal is now transmitted by TV stations once per field in the unviewable part of the scan (called the *vertical blanking interval*). The set detects this reference and automatically corrects for drift, yielding proper color accuracy and stability.

The picture part (i.e., the video) is an analog signal which covers a bandwidth ranging between 20 Hz and 4.2 MHz. As described in [Noll, 1989], the television channel bandwidth is actually 6 MHz, allowing for transmission of a vestigial lower sideband along with a subcarrier frequency at 4.5 MHz that contains the audio portion of the broadcast. This structure is what the typical home television receiver is seeking when connected to an antenna or cable hookup. Domestic and international long-haul transmission involves different signal formats, as discussed subsequently in this chapter. The reason for doing things differently in telecommunication (as opposed to broadcast) is that a broadcast transmitter at a television station actually consists of separate video and audio transmitters. The high level of output power from each is summed at the transmitting antenna. Up to that point, video and audio must be kept separate. Also, bandwidth in the broadcast channel is very valuable and the audio must be squeezed as closely as possible to the video baseband. In contrast, telecommunication systems usually have extra bandwidth with which to play.

7.1.1.2 PAL (625 line, 60 Hz)

The PAL system, developed by AEG Telefunken and first officially used in West Germany in 1967, was a decided improvement over earlier implementations of NTSC. Color instability was eliminated as a problem because the receiver did not use an internal phase reference. Rather, the phase of the color subcarrier was reversed from line to line so that the relative difference could be used (hence, we have the name, *phase alternation line*). The first thing you notice in PAL reception is the accurate representation of color, particularly the green of grass and leaves, and proper flesh tones. Another advantage of PAL over NTSC is the improved resolution in having 625 lines instead of 525. Someone accustomed to the 60 Hz field rate, however, would probably detect the flicker of the picture, a consequence of using a 50 Hz field rate.

Many countries of the world which were using 50 Hz black-and-white television adopted PAL as their standard. This gives PAL a strong position in world markets. The broadcast television channel occupies 6 MHz in all countries except the United Kingdom (plus Hong Kong, Ireland, South Africa, and a few others), where 8 MHz is allocated. The intention was to reduce the interference between the video baseband and the sound by placing the sound subcarrier higher in the baseband in the U.K. system. As with NTSC, the transmission format in long-haul systems is different to maintain the sound separate from the video.

7.1.1.3 SECAM (625 line, 60 Hz)

The third and final color system is the French-developed SECAM standard first used in 1967. French-speaking countries generally employ SECAM and it is common in the Middle East, Asia Minor, and much of central Africa. Interestingly, the part of the world where SECAM is the most consistently used is within Eastern Europe. To eliminate the problem with color drift, the hue information is transmitted using an FM technique. The term "sequential color" describes how the two signals which determine the color are transmitted. That is, they alternate in time from line to line. As with the U.K. system, the broadcast channel occupies 8 MHz.

7.1.2 Analog Transmission on International Links

The dominant mode of transmission of color television signals on domestic and international radio links (terrestrial and satellite) is in an analog baseband format. The actual transmission uses frequency modulation of an RF carrier, producing a wideband signal which occupies a full satellite transponder. Figure 7.1 indicates the general arrangement of the baseband used to modulate an RF carrier. Not unlike the format of the ultimate broadcast, analog transmission arranges the video portion of the signal in its normal range of approximately 20 Hz up to a maximum frequency which depends on the particular system (NTSC, PAL, or SECAM). The vestigial lower sideband used in broadcasting is not transmitted in this case since it contains redundant information. The sound portion of the signal is prepared in a much different way, using a higher than standard subcarrier frequency which again is frequency modulated. The portion of the video channel that ultimately conveys the sound to the home receiver is vacant. This facilitates the complete

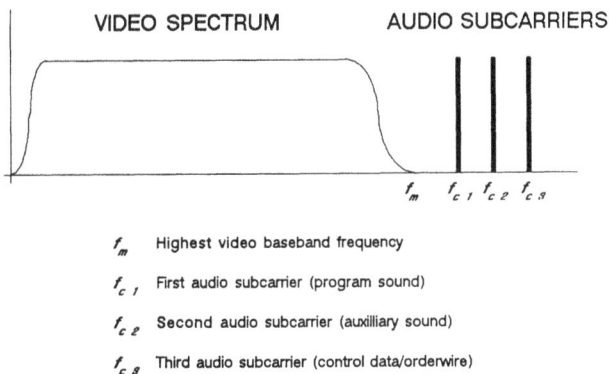

f_m	Highest video baseband frequency
f_{c1}	First audio subcarrier (program sound)
f_{c2}	Second audio subcarrier (auxiliary sound)
f_{c3}	Third audio subcarrier (control data/orderwire)

Figure 7.1 Arrangement of the video baseband used in analog FM transmission over terrestrial radio and satellite links.

removal of the sound portion at the broadcast station, which must apply it to a separate transmitter from that used for video. As shown in Figure 7.1, several sound subcarriers usually may be added to the signal, occupying baseband frequencies starting about 0.5 MHz above the video spectrum and covering a bandwidth of 2 or 3 MHz. These subcarriers can carry stereo sound, multiple language channels, data for broadcast network control, and a "shout down" voice order wire.

Some examples of the baseband and subcarrier frequencies in use around the world are provided in Table 7.2. Aside from the video baseband, there are no accepted international standards for the location of audio subcarriers nor for the FM transmission parameters. INTELSAT specifies these characteristics of the international satellite network for point-to-point and point-to-multipoint video transmission. Likewise, cable television programming which is distributed on U.S. domestic satellites generally follows the frequency assignments given in the table.

Most cable program services employ the Videocipher II scrambling system which encrypts the stereo audio and applies the resulting data to the vertical blanking interval of the video waveform. More specialized or private networks can adopt their own unique set of subcarrier frequencies and modulation parameters. In addition to the frequency itself, the detailed characteristics of audio information can vary greatly. The audio, which is frequency modulated on the subcarrier, can comprise multiple channels. Furthermore, users can control the audio volume, and have a tendency to increase this volume so that their particular information punches through. Anyone with a home dish in the United States can appreciate this problem, because it is obvious as one tunes across the different transponders on a satellite.

Table 7.2 Recommended Baseband Subcarrier Frequencies Used in Satellite Video Transmission to Convey Program Audio, Stereo, and Data (All Frequencies in MHz)

Frequency (MHz)	NTSC	PAL	SECAM
Highest baseband frequency, f_m	4.2	5.0	6.0
TV channel width	6.0	6.0	8.0
Primary audio, f_{c1}			
U.S. cable TV	6.8	—	—
U.S. broadcast	6.2/6.8	—	—
SES Astra	—	6.5	—
EUTELSAT	6.6	6.6	6.6
INTELSAT	6.6	6.6	6.6
Stereo audio, f_{c2}	(left/right)		
U.S. cable TV	(encrypted)	—	—
U.S. broadcast	5.94/6.12	—	—
SES Astra	—	7.02/7.20	—
EUTELSAT	—	—	—
INTELSAT	6.65	6.65	6.65

Source: Norman Weinhouse Associates.

If multiple-transponder operation is not contemplated, then audio volume on sub-carriers can be set to an acceptable level for the particular service.

Many TV networks have automated their operations and are using a sub-carrier channel as a convenient data broadcasting medium. A data stream at 9600 b/s or 56 kb/s can be received with a simple subcarrier demodulator that detects the data and connects it to a computer device such as a PC. This broadcast data channel carries instructions for the video equipment and for the receiving antenna itself. For example, a complete program schedule can be downloaded into the PC, calling for the antenna to be repointed to a different satellite at various times during the day. Other applications of the data channel include newswire services, E-mail, and future schedules. Any return data can be sent over a dial-up connection using a standard voice-band modem. The disadvantage of this approach is that the data reception is interrupted when the ground antenna is moved off the specified satellite. This may be overcome either by transmitting the data broadcast through all satellites of interest (which could be prohibitively expensive) or by dedicating an antenna for this purpose at each receiving site. NBC adopted the latter approach and, in fact, has a separate VSAT network for two-way inter-active data. Additional information on VSATs can be found in Chapter 6.

With regard to satellite transmission characteristics, the luminance portion of the video signal determines the RF bandwidth. Frequency deviation is the term which describes the amount that the RF carrier is swept to one side of the transponder by the luminance signal, with 11 MHz being the typical maximum (peak) value. According to FM theory, the RF bandwidth is approximately equal to two times the sum of the peak deviation and the highest modulating frequency (*Carson's rule*). For NTSC, this amounts to $2 \times (11 + 4.2) = 30.2$ MHz. Additional deviation of the carrier in the transponder is needed to allow for modulation of the carrier by the audio subcarriers; normally, a total RF bandwidth of 32 MHz is sufficient. This is well within the usable bandwidth of a 36 MHz C-band transponder. Two of these carriers can be placed within a 72 MHz transponder. A lesser peak deviation of 9 MHz is also used in special situations like satellite news gathering and direct broadcast, which produces a Carson's rule bandwidth of 26.4 MHz. In this case, an RF bandwidth of 27 MHz can suffice, either as a single carrier or as a pair of carriers in 54 MHz.

Bandwidth is one of the critical parameters in the reception of satellite FM television signals because of its interaction with the earth station receiver demodulator. FM demodulation is a well understood process, yet receiver design is still thought to be a "black art." The key demodulator performance parameter is the FM threshold — the weakened RF carrier level (coming from the receiving antenna), where reception is visually degraded but still acceptable. Degradation in the picture at the threshold point takes the form of small black-and-white horizontal streaks, called "speckles" in the trade. When the signal strength is reduced below threshold, the quality is usually unacceptable for commercial viewing purposes. On a well designed and tuned demodulator, the threshold point occurs at a ratio of *carrier power to noise power (C/N)* of approximately 8 dB, while a conventional demodulator might perform at the 11 dB level. This 3 dB difference equates to a factor of two in satellite power, making receiver threshold performance an important parameter in satellite link design. Oddly, the cost of the receiver does not correlate with the threshold performance. We have found that the only way to know how well the receiver is designed is to test it in the laboratory or on an actual satellite link.

7.1.3 Digital Television Transmission

The television baseband signal can, like any analog signal type, be converted into a digital format. Applying the PCM process commonly used in voice communication, the required bit rate would be twice the highest baseband frequency times the number of bits per information sample. Because of the complexity of the video baseband, there are twelve or fourteen bits per sample, producing a bit rate of approximately 120 Mb/s. As discussed in Chapter 5, this data rate can be compressed down by factors ranging from three to one hundred or more by removing redundancy in the bit stream. A more probable value of three to four has received acceptance by broadcast engineers as this will not cause significant degradation in critical motion performance.

Equipment manufacturers and common carriers have worked to reduce the speed needed for an acceptable broadcast quality signal. As of 1990, an algorithm at 45 Mb/s implemented in an NEC codec is generally accepted in the industry. Audio quality is excellent as a digital format is provided and other capacity is available for data broadcasting to network stations. Due to the relatively high cost of codecs and transmission capacity, 45 Mb/s is primarily used in regional point-to-point links for backhaul (which is discussed later in the chapter).

Bandwidth on international links can cost twice as much or more than on domestic links, so there is motivation to decrease the bit rate further. An experimental codec operating at 30 Mb/s was demonstrated by KDD in 1988. That 30 Mb/s or even 15 Mb/s codecs could met the needs and expectations of broadcasters is entirely plausible, particularly when one considers the improvements in highly compressed codecs used for teleconferencing. Television transmission at 15 Mb/s over fiber optic cables is attractive from an economical standpoint. The added complexity of digital conversion and data compression, however, again make the equipment element of video transmission on fiber costly. Top quality analog FM television transmission equipment is only a fraction of the cost of even the most common 45 Mb/s codecs. As long as the objective is to minimize the transmission cost and as long as the number of receive points is modest, digital transmission can be beneficial.

Satellite transmission of digital video employs *quadrature phase-shift keying* (QPSK), the most common digital modulation method. The occupied bandwidth is slightly greater than half of the bit rate, amounting to 27 MHz for the 45 Mb/s standard. The bandwidth of the carrier increases when error correction coding is used to compensate for operation at reduced C/N. For example, the bandwidth required for a rate 3/4 codec is 36 MHz. More complex modulation such as QAM as used in voice-band modems is generally avoided because of the nonlinearity of the satellite transponder. Ideally, there would be a single carrier in each transponder, but multiple channels can be multiplexed using TDM. INTELSAT does allow dual television carriers per transponder, as discussed previously for analog transmission.

Analog video can be transmitted through a fiber optic cable using phase modulation where each video channel requires its own fiber. Because this is inefficient, digital television is a better match to fiber because several channels can be multiplexed together. The number of digital television channels depends on the combined transmission speed that the fiber optic transmission equipment can handle. Operating at approximately 500 Mb/s, which is typical for transoceanic cables, one fiber can support ten channels. Improvements in compression and manufacturing economics are on the horizon, but extensive deployment on fiber will probably be delayed until the late 1990s.

7.1.4 Multiplex Analog Components

A set of color transmission techniques under the general designation of *multiplex analog components* (MAC) has been developed specifically for use on satellite links. These MAC systems replace NTSC, PAL, and SECAM, and encode the luminance and hue information in an altogether different manner. From a technical standpoint, MAC has superior characteristics over frequency modulation for satellite transmission for the three standards. Another important attribute is that MAC provides encryption of the video and audio so that reception by the general public can be controlled. This is particularly important when considering the possibilities of using direct broadcasting of special television programming such as first-run movies and premium sporting events. Business video also benefits from the security feature of MAC.

As is all too often the case, there are several MAC standards in use. The *de facto* North American standard is B-MAC, which is offered by Scientific Atlanta. Cable TV satellite transmission with FM employs the Videocipher II scrambling system, so the principal application of B-MAC is for private networks. Corporate teleconferencing, pay-per-view events like major boxing matches, and horse racing for off-track betting are heavy users of B-MAC. The system was adopted in Australia as the national standard for all satellite delivered video. In Europe, the basic version is D-MAC with two versions of the standard competing with each other. NV Philips of Holland established a system called D2-MAC, which provides a great deal of flexibility for the broadcaster. In addition to the normal audio channels, which are digitally encoded, the system incorporates several additional channels for high quality audio services and data broadcasting.

The MAC approach has proved itself to be a high quality satellite delivery vehicle with good security capabilities. The cost of receiving equipment is relatively high because of the thin market. As direct broadcasting becomes more established in the United Kingdom and Europe, MAC systems could achieve critical mass and yield unit costs which are attractive for consumers. The basic receiving equipment could easily be built into the television receiver itself. Another advantage of MAC is that improved definition is possible at relatively low cost. The topic of HDTV is explored in the next section.

7.1.5 High Definition Television

Because HDTV is still very controversial, we simply wish to identify the capabilities and issues as they relate to international telecommunication management. HDTV has become highly politicized in the United States, Japan, and Europe, with governments and large corporations exploring how the technology can be exploited (and how a given country itself can avoid being exploited). Most attention is paid to the possible standards for HDTV receivers and the broadcast transmission format.

HDTV demonstrations have been conducted regularly in the United States at television trade shows and by broadcasters. Because of the head start of NHK of Japan, the system most often seen is the MUSE HDTV technique (also referred to as HI-VISION). MUSE stands for *multiple sub-Nyquist sample encoding*. Underlying MUSE is a 1125 line 60 Hz basic television system, doubling the intrinsic resolution of NTSC. The baseband bandwidth of the unencoded signal is essentially four times that of NTSC, amounting to approximately 30 MHz. The MUSE encoding process is able to compress this bandwidth by a factor of nearly four, from 30 MHz down to only 8 MHz. The composite signal includes stereo audio which is also digitally encoded. At 8 MHz, the signal is wider than most standard television signals. Further processing can reduce the bandwidth to fit within the 6 MHz channel used in most countries, an approach which NHK is taking in the 1990s.

NHK and its supporting contractors have created a working system with excellent technical quality. In addition to the improved definition, the shape of the picture is expanded horizontally to give the feeling of a movie screen. Specifically, the aspect ratio, which is the ratio of the horizontal dimension to the vertical dimension, is increased from 4/3 to 5/3. This gives MUSE a presence not possible with regular television. Most HDTV systems have copied the aspect ratio of MUSE.

The MUSE system has a few well-publicized flaws that are a consequence of its position as a pioneering system. Other approaches, many of which are essentially proposals or demonstrations, have responded to these apparent flaws. One problem is that MUSE is not compatible with any existing color television standard (i.e., the MUSE signal cannot be received directly on a conventional receiver). Converter boxes, which are familiar to cable television subscribers, could be made available at modest cost during a transition from normal definition to HDTV. NHK has also responded to this criticism by introducing compatible versions of MUSE, but much attention is still directed toward the system that already works. A second area of concern is that MUSE operates with a 60 Hz field rate, the same as NTSC. European interests point to this difference from both PAL and SECAM.

Several research organizations in North America are pursuing other HDTV technologies. A consortium comprised of David Sarnoff Laboratories of Stanford Research Incorporated, RCA-Thomson, and NBC have proposed the *advanced compatible television* (ACT) standard. In what would be a two-stage evolution toward HDTV, the ACT concept is appealing because current television sets would be completely compatible. The first stage, called ACT-I, gives the wider aspect ratio and improved definition to HDTV receivers, while ACT-II in the second phase would add the bandwidth necessary to provide the full properties of HDTV. Other organizations are pursuing an approach which uses two channels per HDTV signal. The first contains a base color television signal which is compatible with existing receivers, while the second channel is modulated by enhancement information to add the greater definition. More recently, the simulcasting scheme has been advanced wherein the same programming is sent over a normal NTSC channel and a separate incompatible HDTV channel (which, interestingly, may employ MUSE).

European initiatives are also the norm in the realm of HDTV. NV Philips has already developed an improved definition system based on D-MAC; demonstrations of the technology were made in 1989. A cooperative effort on the continent has been mustered under a project called Eureka. The approach of strong government backing, which is followed in Japan and Europe, has also been advocated for the United States. The U.S. Department of Defense initiated an HDTV research program to develop displays and techniques for signal processing. Perhaps some of this work can be applied for commercial development.

International television transmissions of HDTV are just beginning to appear. In North America, PBS, Telesat Canada, Hughes Communications, and GE Americom arranged the transmission of MUSE over the Canadian Anik C satellite. The mode used was FM, occupying 27 MHz of the bandwidth of a 54 MHz transponder. This supported the notion that an HDTV transmission can be accomplished with a standard transponder available on a domestic satellite. In April 1989, KDD, AT&T, NHK, and COMSAT cooperated in a pioneering demonstration of MUSE transmission from the United States to Japan over INTELSAT. A unique feature of the test was the use of a 140 Mb/s codec developed by KDD Laboratories. Digital compression was necessary to provide a broadcast quality picture at this rate. The input rate to the compression algorithm was 648 Mb/s. The bandwidth of the transponder used for the transmission was 72 MHz, which is twice that required for analog FM-TV.

For both these demonstrations, compatible codecs were used on both ends of the link. The encoding equipment for HDTV is very complex and represents a custom installation. It will be several years before the selection of a standard, which will probably be different for each region of the world. After that, equipment development will probably move forward quickly. In 1990, NHK began conducting regular broadcasts with MUSE through their direct broadcast channel in Japan. That system already has approximately two million subscribers who purchased the necessary TVRO dish and electronics to receive the service by using NTSC. Consumers can view HDTV at central locations set up by NHK, but home receivers have appeared on the market. This is the first true market test of the system.

While a lot of attention (both in a marketing and a political sense) is being focused on HDTV standards, the technical performance of NTSC, PAL, and SECAM is improving markedly due to new digital processing and storage technology. For example, a device called a frame store can hold both fields of a frame and then display the frame continuously for the full 1/50th or 1/60th of a second, as appropriate. Further, the apparent number of lines of resolution can be doubled by displaying the mathematical average of the information on adjacent lines. This sophistication is provided by digital signal processing chips within the receiver. Under typical viewing conditions, this processed standard video looks very much like HDTV (although the close-up quality of true HDTV is still better).

7.2 RADIO AND AUDIO COMMUNICATION TECHNOLOGY

Audio channels transmit programming for both AM and FM radio station broadcasts. The technology is essentially the same as for the audio part of television. In fact, it is common for one or more audio channels on subcarriers above a video baseband to be resold to radio networks. The modulation can be either analog or digital.

7.2.1 AM Radio Services

AM radio broadcasting has the least demanding requirements for international transmission because a single 5 kHz channel suffices for the program information. There is a trend in the United States toward AM stereo, but this usually is not required for news and sports. A clean telephone channel with low background noise can be employed, but higher quality may be demanded. In this case, two voice channels can be combined.

A satellite transmission format which is very popular within the United States for AM radio programming distribution is *single channel per carrier* (SCPC). For a complete discussion of this technique, please refer to our earlier work [Elbert, 1987]. Basically, each audio channel is applied to a separate radio frequency carrier and transmitted to an assigned frequency channel within the satellite transponder. There are as many individual carriers operating simultaneously as there are audio channels. The technique has the important benefit that multiple uplinks can be used to access the transponder, each operated by a separate radio network. This provides the independence that many radio networks like to have. In contrast, the subcarrier approach used in television transmission forces all users to share a common uplink earth station, with the added complexity of getting the audio channel to that uplink over some considerable distance.

The modulation on the SCPC signal can be either analog FM or digital PSK. Using FM for an AM radio service may seem contradictory, but there are good technical reasons for doing so. AM is very sensitive to noise and interference within the satellite transponder and from adjacent satellites. FM, however, is very resilient to these impairments and hence is used almost exclusively for analog transmission on commercial satellites.

With suitable encoding, a 64 kb/s circuit can deliver the equivalent of 5 kHz of service bandwidth. Also, the noiseless property makes it very suitable for broadcasting. On top of the basic program channel, broadcasters may require one or more coordination and control channels. Duplex operation through a return SCPC channel would be desirable to aid in coordination of foreign activities. A total of ten digital audio channels can be multiplexed onto a T1 private line or satellite carrier. More information on this transmission system can be found in the following section on FM radio service.

7.2.2 FM Radio Services

Satellite links provide high-quality audio program distribution for FM radio broadcasting networks (i.e., which is delivered over the air locally in the 88 to 108 MHz band). FM radio service has audio quality requirements which are basically the same as the audio associated with television, particularly now that stereo audio has become so popular with television broadcasts. A dual channel configuration is required (i.e., the left and the right), where each has a baseband bandwidth of 15 kHz. The two channels are carried separately even though they are eventually combined within the FM station transmitter. With the advent of compact disc players and broadcasts, the quality of the satellite link has become paramount in FM radio service. The SCPC approach employing FM has been used successfully for several years by National Public Radio on their domestic network. There is the prospect in the 1990s that NPR will use its domestic satellite transponder for limited transborder services. For example, a transportable earth station could be taken to Jamaica in the Caribbean and used to uplink directly to NPR's transponder on a domestic satellite. From a technical standpoint, the SCPC service would work in exactly the same manner as the current service.

Some broadcasters have decided that analog FM SCPC transmission will not provide the quality and clarity that they want under all circumstances, so there is a considerable block of users employing digital audio. The T1 format provides the bandwidth to transport a multiplexed combination of several channels. This can be done on an SCPC basis, where as many as 24 T1 carriers can share a single 36 MHz transponder. The advantage of this is that several networks can uplink their signals directly to the satellite without passing through the same transmitting earth station. In another approach, the T1 channels are multiplexed together with TDM for transmission as a single wideband carrier from a common uplink. The constraint of a single uplink is eliminated if multiple uplinks time-share the transponder by transmitting sequential bursts of audio channels. Such TDMA networks have been deployed over the years, but have not generally been targeted to the audio application.

7.3 INTERNATIONAL TRANSMISSION SERVICES

International television and audio signals are carried across borders and oceans on a daily basis by public and private service organizations throughout the world. We will focus on the long-haul aspect of the problem, involving primarily satellite communication systems. Digital transmission over fiber optic cables is feasible today, but essentially all international transmission is still being conducted by satellite due to a substantial cost advantage.

7.3.1 Video and Audio Transmission Facilities

INTELSAT is the primary carrier of international television and audio services. The reason for this is that nearly every country in the world has an INTELSAT gateway earth station, and some have several. Because major news organizations have worked with carriers and TAs to develop the applications, these INTELSAT earth station operations are accustomed to providing occasional video and audio transmission. COMSAT maintains a directory of the operators throughout the world and their rates, thus greatly simplifying the problem of knowing to whom to talk or how much the service will eventually cost. The process of creating an international network, however, can still be burdensome, whether for a limited time as in the case of an Olympiad, or for a continuous service such as CNN

International. If you do the job yourself, you need to understand the technical details of video communication and how to deal with foreign governments.

A few international carriers package services for broadcasters or others who require temporary links. BTI and France Telecom are strong in this marketplace. Quite obviously, their strengths are in dealing with the end of the service in their home country or region. For example, the U.S. Information Agency chose France Telecom as its main carrier to expand the Worldnet television service. AT&T and KDD are also important international television carriers, but they are more focused on their domestic markets.

The satellite-based video service providers are another important segment of the service market. These are organizations that do not own major transmission facilities, yet provide all of the coordination efforts needed to create a network hookup. The company that defines this market is BrightStar Communications Ltd of the United Kingdom. A subsidiary of Visnews and an affiliate of Reuters, BrightStar has grown steadily over the past ten years to become the leading provider of occasional video and audio links across the Atlantic.

BrightStar has implemented a full-time bidirectional link between the United States and London, where the terrestrial facilities are located at dedicated teleport earth stations. Two transponders have been leased on the Intelsat V satellite located at 332.5° w. The U.S. teleport is provided by Washington International Teleport, Inc., a local satellite earth station operator in Alexandria, Virginia. This gateway to the Atlantic can be reached through a local microwave relay system that is interconnected with Washington, D.C., area broadcasters and carriers. There is a fiber optic connection from the WIT facility to the New York area as well. Alternatively, the international uplink can be reached from any domestic teleport via one of the U.S. domestic satellite systems.

On the European side of the BrightStar satellite network, a full-time facility is maintained in London for east to west and west to east transmission. Local transmission is provided over microwave facilities. Service to the continent is available over the ECS satellites of EUTELSAT. To employ the INTELSAT transponders directly from TA-owned earth stations in France, West Germany, Italy, and Spain would be possible, and mobile uplinks could be arranged on occasion.

Another important international service provider is IDB Communications Group, Inc., of Culver City, California. A relative newcomer to the international carrier business, IDB's capabilities and accomplishments are nonetheless impressive. They have absorbed the operations of several other companies, some larger than their own. One of their early accomplishments was the creation of a domestic network of 20 earth stations to deliver away-from-home baseball game coverage (backhaul) from essentially every stadium in the United States. Today, IDB is the leading U.S. transmission company for the FM radio networks, relying primarily on T1 transmission. Both the SCPC and TDM approaches are followed, where the configuration depends on the application. TDM is used for distribution of FM programming within the United States, while SCPC better fits the bill for international services. IDB's capabilities in international audio, particularly for occasional or one-time services, are unrivaled in the world.

IDB developed an extensive video capability over the last few years, including international gateway earth station facilities on the East and West Coasts of the United States. Through a joint venture with BTI, a full-time video and audio link has been established between the United Kingdom and their teleport on Staten Island, New York. Other facilities are maintained by IDB in Moscow, following the installation of the Bear Lake earth station for the Moscow Summit meeting in 1987. Being on the West Coast, IDB has expanded its international operations into the Pacific Basin. They purchased the earth station facilities of Contel/ASC located in Santa Paula, California. Another site is situated near San Francisco in Valejo, California. IDB has moved very aggressively to become a leading international video and audio transmission services company, with a heavy investment in earth station facilities on both coasts.

7.3.2 Service Capabilities Available to Users

The international service providers reviewed in the previous sections can be used to create temporary or permanent video and audio transmission systems for use by broadcasters and other strategic units. In areas where these providers have facilities or the necessary relationships, it probably behooves users to take this particular route. There will be instances where the user would take responsibility into its own hands and create a private network. To do this, the user should be prepared to invest a great deal of time and effort, including the associated travel, to arrange for the facilities and to provide the necessary on-site coordination. Leading users, such as Turner Broadcasting, ABC, CBS, and NHK, have dedicated the necessary personnel and logistical resources. Discussions of what some of these organizations have done in international television are provided later in the chapter. Strategies for dealing with the foreign PTTs, which can make or break an international video or audio service, can be found in Chapter 8.

7.3.2.1 Backhaul of Event Coverage

The majority of international video and audio transmission is used to carry back the coverage of an event for use as a program element in a domestic broadcast. Known as backhaul, the service was pioneered by U.S. networks particularly for the major events like Olympic Games. Experimental transmissions on the low orbit Telstar satellite caught the public's eye in the late 1950s. The first continuous international satellite transmission, however, was in fact the Tokyo Olympiad in

1962, employing the entire bandwidth of the newly launched Syncom satellite from geosynchronous orbit. Prior to satellites, international television was severely restricted and in fact the only significant event which was sent electronically was the coronation of Queen Elizabeth. The narrow bandwidth of a transatlantic telephone cable was employed by playing back a recording of the telecast at a slow enough speed to permit the signal to pass through the limited bandwidth of the cable.

Backhaul naturally involves significant logistical operations and lots of coordination for a transmission that may last a few hours or at most a few weeks. The elements in a backhaul, shown in Figure 7.2, include production crews, a studio to assemble the material for transmission, a terrestrial link from studio to uplink, and the uplink earth station itself. Fortunately, international satellite earth stations maintained by carriers and TAs are interconnected with cities through dedicated microwave facilities. The difficulty would come when the event site or studio is located some distance from the access point to the existing microwave link. A TA may be willing to install such temporary microwave systems, and, in fact, may hold a total monopoly on such services. The cost of constructing and

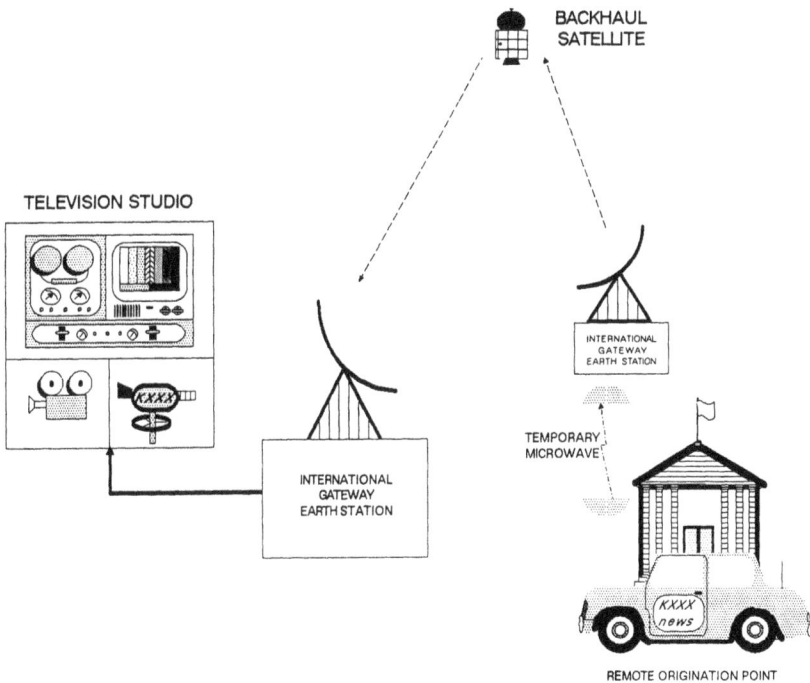

Figure 7.2 Backhaul of video or audio signals for remote event coverage, using point-to-point satellite transmission.

using temporary microwave can be substantial, and it may be more productive to use an on-site earth station.

Domestic and regional satellite systems offer the opportunity to bypass the domestic transmission system and link up with the international gateway. Video and audio services are primarily one-way, so a double satellite hop does not pose much of a problem. In some news and sports broadcasts, interviews are conducted remotely over a satellite link, so the on-air personalities must be alerted not to be confused by the one-second, round-trip time delay.

Backhaul services can be very complicated and expensive to set up because of the potential number of operating entities that must be managed. The service may only last a few hours even though many months of effort may have been involved. A dedicated international link could be installed if the service is repetitive in nature, such as the backhaul used by a consortium of foreign broadcasters. For example, the JISO group which is discussed later in this chapter leases a full-time video channel to Tokyo from a shared studio in New York City. American broadcasters like CBS and ABC do the same for a link between the United Kingdom and New York. To justify such an expenditure, there should be at least four to eight hours of daily usage. Otherwise, occasional services from an entity like BrightStar or IDB would be more economical (and simpler to manage).

7.3.2.2 Program Distribution

International distribution of television and radio programs is just coming into its own in the late 1980s and early 1990s. The medium that many industry observers expected to provide the impetus was direct broadcasting from satellites. Innovators like Turner Broadcasting, NHK, and MTV Network, however, took early steps to create international distribution networks using existing low- and medium-power satellites. The object is to deliver the signal to points of further distribution on a distant continent or region.

The age-old problem with international broadcasting is how to find a program format that appeals to a wide audience considering the language to be used. MTV Network believes that this problem does not exist for them as their service appeals to rock music–loving teenagers and young adults. CNN International is able to play upon the preeminence of American English as a *de facto* international language of business, obtaining a position on cable television systems at hotels and in countries where a significant percentage of the population can understand the language.

Our purpose in this section is not to find a formula for a successful broadcasting format, but rather to review the approaches for interconnecting such a network (assuming that a format has been selected). Like many telecommunication networks, a distribution system for video or audio is built up from individual links. It evolves over time, growing as new markets are located. The key is to identify the locations and the facilities that are potentially available, a process that justifies having a full-time telecommunication staff.

There are two basic options open to the network implementer. The first is to locate a remote studio in the host country. This simplifies international transmission, reducing it essentially to the backhaul problem. Occasional transmission over an international link would probably suffice. The approach has the added advantage that the programming can be tailored to the needs and tastes of the local market.

In the second approach, all studio facilities are centralized in the home country and international links used to deliver the programming to each destination. Remote studios are not required. The problem is reduced to a pure telecommunication networking operation, where a star network must be established. Personnel costs are minimized and much of the operation can be automated. The transmission costs of this technique are relatively high, however, because of the need for multiple international links.

Turner Broadcasting uses a combination of these approaches for CNN International. There is a remote studio set up in London, but several satellite links are used to reach countries throughout the world. CNN gets double usage because the London studio originates daily broadcast transmissions to the United States. The international satellite network is very extensive, with transponders leased on a full-time basis from INTELSAT, EUTELSAT, PanAmSat, and even INTERSPUTNIK.

A considerable volume of political water has passed under the bridge of international satellite broadcasting regulation. The concept of European television without borders is coming in the early 1990s and several satellite systems are going to try to build profitable businesses. The White Paper of the European Commission authorizes satellite broadcasts to reach markets in several countries at the same time, but it is up to the local country's administration to decide if and how the population can take advantage of the service. The complexity of language and cultural differences stands as a very high barrier to a profitable medium. In fact, most of the current efforts in satellite broadcasting across borders are to reach the more liberal markets in the United Kingdom and Belgium.

The most challenging technical problem in program distribution, particularly to cable television systems and the home, has nothing to do with launching spacecraft or generating sufficient power in space. Rather, it is the problem of tightly controlling access to the signal after it leaves the uplink earth station. Scrambling systems have been around for several years and a few are in active use in the United States. The Videocipher II system, used to scramble essentially all cable television programming, is not secure from pirating. The current descrambler units can be unlocked with the simple insertion of a duplicate address key. Approximately one-half of the 800,000 home Videocipher II units have been estimated to have been broken in this manner. The best defense that cable television networks have been able to muster is aggressively to locate and prosecute the sellers of duplicate keys. Whenever pirate descramblers are located, the duplicate key is simply turned off through the encryption data channel which rides over the satellite.

Satellite capacity for program distribution is available in most developing countries, as discussed in Chapter 2. Because satellite operating organizations must invest heavily in the satellites themselves and because the systems have a limited life of from eight to twelve years after launch, satellite operators are motivated to convert to revenue as much of their capacity as possible. The Turner Broadcasting lease of a Statsionar 10 transponder from INTERSPUTNIK is a case in point. The successful negotiation benefited from a motivated party and the existence of some competition. The INTELSAT system is extensive; the particular type of coverage that the user prefers may not be available, however, due to constraints of the satellite design. Also, some of the INTELSAT birds are filled to capacity and full-time transponders may not be available for video services.

Audio services using the SCPC technique occupy only a portion of a transponder and therefore are more easily acquired even on a crowded INTELSAT primary satellite. Subcarrier capacity can usually be leased from the operator or user of a full-time video uplink. For example, a radio network can extend its audience to subscribers of cable television systems by connecting the audio program channel to a subcarrier on the video uplink. Any TVRO at a cable system can recover the audio channel with an inexpensive subcarrier demodulator and pass the radio network feed over the cable system at a convenient FM band frequency.

Another characteristic of satellite systems worthy of note is the large increment of capacity that is created when a satellite is first launched. It usually takes an operator a few years to sell off as much as half of the capacity. The primary cable television network birds in the United States, which are usually pre-sold before launch, are an exception to this rule. Less popular satellites almost always have ample capacity throughout their operation lifetimes. This gives rise to the principle that is well known to people in real estate, namely that the three most important things to have are "location, location, and location." In the case of satellites, the location relates to the orbital position and the extent to which there is an existing ground infrastructure of TVRO and uplinking antennas pointing at it. Some video network operators use movable antennas and therefore are able to select a satellite based on the lowest cost of use. The same applies to a private video distribution network which does not need to be integrated into an existing system.

7.3.2.3 Private Broadcasting and Video Teleconferencing

Private broadcasting (synonymous with business video) employs the same analog full-motion video technology used by television networks and other commercial broadcasters. The medium is very popular in the United States among large corporations and government agencies to communicate with employees and customers across a widespread region. Some companies even own their own television studios and can produce programs which rival regular commercial television shows. For strategic units to contract with a television production company or specialized teleconferencing company to set up a private broadcast is very common, as this can be cost effective for relatively infrequent events. More discussion on private broadcasting can be found in [Elbert, 1989].

Some organizations have already conducted international private broadcasts, primarily between the United States and Europe. A user wishing to pursue this would follow the outline already given for establishing an international video link. The same service providers can be engaged to assist with arrangements in the various host countries. As an example of an international private broadcast, Hewlett-Packard used satellite links from INTELSAT and the French Telecom 1 satellite to include points in Europe with their normal teleconferences which are carried on a U.S. domestic satellite. With the assistance of France Telecom, HP reached fifteen sites in Europe and encoded the video for security with the B-MAC system. Sites in Canada were also included by virtue of the transborder coverage of the U.S. domestic satellite.

7.4 ARRANGING INTERNATIONAL SATELLITE TRANSMISSION FROM INTELSAT

International television transmissions are carried over satellites primarily provided by INTELSAT. Users can work with a domestic TA or provider which is an authorized user of the INTELSAT system. Fees include the charge for using the uplink, the transponder on the uplink side, the transmission from the satellite on the downlink side, and the use of the receiving earth station or earth stations. To these costs are added charges for local transmission, as discussed previously in this chapter. While the satellite transponder is only used once per transmission, an older INTELSAT fee structure has users paying several times over; once for the ground to space direction, and again for each space to earth path to a receiving earth station. This really makes no sense and INTELSAT has modified its fee structure so that only one downpath is charged regardless of the number of receiving earth stations.

In the United States, any carrier or user can acquire satellite capacity from COMSAT once a standard set of administrative procedures is satisfied. The space segment is, in fact, the easiest part of the linkup to purchase because the process has largely become automated in recent years. The more complicated part is arranging for earth stations and local connections on the ground. Users can, however, solve these additional problems themselves and save considerable money in the process. The trade-off is that service providers like BrightStar and IDB can save the user considerable time and trouble, especially if the transmission is for a one-time event.

7.4.1 Television Service Modes

Transmissions are routed through more than 11 INTELSAT satellites located over the Atlantic, Indian, and Pacific Oceans for distribution to literally hundreds of earth stations around the world. The four types of satellite coverage beams available are global, hemispheric (hemi), zone, and spot beam. Examples are shown in Figures 2.4, 2.5, and 2.6 for the Intelsat V and Intelsat VI series of satellites. For reference, the global beam has full coverage of the portion of the earth visible from the longitude position of the satellite (which is approximately one-third of the total area of the globe). Satellite deployment across the ocean regions changes from time to time as INTELSAT launches new birds and retires those which have reached end of life. Satellites which reach end of life are boosted into a higher orbit and deactivated so that they cannot interfere with operating satellites. At this altitude, the orbit will not decay and there is no danger of the retired satellite ever crashing into the earth.

Occasional video transmissions are carried primarily at C-band on the global beams of designated satellites. The limited radiated power of global beam transponders must be compensated with ground antenna diameters in the range of 15 to 30 m, which correspond to INTELSAT standard A earth stations. These earth stations will also function for the hemi and zone beams which have greater radiated power than the global beam. Ku-band service is offered through the spot beams, which are smaller in geographic dimensions than the C-band coverage beams and produce correspondingly greater radiated power on the earth's surface. As a consequence, Ku-band spot beams can be accessed by smaller earth stations. Ku- and C-band coverages can be cross strapped, allowing a Ku-band uplink to reach a C-band receiving earth station, and *vice versa.*

The following definitions apply to transmission service modes. Simplex transmissions are transmitted in one direction only and are the predominant form of service. In half-duplex transmission, a single transponder allows alternating transmission of a one-way signal by two or more earth stations. The transponder is therefore shared by the transmissions, because only one may be present at a given time. Duplex transmission allows transmission in both directions between two locations and therefore involves two transponders. An exception is where the two carriers share a common wideband transponder, which has double the bandwidth of a conventional transponder.

7.4.2 The Video Service Order Process

The sequence of events which an authorized user must follow to book and use INTELSAT satellite capacity for occasional video service are summarized below for the United States. It is assumed that the service is to be ordered through COMSAT and rendered between the United States and another country. The following paragraphs identify an eight-stage process.

The International Service Order

The user calls the COMSAT International Satellite Services (ISS) to place the booking. An account can be created at this time if one does not already exist. The COMSAT service coordinator will ask for the following details for the service: the broadcast time, date, number of receive sites, the type of programming (news, video conference, sport event, special world event, program syndication, *et cetera*), and the billable party. COMSAT maintains an on-line database of the available satellite capacity and is in direct contact with INTELSAT to ensure that arrangements are properly made. The on-line system is available to users on a dial-up basis and for additional fees.

The Domestic Order

To arrange for domestic uplink or downlink transmission, the user must contract directly with a U.S. earth station operator. Transmission over a domestic terrestrial or satellite link may be required in order to carry the video from or to an international gateway. COMSAT maintains a current listing of international video carrier companies, which is summarized in Table 7.3 as of 1990.

The Path

The COMSAT ISS coordinator will recommend a satellite and transponder (the path), thereby matching the earth stations in the various countries for the desired connectivity (not all earth stations can transmit to or receive from every satellite). When the requirements are complete, the video service order is entered into COMSAT's automated television reservation system.

INTELSAT Coordination

COMSAT is responsible for carrying the booking forward to INTELSAT to reserve the space segment for the particular event. Simultaneously, COMSAT will inform the foreign authorized carrier of the service requirement.

Table 7.3 U.S. International Video Carriers

Carrier	Telephone No.
AT&T	(212) 219-7773
BrightStar Communications, Ltd.	(212) 582-8578
Crescomm Transmission Services	(201) 882-4184
GE Americom	(212) 930-4804
Houston International Teleport	(713) 438-3600
Hughes Communications	(213) 607-4000
IDB Communications Group, Inc.	(213) 870-9000
MCI International	(201) 585-3341
Overseas Telecommunications, Inc.	(703) 764-2829
Turner Teleport International	(404) 827-1543
Washington International Teleport	(703) 734-2724
Keystone Communications, Inc.	(213) 474-3500

Foreign Coordination

The user must now advise its foreign counterpart of the service that they must place a matching order with their authorized carrier (such as the TA or international carrier in the host country). Recall that INTELSAT service always consists of separate parts — one for each country in the connection.

The Match

Once both the end user and the foreign carrier have placed the order with IN-TELSAT, COMSAT ISS returns a written confirmation to the user. COMSAT also sends a confirmation to the gateway earth stations along the transmission path.

Alternative Paths

INTELSAT occasional video capacity is committed on a first come, first served basis, so there will be times when matching orders cannot be made before someone else acquires the desired time slot on the recommended satellite and transponder path. If this happens, COMSAT will advise the user of alternative paths and available times.

The Follow-Up

The COMSAT Operations Bureau will require that earth station operators along the path submit a television completion report when transmission is completed.

This assures that the transmission was successfully completed and is the basis for the submission of accurate billing to the user.

The COMSAT Television Scheduling (CTVS) service mentioned under the first stage above is an important capability which is available to U.S. international television users. Subscribers to CTVS pay an extra fee to have direct access to an on-line computer scheduling database. By dialing into the COMSAT computer, the user can employ nine different information screens to determine available satellite time, to configure the transmission path, and to determine the associated costs. An E-mail capability is provided to permit the user to request that COMSAT convey an appropriate service order with INTELSAT. Because only COMSAT can place such orders, the process requires the intervention of a human operator prior to confirmation. The system has proven so effective that some foreign organizations are subscribers to CVTS.

International video services are readily available in the United States because of the largely free market there. Table 7.3 lists the registered international video carriers as of 1989. Some of these companies simply operate earth station facilities and are really not full-service providers. They can, however, play a role in the combined service.

7.5 COOPERATIVE BROADCASTING ACTIVITIES

Commercial, public, and private broadcasters are not alone when pursuing international video and audio transmission activities. Many of the most basic problems of connecting broadcast networks together have been solved by cooperative groups which seek to establish standards and gain leverage by pooling efforts. Broadcasting has always had international repercussions because radio transmissions naturally cross borders, both potentially causing RF interference (RFI) with legitimate radio receivers and allowing foreigners to listen in on the television and radio shows of another country. Short wave radio is particular effective for international broadcasts, but attention is now focused on satellite transmission because of its technical superiority. The extent of radio communication around the world has made the International Telecommunication Union into an important UN agency for broadcasting operations. Other UN agencies are concerned with issues of copyrights and the rights of individual nations to control what their citizens can see and hear. Beyond the issues of radio spectrum management and domestic politics, we consider such organizations which broadcasters employ to gain leverage with suppliers and governments around the world. Without these efforts, a given broadcaster would be largely on its own in dealing with a potentially much stronger entity.

7.5.1 National Association of Broadcasters

Incorporated in the United States and headquartered in Washington, D.C., the National Association of Broadcasters (NAB) is the leading trade association for companies and other organizations which engage in radio and television broadcasting. While a primary mission is to influence U.S. government broadcasting policy, the NAB's technical and business activities are watched carefully around the world. The NAB is very much an international pace setter, as demonstrated by the popularity of their annual trade show among Japanese, Latin American, and European broadcasters and manufacturers.

The NAB has an emphasis on the technical and administrative operations of networks and broadcast stations. Programming development and distribution (including syndication) are not a primary focus of the NAB. Members include individuals and organizations, making the NAB a particularly diverse and well-represented organization in the industry. Manufacturers of equipment are included, as are telecommunication companies which provide terrestrial and satellite transmission of video and audio services.

Within its engineering segment, the NAB provides a forum for the resolution of many of the important technical issues. One which is getting strong attention is HDTV, both from the standpoints of program production systems and transmission of the HDTV signal itself. The first public demonstrations of HDTV were at the NAB convention. In 1988 and 1989, the NAB reserved an entire room at the convention for HDTV developers to provide actual side-by-side comparisons of the different systems. Researchers can present technical papers at engineering sessions held during the annual convention. Because of the wide attendance at these sessions by engineers from networks, stations, service providers, and manufacturers, the question and answer periods are always very stimulating. Participants from outside of North America add an international flavor to NAB conventions.

Between conventions, the NAB maintains ongoing technical committees which meet periodically. Topics of discussion deal with standards such as the appropriate format for 45 Mb/s encoding and the data communication formats to be used for remote station control. These topics overlap the various member groups within the NAB, which gives considerable credence to any resolutions which could result from the ongoing dialog.

7.5.2 Regional Broadcasting Unions

The broadcasting union is a cooperative grouping of governmental broadcasters in a particular region. One of the most recognized is the EBU, which, in addition to its intergovernment coordination mission, also functions as a regional network. Other broadcasting unions exist in Asia, Africa, Latin America, and the Arab world. A comprehensive list of regional broadcasting unions is provided in Table 7.4; the key unions among them are reviewed in the following paragraphs [Yurow, 1983].

Table 7.4 Regional Broadcasting Unions

Acronym	Organization	Headquarters
ASBU	Arab States Broadcasting Union	Tunis, Tunisia
ABU	Asia Pacific Broadcasting Union	Tokyo, Japan
CBU	Caribbean Broadcasting Union	Kingston, Jamaica
EBU	European Broadcasting Union	Geneva, Switzerland
OIRT	International Radio and Television Organization	Prague, Czechoslovakia
CBA	Commonwealth Broadcasting Association	London, U.K.
OTI	Ibero-American Television Organization	Mexico City, Mexico
IAAB	Inter-American Association of Broadcasters	Montevideo, Uruguay

The EBU has long held a special position in Europe and the world, providing the vehicle to distribute programming which is acceptable to its members. The 38 active members of the EBU reside in Western Europe and around the Mediterranean. High technical standards are maintained by the EBU's full-time staff of engineers and operations professionals. A broadcasting network within Europe is operated for the benefit of the members. This network started out as a hookup of terrestrial microwave radio links provided by the PTTs, but has gravitated to satellite transmission through EUTELSAT. Also, the EBU obtains live coverage of events in other parts of the world through full-time and occasional video links over international satellites.

The EBU once held a near monopoly on program distribution within Europe but this is waning in the face of commercial initiatives such as Sky Television, CNN International, British Satellite Broadcasting, and the SES Astra satellite system in Luxembourg. The opening up of European markets in 1992 is expected to propel this migration even further. In the face of these changes, the EBU still holds a unique position as the carrier of noncommercial television throughout the continent. In particular, the success of commercial broadcasting ventures which can transcend borders is still very much subject to debate.

In recognition of the rapid development of the Pacific Rim, the Asia-Pacific Broadcasting Union (ABU) is the largest and most diverse regional union. Boasting 26 full members and 38 associates, the ABU maintains its headquarters in Tokyo with another office in Kuala Lumpur, Malaysia. A wide diversity of members is indicated by the presence of such countries as Saudi Arabia, New Zealand, Sri Lanka, and China. The Tokyo headquarters is located at the NHK Broadcasting Center, in recognition of the leadership role taken by Japan's government-owned network. The purpose of the ABU is to promote professional interchange of program, technical, and legal matters within the region and within cultural groups. It is expected that the ABU will grow in importance with the continued economic development of the region.

The strength of Spanish-language media activities in Mexico has raised the Ibero-American Television Organization (OTI) to international stature. Members include Mexico, Spain, and Portugal, as well as many Latin American countries. One recent accomplishment was the acquisition by OTI of the rights to carry the 1984 Los Angeles Summer Olympiad to Latin America. Its stated purpose is to work for professional improvement of broadcasting in member organizations through collective study and mutual assistance.

7.5.3 *Ad hoc* Affiliations (Broadcasting Pools)

Broadcasters often approach international television and audio needs by working together in *ad hoc* groups and affiliations. Some of these are essentially permanent with a dedicated staff and set of facilities to deal with long-term requirements. Other activities are set up just for the duration of a particular event, the Olympic Games being a good example. Practicality is the motivator for taking the *ad hoc* approach, because a single broadcast organization may not have the financial, technical, and personnel resources to undertake a major international project on its own. What was once a joint short-term effort could eventually blossom into a separate activity of its own.

Broadcast pools have been used by U.S. networks when covering an event in a foreign country. The pool is formed by the three major television networks with CNN now representing a fourth member. Others may join depending on the nature of the event and its location. One member is designated as the lead for the purposes of negotiating arrangements with the host country. The lead broadcaster may also incur the majority of the expense of the activity. For this reason, the leadership is passed on to the next network in a predetermined order. The advantage of being the lead broadcaster is that you can be first in line for some of the key facilities, which may be in very short supply. The cost and complexity of being leader is why only a few organizations can sustain the role. Recent competitive trends in commercial broadcasting and news are driving the networks apart and consequently the broadcast pool is not as popular as it once was.

The Japanese International Satellite Joint Users Organization (JISO) is a broadcast pool which is essentially permanent in nature. JISO was created to provide a continuous connection from the Western world to Japan. The headquarters of JISO are in New York, where both U.S. and European occasional broadcast services can be coordinated. JISO maintains studio facilities at Rockefeller Center and has a full-time video link from there to Tokyo. This connection is provided as a service by Keystone Communications over a domestic and INTELSAT satellite link. The principle members of JISO include NHK, Tokyo Broadcasting Company (TBC), and Fuji Television. NHK also maintains a separate full-time video link for direct broadcasting service within Japan, as discussed previously in this chapter.

Ad hoc groups of broadcasters have evolved from necessity because of the difficulty and cost of establishing international video and audio links. Perhaps one day these links will be readily available on demand from the international carriers but there will always be a role for *ad hoc* approaches because the nature of events continuously changes as do the players themselves.

Chapter 8
MANAGEMENT AND LOGISTICS

To implement a global network is anything but simple, yet there are certain management principles that can significantly aid the process. International networking, after all, is another form of international business, something that many companies have done successfully for more than a hundred years. The difference here is that telecommunication networks are evolving rapidly, as are the information processing systems that employ them. Political and economic change is also underway, particularly in Europe and Asia, where much of that which has worked in North America can be applied. Telecommunication managers who must proceed into the international environment would do well to study the two aspects: how to do business internationally, and how to apply modern digital technology.

In this chapter, we present network implementation strategies that apply to the international environment. One of the priorities is for the strategic unit to develop a presence in each country which is to be served and to contain a node. There is, of course, the basic trade-off of whether to obtain telecommunication as a service from international and domestic carriers, or to implement a private networking capability using owned facilities. We primarily focus on the latter, although any real network will be a hybrid of public services and private facilities. The traditional network design process is reviewed from the perspective of international capabilities. There will be significantly greater logistical and administrative problems when operating in several countries at the same time. Perhaps the one commodity that the network implementer should maintain in reserve is time because it will almost always take longer to accomplish any task. Very often, costs and time schedules must be doubled or even tripled to account for the complexity and added expense of international operations. Whoever said "patience is a virtue" might well have had the role of the international telecommunication manager in mind.

8.1 INTERNATIONAL NETWORK STRATEGIES

Strategies for telecommunication networks are presented in Chapter 8 of our previous book [Elbert, 1989]. We discussed in that chapter four basic telecommunication network strategies, each representing a particular focus for defining, implementing, and operating the required facilities. The four strategies are *control of destiny, network reselling, use of public networks,* and the fostering of *customer dependence.* These are common themes encountered in the context of private networks for strategic units. These network strategies are applicable to the international situation, but there are several additional considerations. Some of the simplest problems that are easily handled in the domestic context can be "show stoppers" when extending networks overseas. For example, the basic question of "who do you call?" if a circuit fails to perform due to an unseen equipment problem must be addressed. Should you have qualified personnel on every site, or can you rely on nontechnical personnel to stand in? Are there network management systems that can replace people on site? If an internal approach cannot be developed, can the local telco or PTT be expected to assist in times of need? Or, might there be a third-party service organization that can take action? Therefore, while creating an international link is a major challenge in and of itself, maintaining a vital network will raise all kinds of new issues.

In some ways, this is not unlike the post-divestiture environment in the United States, where telecommunication managers must deal with multiple carriers to complete a private line connection. The international environment has its parallels, with international carriers operating long-haul media between countries and TAs providing domestic services on a monopoly common-carrier basis. Adding to this the prospect of deregulation within the countries can cause the problems to increase almost exponentially. For these reasons, a strategic unit needs a separate international network strategy when moving beyond its borders. Examples of such international networking strategies are provided in the following sections.

8.1.1 Balancing Public and Private Facilities

Previous chapters have investigated the opportunities for building private networking capabilities from services offered by carriers (Chapters 2, 3, and 4) and from facilities that a strategic unit can own and operate (Chapters 5 and 6). From the standpoint of "controlling your own destiny," ownership would seem to be the optimum strategy. In fact, many organizations in the United States have sought to build their private networks in this way, relying on carriers only for dedicated links at the T1 and higher rates. Some of these strategic units have embarked on the somewhat risky strategy of reselling excess capacity to other organizations as a way to recover costs and to justify their initial over-buying.

International telecommunication network development poses many additional problems, and a single strategy cannot be applied with ease. Controlling your own destiny might be desirable, but the size of the endeavor could sink the ship in the process of having the basic network work as planned. Of course, if requirements are simple enough, such as a single point-to-point circuit that can be implemented with IBS, the ownership route may be feasible. In most cases, however, the number of points to be reached and the variety of applications to be provided should move the situation toward a combined strategy of some type. The solution process for this type of question has been called a "make *versus* buy" decision; this sounds simple, but the problems are complex enough to demand a thorough investigation. The resulting strategy will not be based on a single approach (e.g., a "make") but rather on how the services and facilities are mixed together. Therefore, make *versus* buy trade-offs are performed on a case-by-case basis, selecting from a predefined menu of choices that the strategic unit has assembled.

A graphical representation in Figure 8.1 defines the relationship among the size of a strategic unit, the degree of use of services from carriers *versus* ownership and a mix thereof, and whether the applications are for domestic or international telecommunication. The largest strategic units can often justify owning domestic networks; even these large units, however, must move toward a mix of public network services and owned facilities when deploying international network

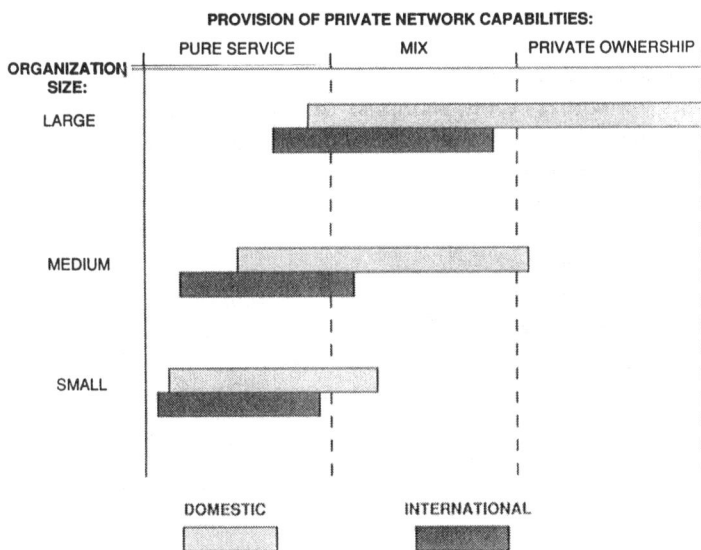

Figure 8.1 Selection of pure telecommunication service *versus* pure ownership of facilities for small, medium, and large strategic units, considering the difference between domestic and international requirements.

capabilities. The role of international carriers in this regard is discussed in the next section. Medium and small strategic units will almost always rely on carriers when doing business internationally, simply because of their reduced business attractiveness for, and leverage with, foreign governments and TAs. This important topic is also reviewed in a subsequent section.

Strategic units engaged in international business have historically relied on the PSTN and the telex network. The PSTN has become even more valuable as the quality of connections has improved greatly, facilitating low- and medium-speed data applications and Group 3 facsimile. Telecommunication managers should consider how good-quality dial-up service can supplant private lines, particularly when several time zones are being traversed. In cases where the predominant application is telephone and fax, dedicated connections may never be sufficiently utilized to justify the cost. For example, the overlap of work hours between Frankfurt and Los Angeles is only about one or two hours in the morning, Frankfurt time. Similarly, the overlap for Los Angeles and Tokyo is also only an hour. Private lines will still be necessary for 24-hour operations and if the circuit can be utilized in "off" hours for unattended operations.

Some have argued that the telex network is outmoded and need not even be considered by international businesses. Others state that telex is still the most important international networking medium, particularly when dealing with developing countries and the socialist nations of the world, who have yet to implement PDNs. Old-fashioned telex machines are probably a relic of the past, particularly now that PDNs and E-mail services connect to telex networks. The user can format a message on a computer terminal, and then employ a mailbox facility, which is, in fact, a telex number. Distant telex users employing traditional terminal equipment can exchange telex messages with distant subscriber numbers that are actually E-mail boxes. Open availability of fax machines within Eastern Europe will hasten the decline of telex, as has already occurred in the West.

In addition to the size and geographic extent of the activities of the strategic unit, the type of activity and the nature of the services must also be considered carefully. For example, a television network wishing to cover events worldwide and having a few fixed bureaus in such places as London and Hong Kong, will not require a dedicated network. They must, however, have relationships with foreign service providers and transportable facilities, which, on a combined basis, permit rapid deployment and quick activation of service. Contrast this with an airline that needs an extensive telecommunication network to connect branch operations in major cities worldwide. Rapid deployment is not as important as consistency because the service to each point must be uniform and reliable.

8.1.2 Role of International Carriers

The leading international carriers have embarked on a strategy of gaining major customers and are attempting to build market share. Among the more familiar of these organizations are AT&T, BTI, France Telecom, and Cable and Wireless. Some of the leading PTTs, who previously concentrated on their domestic monopoly markets, are, however, entering the international marketplace. The Deutsche Bundespost is presenting itself as an international carrier, DB Telekom International, and is willing to solve some of the problems of major customers. Likewise, KDD has recognized that deregulation in Japan and the entry of competitors require that they, too, offer more substance to their customers and even enter new markets in foreign countries. The other two leading U.S. long-haul carriers, US Sprint and MCI, have made strong moves into international telecommunication. All of these companies have in common that their contacts with key foreign service providers as well as their technology and financial bases enable them to implement almost any plan that they may undertake.

As a case in point, in 1988 and 1989, British Telecom and France Telecom were successful in acquiring the international business of major international users like General Electric of the United States, the U.S. Information Agency, Arthur Andersen, and American Express. Furthermore, the European airline reservation consortium, Amadeus, signed an agreement with the Deutsche Bundespost for that PTT to be the single source (i.e., "one-stop shopping") for the first European E1 backbone network. The fact that these sophisticated telecommunication users chose the leading international carriers to provide their international networks is profound.

The major user organizations, called "blue chip customers" in the United States, have the transmission requirements and industrial strength to motivate the international carriers to become more innovative. Competition for these customers' accounts is particularly intense during this period of telecommunication restructuring. How can a smaller strategic unit get this kind of attention from the carriers? The answer for the time being is that they probably cannot, but help is on the way. The smaller user ought to be able to enlist the carriers for some of the traditional leased line services. In this respect, British Telecom is taking steps to improve the management of its domestic and international facilities, which will allow them to be more responsive to user needs. Such steps are particularly important for BT because its chief U.K. rival, C&W's Mercury Communications, is already very customer-oriented. US Sprint, as discussed in Chapter 3, is building an international fiber networking capability with the acquisition of PTAT Systems, Inc., and their relationship with C&W in the Pacific. With carriers such as US Sprint and C&W Mercury in the market, the more established carriers will have no choice but to become more responsive to the mid-sized user.

8.1.3 Role of Public Data Networks and E-Mail Services

International PDN operators like SprintNet, GE Information Services, and Infonet offer unique opportunities to a strategic unit that cannot justify the creation of a private international data network. As discussed in detail in Chapter 4, the PDNs as well as E-mail and information services like Dialcom, MCI Mail, and the IBM Information Network have solved the bulk of the logistical and technical problems encountered in international data networking. What is more, these organizations arrange for local staff around the globe, providing the most convenient answer to the question, "who do you call?"

The PDNs provide a certain set of telecommunications capabilities, primarily in the area of data. As discussed in Chapter 4, these companies have made agreements with PTTs in the countries that they serve. A strategic unit can contract with a PDN for a uniform set of data communication services almost without regard to the foreign telecommunication environment. PDNs either offer information services or allow users to provide access to third-party information service providers, giving the strategic unit the benefit of a much larger information resource. A case in point is Dialcom which evolved out of the information utility business primarily in support of the U.S. government.

A number of the PDNs, Infonet in particular, are in the business of providing facilities management. This means that the strategic unit can install node equipment throughout the world, then contract with the PDN for maintenance and other support. Data communication service could be provided over a combination of the private network and the PDN.

International data companies now offer EDI services through their PDN and other computer facilities. As discussed in Chapter 4, GEIS and IIN are in a position to interconnect suppliers, manufacturers, and retailers around the globe. Incompatible EDI architectures can be supported because the network supplier has the software to make the necessary conversion. For example, X12 standards are in use in North America, whereas Europe is focusing on the UN standard, EDIFACT. The actual document information is forwarded over the proprietary network of the EDI service company because conversion to X12 and EDIFACT occurs prior to delivery to the user. These capabilities put the EDI service company in the position of being able to solve a critical international networking problem for small- and medium-sized strategic units. The large users can justify their own EDI networks, but even the largest will need to interconnect with the public EDI service.

The PDNs are extending the services outside of basic packet switching by offering facsimile transmission, bypassing the international PSTN. In addition to the normal kinds of dial-up access to fax machines, the PDNs allow users to broadcast messages to several machines throughout the world. Also, E-mail messages can be generated on computers and terminals, then delivered via fax to machines connected either to the PDN or to the PSTN. By using the facsimile networking services of an international PDN, an organization can deal effectively with time zone differences, incompatible E-mail systems, and the vagaries of the postal system.

8.1.4 Role of the System Integrator

A strategic unit which has decided to install some of its own facilities around the world can employ the services of a system integrator to deliver the network capability on what is referred to as a "turn-key" basis (meaning that after the network is complete, operations are initiated by the system integrator, and the "keys" to the facilities are given to the customer, who takes operational control). Many of the major telecommunication and computer equipment manufacturers have entered this business area, and the topic of system integration has received much attention in the trade press. Furthermore, there are several large system integration houses, such as Electronic Data Systems Corporation and Martin Marietta, which themselves do not manufacture the equipment that they install.

The system integrator's task is to perform nearly all of the system design and installation functions, relying on its own resources instead of those of the strategic unit. These functions are reviewed in subsequent sections of this chapter. Importantly, the system integrator takes responsibility that the network will perform as specified. The significance of this aspect should not be overlooked, and often it would be the justification for deciding to contract the system integration role.

System integrators can also operate the network for the strategic unit after installation is complete. This approach was taken in some large projects for the U.S. government, such as the Distant Early Warning (DEW) Line radar detection system. More recently, EDS developed a substantial business managing and operating the data processing and telecommunication facilities of major U.S. and multinational firms. Under this basic scenario, the system integrator takes control of data processing and telecommunication resources (involving possibly sharing facilities with other customers), and then charges a fixed annual fee for all services. The facilities are operated by personnel who are either brought in from the outside or are hired away from the customer's former data-processing or telecommunication staff. Although EDS has the largest market share, IBM and Computer Sciences Corporation have each gained important customers and can be expected to provide EDS with some tough international competition.

8.1.5 Strategies for Operations in Foreign Countries

We assume from this point in the discussion that the telecommunication manager has taken responsibility for the design, installation, and operation of network facilities in one or more of the foreign countries to be served by the network. He or she, therefore, is to take on the role of the system integrator. We have made the point that international networking is just another form of international business, wherein a strategic unit develops the processes for dealing in several countries at the same time. Large corporations and government agencies, particularly the foreign service and military, are very experienced at setting up foreign operations in developed and developing countries alike.

These network strategies have in common that their primary purpose is to gain leverage with the local infrastructure of the foreign country. A strategic unit thus has numerous ways to develop and apply leverage on governments, financial institutions, customers, suppliers, and other foreign corporations. The telecommunication manager should consider how to take advantage of any leverage that can be developed by members of the strategic unit or its support groups. The key point is that whoever has the leverage ought to use it. As an example, a satellite transmission company operating in the United States had a customer who wanted to transmit a closed-circuit television signal into northern Mexico. The operator of the satellite itself, an American company, was unable to make arrangements with the appropriate Mexican government authorities simply because the amount of business involved was minuscule in relation to that company's overall domestic activities. The customer was highly motivated to obtain the required permissions, however, so he personally visited the agencies in Mexico City and eventually made the arrangements himself.

8.1.5.1 In-Country Consultant

The single most effective way to evaluate the environment in a new country is to create an individual presence inside its borders. Larger organizations have local representatives already on site for the purpose of setting up or maintaining business operations. Because these people are already members of the same strategic unit, they can probably be counted upon for reliable advice and assistance. Unfortunately, they could be so occupied with their own work and problems that they might provide little more than a temporary address for the telecommunication manager to use. The alternative, short of specifically opening up an office for the purpose, is to engage the services of a reliable consultant already familiar with many of the situations that the telecommunication manager may encounter. Even if the consultant does not already know the answers, he or she can rapidly do the research and legwork.

In the following discussion, we will take the position that local representation is to be provided in one of the ways discussed above. Telecommunication managers who proceed by trying to deal over the telephone or in writing will quickly discover that answers will not be forthcoming in the desired time frame. One way or another, someone will have to place their feet on foreign soil. We also assume that the network design (which is discussed later in the chapter) has already been determined for the most part.

The consultant's main job is to discover and help remove the hurdles that must be overcome before the network facility can be installed and operated. While it may be possible to identify the general class of problems, they will no doubt be different from country to country. Locating such a consultant is no easy task; if the organization already has a local office, they can assist with the selection of such an individual or company. If not, the search will be more difficult. The local office of a major international bank could be of assistance and possibly suggest a number of candidates. Of course, the bank may not understand your requirements well enough to be very specific, but at least can provide a starting point. The time spent locating a consultant and verifying the qualifications and integrity of the individuals concerned will be well worth the effort. In the following discussion, under the term "consultant" we group individuals, consulting firms, and members of the strategic unit's on-site staff, if available.

A first priority is to establish relationships with the appropriate government or telecommunication authorities to "learn the ropes" of doing business in the particular country. Depending on the extent of the facilities to be installed, the telecommunication manager could be required to attend some of the meetings with government officials. As has been proven time and again, such contacts are vital and usually lead to the learning of the ways of preventing hurdles and even gaining an advantage. Readers should be advised, however, that more time would be taken than is customary in the United States, often because of the work load of the individuals who must be contacted. This author has spent many hours in lobbies of government buildings in developing countries waiting for meetings to start (some of which never occurred as planned).

What is surprising is that just asking nicely and being patient can work miracles. For many years, no Western carrier was able to establish a wideband telecommunication circuit into the Soviet Union. IDB Communications Group, however, not knowing that it was impossible, approached the Soviet authorities in Moscow and asked how they could install an earth station for a simple T1 IBS circuit to the United States. What IDB found was that by listening to the Soviet government people and being sensitive to their needs, it was able to obtain the necessary permission to install the facility. While originally intended for a short-term operation in support of the Moscow summit meeting in 1987, the IDB-installed earth station has remained in operation for an indefinite period.

8.1.5.2 Transfer of Technology

Modern telecommunication networks often employ technology which has not de-fused around the world. If this is the case in the instance of the network to be developed, then the telecommunication manager may be able to employ it for negotiating leverage with a less than cooperative foreign government or telecom-munication authority. The country may benefit from direct application of the tech-nology in the existing telecommunication infrastructure.

For example, back in the mid-1970s, a data communication link was needed in the Republic of Indonesia between Jakarta, the capital and largest city, and a distant site. The only means of communication at that time was via a newly installed microwave link, which was configured for normal analog telephone service. We were able to install 2400 b/s duplex modems at the computer center in the city and at the remote site; we needed, however, the cooperation of PERUMTEL, the PTT, to acquire decent wireline circuits between the computer center and the telephone exchange where the microwave system terminated. Because the computer center had been operating for walk-in customers only, no data lines had ever been installed and used. The existing telephone wiring was totally inadequate, particularly when it rained (which it did very frequently in that tropical country). The PTT cooperated by installing new cable in a relatively short time span. With the data circuit in operation, the PTT was then familiar with this form of service and could respond more effectively to new requirements from other organizations needing data lines.

European PTTs are more willing than ever to accept new technologies and allow intelligent devices to be connected to public networks. As discussed in Chapter 5, the key is to obtain type acceptance of the device prior to installation. The homologation process is an accepted principle in Europe now that networks are being opened up for access by strategic units wishing to implement private networking capabilities. The manufacturer is normally expected to work directly with the government to have their device accepted. There is the advantage of allowing new technologies into the network because of the potential for increased usage and revenues. This is particularly the case for international telecommunication services, where major PTTs and international carriers have invested billions of dollars on fiber optic cables and satellite systems.

In some nations, data processing and telecommunication systems should partially or wholly be supplied by local manufacturers under domestic content rules currently in force. This can be a very troublesome obstacle for network development using some of the more advanced techniques such as compressed video teleconferencing and narrow-band voice multiplexing (see Chapters 5 and 6). Major telecommunication projects, such as the Brazilian satellite communication system installed in the early 1980s, are of such magnitude and duration that local manufacture is potentially viable. Manufacture in Brazil of the satellite itself was not feasible, but the minicomputers and some of the earth station hardware could come from local industry. The kind of problem that one encounters, however, is that the local product may be incompatible with the network operation or interfaces. In such cases, one needs to modify the local product or encourage the domestic manufacturer to produce a new, compatible product. While not a general solution to all problems, local manufacture has been applied successfully in some critical cases in the past.

8.1.5.3 *Employment Opportunities for Local Citizens*

Almost without exception, the governments of the world have a policy of promoting full employment of citizens. This domestic objective can be used to advantage by the telecommunication manager when dealing with an uncooperative bureaucracy. Perhaps the overall business activity of the strategic unit will employ many local citizens in manufacturing jobs. Automobile assembly plants, oil exploration and mining operations, airlines, and electronics manufacturers are in an excellent position to use this type of leverage. Obviously, the new network capability is an important aspect of developing the local operation and therefore can be related to employment. On a smaller scale, the telecommunication facility to be installed will likely require on-site manpower in the form of construction work, installation of equipment, testing, and eventually operations and maintenance support. These areas are treated later in this chapter. Usually, the domestic economy has within it the ability to supply these services, either in the form of skilled individuals or from one or more local contractors.

Locating such a work force can be difficult, but may be the key to obtaining the actual approval for the project. This is where the local consultant can again show his or her worth. What is essential is that any person or contractor so employed on the project be capable of performing the required technical tasks. One way to ensure this is to provide training on the systems to be installed. In the instance of a major telecommunication project in Indonesia, local technicians were actually transferred to the United States, Belgium, and France for training programs provided by the equipment manufacturers. These individuals already had undergone basic technical training (which was available at government technical schools in Indonesia) and could speak English. Fortunately, anyone who can speak a foreign language usually can comprehend the technical training as well.

8.2 LAYING THE GROUNDWORK

Entry into a foreign country for the purpose of extending an international network is not an impossible task, provided that you have done your homework and a certain groundwork is laid. This concerns the technical issues previously raised in this book, but the real-world problems of dealing with a different culture and physical infrastructure are vital concerns. Subject to regulatory restrictions, a practical system design which is simple to understand and which has already worked in another place will relieve the strategic unit of a significant part of the risk. In addition, success will depend greatly on having the support of people and in effectively coordinating their activities. With thousands of miles of separation and potentially poor communication (until the network is in operation), these needs will not be simple to fill.

8.2.1 Realistic System Engineering

The system engineering approach would dictate that the requirements for any new system should be carefully defined before any concrete steps were taken toward implementation. A thorough analysis and series of trade-off studies are usually performed to identify an optimum configuration, considering criteria such as minimum total cost, minimum response time, and maximum data throughput. One should have no preconceived notions about how the system would be constructed, what technology should be used, or which specific pieces of equipment should be bought. This approach is too idealistic to be practical in international telecommunication because of the reduced number of options available and the difficulty of implementing any "optimum" solution with precision. Instead, one needs to be pragmatic and have as many contacts as possible so that whatever is selected can be accomplished within some realistic time frame.

While the telecommunication manager is unlikely to have the luxury of designing a unique network solution for the international requirement, there still are planning steps that ought to be taken. When new requirements suddenly appear, there simply is not enough time to perform numerous trade-off studies. A basic network architecture should be selected early based on the principle that requirements can be attached from time to time. In general, the telecommunication manager should follow a logical process tempered by a sense of humor.

8.2.2 Management Commitment

A statement which has become very popular is that a telecommunication manager needs the commitment of the general manager whenever embarking on a major network project. The situation is no different for international telecommunication, and, in fact, management commitment may be even more essential than in the domestic case. The key point for management to understand is that the job may not be as simple as it is in the home country. As stated earlier in this chapter, time schedules become stretched out from weeks to months or even a year or more. Also, costs can be several times those of an equivalent domestic service. A general manager of an international division of the strategic unit no doubt would want communication operating as soon as business operations are initiated. This person must be made to understand the realities and to allow the time and resources to be committed.

The project plan will need to be presented for approval by the overall management of the strategic unit. If the organization is fortunate enough to have a chief information officer or equivalent on the executive staff, the coordination and approval process can be simplified. Otherwise, an executive less familiar with the technology will need to do the presenting. This person ought to be familiar with the difficulties of international business, so that he or she could explain some of the more critical aspects of the plan.

Extending a telecommunication network into a foreign country requires a team effort. Upper management aids this effort by including telecommunication requirements with the overall business plan. Then, coordination with the in-country elements (other than telecommunication) will be easier to accomplish. Some leverage is gained by simply having the general manager point out to suppliers, customers, and government contacts the importance of telecommunication in achieving the strategic goals.

A steering committee of telecommunication management, key users, and operations personnel could be created to facilitate the gathering of requirements and implementation of the new network capability. This enhances the commitment of upper management as well as others who will come into direct contact with the facilities after they are installed both locally and overseas. The committee group identifies potential problems because of the greater experience base that can be brought to bear on issues. People come to know each other better, which will prove useful after the network is deployed and face-to-face contact is infrequent.

8.2.3 Contingency Plans

Most of us are accustomed to reliable public telecommunication networks, so when basic telephone or fax service is limited can be disheartening. This becomes more critical when information networks are to be extended to a foreign location, where data communication capabilities may be unavailable or incompatible with information systems. The telecommunication manager needs to educate upper management as to what capabilities can be expected and how reliable (or unreliable) they could be. Then, managers will be better able to cope with disruption when it occurs. Contingency plans become more important than ever. For example, a PDN access line could fail, but operations may not be hampered if, in anticipation of such problems, a dial-up backup system is provided.

To prepare contingency plans does little good unless they are distributed to all locations and key people ahead of time. A failed telecommunication link will make direct communication difficult and could preclude coordination of restoral. It becomes very basic under such circumstances — a simple telex message may be all that you can send to inform the other end that the contingency plan is to be activated. Dial backup may be the solution, provided that you can rely on the PSTN to provide the necessary circuit. In areas where satellite communication links formed the basic network capability, the only alternative could be high frequency radio or even messenger. Stating the obvious, these alternatives must be checked regularly to ensure that they indeed work. Before the disaster happens, telecommunication managers need to identify these alternatives and have means to employ them.

8.3 NETWORK DESIGN

Network design is really an ongoing process by which the telecommunication needs of a strategic unit are converted into a clearly defined network configuration. From that point, the telecommunication engineer can implement the network by procuring the necessary elements, being combinations of owned facilities and services from common carriers. This process continues into the future, as requirements shift due to additional deployment of people and locations, and more significantly as the strategic unit's business or functions change.

This section reviews considerations in the design of international networks and is not intended as a tutorial. A more complete discussion of network design, considering economic and technical aspects, is found in Chapter 6 of our prior work [Elbert, 1989]. Economics is important because capital and operating costs need to be recovered. Unless the strategic unit is actually in the telecommunication or information processing business, relating the cost of operating the network to the financial performance of the business activity will be difficult. Therefore, the primary approach will be to select the most cost-effective way of providing the needed service capabilities.

Extending the process to the international network problem is not particularly straightforward because the standard collection of techniques cannot be applied with ease. Certainly, the telecommunication manager can compare the cost of dial-up telephone service with that of an international private line. Basic information on such tariffs in Europe is provided in Chapter 3. These studies are easy enough to perform because the analyst requires only a limited set of data. Yet how would you analyze a truly global network that extends to dozens of countries today and could be required to reach dozens more in a matter of years? By 1992, a uniform service offering will apply throughout Europe, but tariff structures today have a high statistical variance. Moreover, a critical service such as T1 or E1 private lines may not be available to a country or city of particular importance to the strategic unit.

The key to a successful network strategy is flexibility. We have already discussed the importance of having a reliable in-country presence. Flexibility in terms of the network itself will allow the telecommunication manager to respond to a changing environment. The principle of combining international public services from the PSTN and PDNs with capabilities from owned facilities is central to a successful strategy as well. Network design, in this case, means finding the correct mix and selecting an architecture most appropriate to the strategic unit, both in terms of applications and the financial resources that the organization can reasonably afford to apply. In the following paragraphs, we provide some additional thoughts on the network design process with regard to international applications.

8.3.1 Requirements Definition

A telecommunication network design (or *any* system design) is only as good as the accuracy with which its requirements are specified. There is, after all, the old principle of "garbage in, garbage out." The problem for the international telecommunication manager is that the requirements are so obscure and subject to change. To separate requirements for international services from the capabilities that exist at any given time is extremely difficult. The more a designer can abstract the requirements from the systems that will ultimately be employed, the better will be the eventual evaluation of options.

Requirements are typically specified in terms of the number of circuits needed for telephone, data, analog or digital video, and facsimile transmission. In the world of digital communication using T1 and E1 backbones, the individual circuit requirements are aggregated. The time-varying nature of such needs as video teleconferencing, telephone calling patterns between countries, fax traffic, and data transmission demand are incorporated to permit further compacting of information in multiplexing devices with dynamic bandwidth allocation and statistical multiplexing. This is conceptually simple, but difficult to predict with any precision.

Ideally, circuit quantity specifications are themselves determined by an analysis of existing traffic on an operating network. When implementing a new international network, however, there will not be any statistics for use in the analysis. In this case, the designer can postulate traffic transfer requirements as inputs for the analysis. Fortunately, intelligent nodal devices are flexible in terms of bandwidth allocation and modularity of design. Circuit configurations can be adapted after services have operated for some time. A rough estimate is still needed so that initial capacity is neither grossly excessive nor insufficient. The analysis tools discussed in the next section can be used to generate "first cut" circuit requirements.

8.3.2 Analysis Tools

The network design process has evolved to fairly high levels of sophistication in the United States, where telecommunication managers can take advantage of software-based network design tools. If the problem consists of a few conventional telephone circuits, standard traffic engineering tables can be used. More complicated networks, including those with data communication services, are more readily evaluated with commercial software packages. As presented elsewhere [Elbert,

1989], such software packages are available in the United States to run on a personal computer or a much larger machine, such as a minicomputer or a mainframe. The advantage of the PC version is low cost of hardware and software, and the entire package of hardware and software can be transported to another country (local import restrictions notwithstanding). Because of limitations on storage capacity and execution speed, the package may only provide a fraction of the capability of the mainframe version. The version for a minicomputer or mainframe could handle the largest network and would run the program at much greater speed. A strategic unit may already have such a software system in operation and be able to add the international services and facilities to the existing database.

To simplify the tariff research problem, on-line computer services allow the network designer to enter a hypothetical wide-area network configuration and to see results of the analysis in terms of the detailed circuit routings and costs. Introduction of owned facilities brings additional complexities, but several manufacturers of node equipment, such as packet switches and high-level multiplexers, provide their sales staffs with PC-based software packages, which can configure a network for the customer virtually overnight. Still lacking in 1990, however, is a generic network design package with a detailed database describing all vendors' equipment.

Users employ other software packages for network traffic estimation and circuit sizing if the existing network is maintained in a relatively fixed database. From a technical standpoint, the algorithms within the program could either directly simulate the traffic going through the network or employ simplified analytical models from queuing theory. Experience with the particular network will show which kind of data is necessary. In the most complete programs, actual traffic statistics are downloaded into the computer from PBXs and other intelligent devices. Such analysis packages can be used almost anywhere in the world because they presuppose the network configuration rather than synthesize a new design. Creating the design is such a complex problem that practical software has yet to appear on the market.

8.3.3 Technology Assessment

Telecommunication managers and engineers continuously need to assess the technologies that are available in the marketplace. The information contained in this book and that of its references categorizes the technologies and services, and identifies some of the more established sources of supply. This information should be supplemented with sources that are absolutely current at the time of need. Most telecommunication managers make a point of reading the current trade literature and usually will attend one or more of the leading annual trade shows. Publications in the United States such as *Network World, Communications Week, Data Communications, Telecommunications,* and *Business Communications Review* do a good job of publicizing new products and reviewing their capabilities. (Of these, *Tele-communications* and *Communications Week* have international editions available in Europe.) Stories also cover how strategic units have applied some of the latest technologies in building or expanding their networks. Among the best of the annual trade shows are Communications Networks, Washington, D.C., the Tele-Communications Association (TCA) show, San Diego, and the International Communications Association (ICA) meeting. The ITU holds biannual conventions in Geneva, Singapore, and other cities, giving international vendors an opportunity to display and demonstrate their equipment and network capabilities. These generalized shows are supplemented by numerous specialized gatherings on data networks, satellite communications, teleconferencing, and television and radio broadcasting. Most of the newer technologies and products are exhibited at these shows.

Major telecommunication carriers, equipment manufacturers, and resellers maintain direct sales forces for the purpose of locating customers and closing deals. Sales people and application engineers usually cooperate by providing brochures, specification sheets, and operation manuals. The telecommunication manager can request that the vendor prepare a network design that uses their particular type of equipment or network capability. This service would probably be provided at no cost. The marketplace is very active in North America, Europe, and Japan because of the size of the client companies and the potential sales that can be garnered. Included in this market are those services and facilities that can be applied to international telecommunication. Quite obviously, if the international carrier or equipment vendor can establish a close working relationship with the telecommunication manager in the home country, subsequent sales for international use will be greatly facilitated.

Employing these sources of information, the telecommunication manager can maintain a personal file of hardware, software, and services that may prove useful in future applications. A file cabinet full of articles, brochures, and even equipment manuals will be a valuable repository of data. Keeping this information up to date will require continual review of these information sources. What will always be advantageous is to ask vendor representatives for information and to consider the possibility of experimenting with new devices and services. The procurement of equipment and services would follow the *request for proposal* (or *request for tender*) process discussed later in this chapter. For greater detail, please see [Muller, 1989] and [Elbert, 1989].

8.3.4 Topology Decisions

The concept of network architecture for domestic and international networks was briefly considered in Chapter 6, where we showed how nodal devices can provide packet-switched, circuit-switched, and dedicated circuit services. Our continuing emphasis is on integrated digital networks, using 56 kb/s and higher rates on the transmission lines as well as dynamic bandwidth allocation and statistical multiplexing. Such capabilities are now available within continental regions and across oceans via fiber optic cables and INTELSAT Business Service earth stations.

Terrestrial backbone networks in North America employ a ring or mesh architecture to provide alternate routing. International networks are more costly to implement and to operate, however, forcing one to use an interconnected star topology (see Figures 6.1 and 6.2). The advantage of this approach is that the number of dedicated international links is reduced and the bandwidth utilization thereon is maximized. Potential failure of the intercontinental long-haul connection, however, increases the risk of disruption of the entire network. Local distribution within a continent or region is by way of a star topology with a major foreign city located at the hub. VSAT networks, if available in the foreign region, could fit this need particularly well.

The topology question then becomes one of locating the hub or hubs of the network. Location can be determined by considering a number of factors, with a key one being where the strategic unit already has a significant presence. To establish and maintain a hub in a city and country that are familiar and where the organization is already recognized definitely will be simpler. Locating a suitable consultant may be as simple as calling your representative office in the particular country.

Two other factors hence become important: the availability of modern digital telecommunication facilities and the tariff structures under which services are offered. The facilities need to extend beyond the borders for other countries that require connection to the private network. Within a few years after 1992 (or perhaps even by that year), major European telecommunication construction projects will be complete and networks interconnected so that 64 kb/s and wider bandwidth can be nearly uniformly obtained. The TAs on the continent follow CCITT recommendations with regard to the organization of tariffs, but there are wide disparities among countries. This disparity, illustrated in Figures 3.2 and 3.3, may be a basis for selecting one city rather than another for the primary European hub. One of the objectives of an open Europe is to eliminate or reduce this wide variation so that firms can proceed on the basis of business criteria and not circuit costs.

Asia and Latin America, two other important economic regions, are not as simple as is Europe to evaluate for hub locations. In fact, the hub concept may not apply at all, considering the greater distances between centers of economic activity. Mexico and Central America can be viewed as an extension of a North American private network; the nature of telecommunication services to these places, however, is dramatically different. While the problem might be resolved with transborder links, the typical Latin American TA adopts the posture of a totally foreign country, forcing strategic units to transmit information over international gateways to the INTELSAT system and through limited underwater cables. An exception is IBS, which is available to certain cities in Mexico and South America.

Japan can be exploited as a hub location, but connections from that country to centers like Hong Kong, the People's Republic of China, and the Republic of Korea are often no less costly than if direct links are established. Singapore, however, has initiated a strategy to be the Asian hub of choice for strategic units engaged in some business activity in that Asian city-state. In addition to excellent public network services, Singapore Telecom has investment shares in all of the Pacific and Indian Ocean fiber optic cable systems. Singapore is very competitive with respect to enhancing its position as a hub for telecommunication, just as it is with regard to manufacturing and transportation as a whole.

8.4 IMPLEMENTATION PLANNING

The implementation phase begins after the network has been designed and choices have been made about the facilities to be deployed. An international network implementation project assumes new dimensions as compared to a comparable activity within a single industrialized country. Logistical problems are considerably more difficult because so much of what is needed must be transported a considerable distance, and then all operations must be conducted through a long and relatively inefficient supply line. The first principle is to allow extra time for the entire project, and the second is to have modest expectations about the nature of the result. What will be best is to include support personnel and organizations on the implementation team who are familiar with international projects.

Team relationships must be much stronger and cohesive in the international environment. Interpersonal and interorganizational communication is quite difficult, as barriers of distance and time zones must be overcome. Much more must be committed to writing so that correct (and expected) actions can be taken. Every element of the organization should have a good understanding of its respective role. For this purpose, an organizational chart, showing international staff as well as outside subcontractors and telecommunication carriers, needs to be constructed, distributed, and maintained. Telephone, telex, and fax numbers should be included.

If the decision has been made to use a system integration contractor or international carrier to coordinate the project, the strategic unit must still identify a program manager to work in concert with his or her counterpart. Ideally, these individuals should have been involved in contract negotiation so that they would have the fullest possible understanding of what to include and to exclude. Trying to discern the intent of the parties from only the contract, purchase order, or tariff can waste time and cause all kinds of problems later.

The node equipment and transmission services may be procured through a request for proposal (RFP) process, also called a request for tender (RFT). The equipment to be obtained is mainly described in terms of the functions that it must provide and the technical specifications that it must satisfy. Bidders (tenderers) submit proposals to the buying authority for evaluation and selection. The RFP should state where the equipment is to be delivered and whether the supplier is responsible for installation and checking. A system integration contractor may actually conduct the procurement as well as installation. If not, the strategic unit must decide how to complete the project after the equipment is delivered to a staging area or the final destination.

8.5 SHIPPING, INSTALLATION, AND CHECKING

The topics of shipping, installation, and checking would appear to be mundane and unworthy of specific mention. For international network implementation, nothing could be further from the truth. We must keep in mind that at least some of the equipment to be installed in the foreign country will not be available on the local market. For example, simple devices like cable connectors and test meters could be vital to timely operation of the network, but if these were left behind, days or even weeks could be taken to resupply. Overseas operations also dictate that the equipment be fully tested under realistic conditions before being packed for shipment. To send critical elements overseas is inadvisable without first checking to see that they function properly, in a place where corrective action can still be taken.

The most important part of the installation phase actually occurs before the equipment is even sent to the final destination. This is the *factory acceptance test* (FAT) phase, which is conducted by each vendor on its portion of the system. Usually, each manufacturer conducts a FAT on its segment of the project so that every device meets its performance requirements and is in the correct configuration for the network. Representatives of the strategic unit, such as in-house engineers or consultants, may be present during key tests. In a large system, to conduct an integrated FAT of overall operability is desirable prior to deployment in the foreign environment. This would involve connection to existing network facilities and the use of dummy traffic to load the new facility. The more of this type of pretesting that can be done, the more likely the network capability is to function as expected.

Shipment of equipment to a foreign country is a complex subject and worthy of its own chapter. People who are very familiar with international logistics are essential to have involved in this phase. International freight forwarding companies can be used to advise and even conduct the entire logistical effort (but at a price). Generally speaking, air shipment is the most reliable, and it is certainly quicker than transport over sea or land. Also, airport facilities throughout the world are easier to access and tend to be managed by younger, more flexible people. The cost of air shipment can be more than compensated by the reliability of delivery and confidence in time schedules. Of course, if we are talking about shipping large towers, generators, and building components, sea transportation will be necessary. In such cases, allow sufficient time for the shipment to arrive at its destination and to clear the customs bottleneck. Not uncommonly, shipments are misrouted or lost, arriving months after the expected delivery date. A potential worst-case situation would be that the shipment never arrives at all. The on-site manager or consultant will be particularly valuable in locating shipments and facilitating the clearance through governmental or other roadblocks. A safe storage facility should also be arranged so that the equipment could be housed prior to its installation.

Telecommunication facilities that have been properly configured for the particular purpose must still undergo certain on-site processes before services can be provided. This ensures that nothing has been broken in shipment and that interfaces are properly configured. The requisite equipment is typically installed by a subcontractor or telecommunication service company working under an agreement of some type. Facilities which are leased from a service provider such as a telco or PTT may still need to be installed prior to use, which effectively results in the same set of circumstances. The telecommunication manager needs to monitor the progress of installation and the installer must be held accountable for schedule and quality. One would hope that the installer could be held liable for any disruption of ongoing activity or damage to existing facilities. The question of liability is a complex one, and depends on the nature of contracting in the particularly country. Most telecommunication service providers attempt to shield themselves from liability beyond the direct expense of correcting a fault that any of their installation people may have caused.

Assuming that installation is complete and acceptable, the next order of business is to exercise the network capability to ensure that it will perform correctly. Typically, an acceptance test procedure is written and reviewed by the buyer prior to testing. The testing is often a joint effort of buyer and seller because the buyer's applications ought to exercise the network and the seller's personnel are needed to operate and troubleshoot the new equipment and transmission facilities.

The installation and checking phase seems to exhibit the best and the worst in people. Installers in foreign countries often work under difficult circumstances and on tight schedules. Therefore they must move quickly, and may need to improvise when parts are missing or instructions are fragmentary. Teams of installers commonly miss making a connection with each other, causing unnecessary delays and even costly reworking. Telecommunication managers should try to develop procedures for tracking progress during installation because subcontractors at times could be deficient in internal coordination. For the subcontractor, a delay may cost a little money, but the effect on the strategic unit of not having the network capability in place is much more dramatic.

8.6 OPERATIONS AND MAINTENANCE

Once the network capability is in operation, to maintain it in working order becomes the responsibility of the telecommunication manager. This is *operations and maintenance* (O&M), a function traditionally handled by telephone service organizations such as telcos and PTTs. According to Bell System practice [Bell, 1983], O&M is divided into three categories:

Provisioning is the process of activating particular telecommunication resources such as switch ports, terminal devices, trunks, and circuits. This process could include forecasting short-term demand to determine the inventory of available facilities that should be maintained. In the modern private telecommunication network context, provisioning is often accomplished nearly automatically through network management facilities, as discussed in the following section.

Administration of the operating network is subdivided into *network administration* and *service administration*. The former monitors network performance and traffic flows to ensure that grade-of-service objectives are maintained. Integrated network management systems again play a key role in gathering such information and allowing it to be used to adjust network resource allocation. Service administration is identical to administrative operations, discussed at the end of this chapter.

Maintenance operations ensure that all facilities work properly and continue to serve users. Functions include testing and repair activities to correct malfunctions, which is called *corrective maintenance,* and those activities that prevent a malfunction from occurring (*preventive maintenance*).

Based on decades of experience with large networks serving millions of users, the Bell System philosophy is worth following. An average strategic unit will not have resources in proportion to that which a major TA can allocate to O&M; therefore, one will need to chose activities carefully. The role of a viable monitor and control system, which can feed status information to a network management center of some type, is clearly established. In the international context, O&M is considerably more difficult to implement and telecommunication managers will need to build upon whatever resources can be found and applied to the problem.

Telecommunication service organizations can be contracted to provide O&M services both within the home country and at numerous foreign locations as well. We have already mentioned that some international carriers and PDNs are in the business of providing this type of support. International carriers and some TAs are beginning to pursue O&M services as a business opportunity and may also be enlisted. Before a contract is signed, however, the telecommunication manager and O&M service provider must have negotiated a clearly defined statement of work. The kind of items to be included are:

The responsibilities of the buyer (the strategic unit), including what functions the buyer's personnel will perform in support of O&M, what access the service provider will have to operating facilities, payment terms (including whether the buyer must provide spare parts), and rules for notification of problems and recourse if the problems are not corrected.

The responsibilities of the provider (the O&M service organization), including the point of contact for problem reporting and the response time for corrective action to start (4 hours, 24 hours, *et cetera*).

Penalties on either party for not performing. The provider may need to refund part of the fee for service if the network is down for more than a certain period. This provision may only apply where actual transmission service is included as part of the package.

In the new telecommunication environment, users take more responsibility for the O&M of devices that are connected to the network. Therefore, for the telecommunication manager to expect users to troubleshoot problems on terminals, modems, telephones, and fax machines would not be unreasonable. These devices can be repaired by commercial repair shops which are generally available in major cities around the world. The area where the telecommunication manager is very much the focal point is network services. This is where network management plays the vital role of keeping the telecommunication manager in control of resources and allowing him or her to perform basic O&M from a centralized location.

8.7 INTEGRATED NETWORK MANAGEMENT SYSTEMS

Modern telecommunication networks, although advanced in terms of technical capabilities, are still difficult to manage in an operational sense because of the diversity of vendor equipment and network technologies employed. Large corporations that operate private networks must also deal with an amalgamation of public network facilities, particularly if the network is international in scope. The basic network management capabilities that telecommunication managers require include monitoring and control; configuration and change management; problem management; and network performance analysis. This challenge is increased by a typical shortage of skilled technical personnel within the strategic unit so that the ideal network management solution is one that provides sufficient automation and reduced training time.

In Chapter 6 of our previous book [Elbert, 1989], we define and illuminate the network management problem along several lines of discussion. We can view it as an organizational problem with requirements for operating the network as a business. This is comparable to the needs of a small telephone company, which must connect new users to the network, arrange for collection of charges and billing, and maintain network facilities by repair personnel. The other context deals with monitor and control of network switching and transmission facilities. In this chapter, we focus primarily on the latter aspect of the network management problem because this will be more difficult in the international situation.

The primary reason network management is so difficult is that most vendors of communication equipment have addressed the basic operations of their equipment, leaving the issues relating to interconnection with other vendor's devices unaddressed. There are efforts by major computer and telecommunication equipment manufacturers to create integrated multivendor environments and the international standards organizations are very active as well. Few such approaches, however, are immediately available on the market. The two primary thrusts in this area are discussed in the following paragraphs.

8.7.1 Network Monitoring and Control

Integration of monitoring and control is the primary objective of standardization and implementation efforts. The three aspects that must be controlled are the physical devices, the logical devices (e.g., applications), and services provided by third parties. Strategic units indeed share many of the same needs and problems as telecommunication carriers and information networks with regard to dealing with these three elements. Network management systems have primarily been created for dedicated networks with common interfaces. For example, Codex offers an integrated NM solution for their modem and statistical multiplexer products. Packet switches are offered with network management software and interfaces to allow the public or private network operator to control the traffic and facilities. Likewise, high-level multiplexer products from NET, Stratacom, General DataCom, and others, have elaborate NM background environments to centralize the operation of this type of backbone network. If the international network is devised for a single application such as a computer reservation system involving only low- and medium-speed data transmission or packet switching, a proprietary NM approach is entirely appropriate.

The real challenge, however, is to take such proprietary NM systems with other devices having more rudimentary M&C capability and to combine them into a cohesive whole. Telecommunication network managers often find themselves with a seemingly incompatible inventory of circuits and equipment. This raises the practical problem of how to integrate NM without discarding the entire network and starting anew with the latest in industry standards. There are *ad hoc* approaches for overlaying a workable NM system, using interface boxes and specially programmed microcomputers. As a case in point, Teknekron Telecommunications Systems of Berkeley, California, has developed this type of approach to a particularly effective level. Other vendors of NM workstations also have good capability to improve the responsiveness of NM systems in today's environment. Two leaders in the field are AT&T and IBM, with the former pursuing an open architecture and the latter emphasizing the use of a tightly designed proprietary M&C structure. As the two fundamentally different structures under development in 1990, AT&T's Universal Network Management Architecture (UNMA) and IBM's NetView present a basic dilemma for the telecommunication manager who wishes to create an extensive integrated network management system for worldwide applications.

The AT&T UNMA formulation is conceived in the context of international standards conforming to the Open Systems Interconnection model. AT&T's desire is to divide the NM problem into the following four levels [Sloane, 1989]:

1. The on-premises elements, such as PBXs, PCs, LANs, modems, wiring, mainframe computers, and multiplexers;
2. The local exchange carriers, who provide switched and private line services and access to other long-haul carriers;
3. The interexchange (domestic long-distance) portion of the network, typically provided in the United States by AT&T, MCI, and US Sprint, and in other countries by a single governmentally owned or authorized carrier;
4. The international carrier, which maintains international long-haul circuits and has a relationship with foreign service providers.

This model is only now being expanded to the fourth level and much attention will continue to be focused on this area. AT&T and the other leading international carriers have a sizable stake in being able to solve the global networking problems of large strategic units that demand integrated NM approaches.

The drive toward integration of different networks and NM systems through standards is accelerating among equipment and software suppliers, telecommunication carriers, and major users. In the United States, much of this standards-making activity is by two OSI-based groups: the National Institutes of Science and Technology (NIST) OSI Implementors Workshop, and the OSI/Network Management Forum. As discussed in [CSI, 1989], these two groups do not implement standards, but rather lay the conceptual groundwork so that implementation is possible by software developers, equipment manufacturers, and service providers. Although the agreements are not specific enough to build hardware or write software, they promote compatible structures among NM systems of different suppliers that choose to follow the approach.

The particular NIST committee involved with NM is the NM Special Interest Group (NMSIG), which includes all interested parties. With government oversight, there is a general impression of fairness so that those who attend will accept the standards. Communications Systems and Information Group is of the opinion that government involvement is problematic insofar as other countries may view the United States as having too great an influence. In contrast, OSI/NM Forum is made up of vendors and telecommunication carriers, giving it a decided bias for action. These members need solutions quickly so that they can better manage their networks and sell products and services. In this case, the potential criticism is that membership is more restrictive, precluding users in particular.

The OSI approach to NM is clearly attractive for telecommunication managers needing solutions that can be applied for diverse networks. If major equipment vendors along with international carriers and PTTs can adopt OSI NM linkages, network management can become a centralized process for strategic units. This process will obviously take time. The prudent telecommunication manager will follow these developments and work with those vendors that also are actively pursuing OSI standards. Data networks for the U.S. government are to employ OSI-compliant protocol sets, a policy intended to reduce the grasp of proprietary systems and to increase economy of scale. The European TAs within the EC are also supporting OSI applications as a means to simplify network integration and management, which tends to create a more open competitive environment.

IBM NetView, discussed in [Elbert, 1989], is an important facility for the IBM brand of information processing in host-based networks. With increasing effort, IBM has extended the reach of NetView so that peer-to-peer applications can be provided over a well managed network. One of the interesting innovations is the use of one host node as the NetView server (on a grand scale), providing the intelligence to control, monitor, configure, and guide the entire information processing environment. Reliance on a single host for NetView may seem risky, as IBM has arranged it so that the host assists in establishing sessions; once established, however, the NetView host can disappear entirely. Reports of session completion, for example, are sent to the host over the SNA network for information purposes, but are not actually required to maintain the session. New sessions, however, cannot be established until the NetView host is reactivated.

Whereas standards are developing somewhat slowly, suppliers of telecommunication equipment have taken concrete steps to enhance network management. We have mentioned how intelligent multiplexers, packet switches, PBXs, and the like allow network managers to access the "brains" of the system for NM purposes. This type of capability is inherently usable in a future integrated NM environment. The only problem is that a conversion process must be added to the particular system so that it can be controlled through the standard OSI-based NM system. Looking at NetView, IBM has already created such a conversion system, called NetView/PC. The concept is that a dedicated IBM PC-AT provides the conversion of the proprietary NM link into a NetView compatible format. A unique set of routines must be developed by the supplier and subsequently accepted by IBM. NetView/PC has been deployed in SNA networks of major organizations like American Express and EDS, and much of the original promise has been delivered. There is much that needs to be done with the adaptation approach, however, and there is vocal displeasure among equipment vendors and users alike. One of the early shortcomings of NetView/PC is that it conveys status and performance information while limiting the ability to control remote non-IBM facilities.

8.7.2 Administrative Operations

The administrative side of NM, which we call administrative operations (AO), is an important telecommunication management function for strategic units. AO comprises service provision, service maintenance and quality, and money. The network is a key resource and, once users receive services, breakdowns are unacceptable. Furthermore, the costs of creating and operating the network must be covered through revenues or economies achieved.

Unfortunately, international telecommunication management is often unmanageable. There is a concept that can easily apply: when situations get very complicated (as they often do in international business), keep things simple! As a telecommunication manager, you probably should not try to "micro-manage" an international network. Rather, a straightforward approach to handling service installation, maintenance, and changes should be followed. Use operational networks that are well maintained and reliable for vital traffic. Have customers pay for what they use — perhaps even by direct billing from the foreign TA or service provider. Telecommunication management would appear to be more like a consulting service for remote business operations as opposed to an international utility.

Chapter 9
INTERNATIONAL TELECOMMUNICATION REGULATION AND TRANSBORDER DATA FLOWS

A concern of telecommunication managers is that international rules and regulations can cause major difficulties when planning and implementing a telecommunication network project. Telecommunication law is a very complex subject when individually looking at the United States, Western Europe, or Japan; viewed as a whole, there is little consistency from region to region or even country to country within a given region. Fortunately, certain basic principles can be applied, although all will not apply uniformly across the globe.

Organizations such as the United Nations (UN), the International Telecommunication Union (ITU), the International Organization for Standardization (ISO), and the Commission of the European Communities promote the standardization of networks and services by holding conferences and publishing recommendations. Among industrialized countries, many of the obstacles and barriers are being removed through these efforts, and telecommunication management is becoming somewhat simpler. Developing countries still pursue more parochial policies of protecting domestic industry, however, and even attempt to use the international regulatory process to divert resources in their direction. The telecommunication manager must allow for these problem areas and maintain a watchful eye on developments in international regulation.

To aid the reader in the area of regulation, we provide a review of the international organizations that are relevant to network development and operation. This review is not exhaustive, and readers are encouraged to pursue additional study by using the references. One actually needs to investigate some particular problems to gain a feel for the severity of a particular regulatory obstacle in a particular country. There is, however, a methodology for proceeding in a new country, as discussed later in the chapter.

9.1 THE INTERNATIONAL TELECOMMUNICATION UNION

Founded in 1865, the ITU is the oldest and most prominent of the international organizations in the field of telecommunication regulation. As a specialized agency of the United Nations and with its origins dating back before the age of radio, the ITU has proved itself in a changing and turbulent world. Today, the union has more than 150 member nations and its headquarters is in Geneva, Switzerland. Membership is restricted to government telecommunication agencies, called *Administrations;* each is required to contribute to the union's financial support more or less in proportion to its economic strength. Telecommunication carriers, equipment operators, and users are represented by their respective governments, and often attend meetings together with their governmental counterparts.

The objectives of the ITU, reflecting the shared purposes of the Administrations, are:

1. To maintain and extend international cooperation for the improvement and rational use of telecommunication of all kinds;
2. To promote development of technical facilities and their efficient operation, so as to improve telecommunication services and increase their usefulness and availability;
3. To harmonize the actions of nationals in the attainment of these goals.

Generally speaking, the ITU does not have direct power to regulate telecommunication and radio transmission, as these are considered to be the right of sovereign states. Because the ITU is a common body among governments, however, its regulations and recommendations are followed by the member nations due to the practical need for a realistic framework. There have been instances where politics has entered into the process when particular groups, notably the developing countries, see the opportunity to work as a bloc and pursue parochial goals. Developing countries have a special place in the ITU because of the shared objective of improving worldwide telecommunication services. A country with a poor telecommunication infrastructure cannot provide international telecommunication services of good quality. Furthermore, there is the more general UN mission of improving the living conditions of the world's population, for which the ITU implements the telecommunication aid program.

There are two basic areas where the ITU regulates international telecommunication. The first principal area is management of the radio frequency spectrum, a critical role because individual countries and operators of radio transmitters need "rules of the road." Otherwise, radio stations would transmit on any frequency that they chose, and radio frequency interference would be very common; therefore, no one would be able to expect reliable radio broadcasting and point-to-point radio communication. Through international conferences and the *Radio Regulations,* the ITU oversees the use of radio frequencies and provides an effective forum for the resolution of interference difficulties. The union does not, however, actually police the airwaves.

The second principal area is the interface of telecommunication networks with one another. In Chapter 3, we discuss the *Recommendations,* which define interfaces between telephone, data, and other networks. Without these definitions, telephone calls could not be completed automatically across borders. With the digitizing of public networks and expansion of new data communication services, the importance of ITU activity has increased significantly in the past few years. For example, the newly adopted (1989) International Telecommunication Regulations officially recognize "special arrangements" for innovative services and applications not generally available. Subject to national laws, member nations may allow Administrations and private organizations to enter into special arrangements with entities in other countries to establish, operate, and employ specialized networks and services within or between the territories of those involved.

In terms of the ITU's structure, the General Secretariat is the administrative arm, which organizes the operations of the union. The real regulatory functions are implemented by the *International Radio Consultative Committee* (CCIR), the *International Telegraph and Telephone Consultative Committee* (CCITT), the *World Administrative Radio Conferences* (WARCs), the *World Administrative Telegraph and Telephone Conferences* (WATTCs), and the *International Frequency Registration Board* (IFRB). To facilitate international cooperation, the ITU operates as does any UN agency, having simultaneous translation into the major languages of the world. Critical aspects of the regulatory functions of the ITU committees and conferences are reviewed in the following paragraphs.

The charter and administrative operations of the ITU are established by international treaty. Periodically, these provisions are modified by a Plenipotentiary Conference (the *"Plenipot"*), where new officers are elected and the budget and agenda for the union are set for the future. A recent Plenipotentiary was held in June 1989, in Madrid, Spain. While the international conferences have become increasingly important over time, the cost of ITU operation has continued to escalate. Because the Administrations ultimately pay for the operation of the union through allocations (the United States having the biggest), there is continued pressure to reduce the number and duration of the conferences and to reduce the full-time staff. The General Secretariat in Geneva runs the ITU on a day-to-day basis, maintaining a permanent staff and arranging future conferences. All of the regulations, reports, and supporting documents are published and distributed to members. Copies of documents, which are quoted in Swiss francs (and are expensive), can be obtained from the ITU's publications division in Geneva.

Another function of the ITU is to further world telecommunication development by assisting developing countries with the modernization of their telecommunication infrastructures. The UN Development Program (UNDP) and its telecommunication activities are managed by the ITU, consisting of projects for telecommunication systems and training; maintaining a fellowship program that allows candidates in industrialized countries to obtain training; and offering administrative assistance to developing countries by conducting seminars, studies, and training programs. For example, the ITU supports the Telecommunication Training Institute in Indonesia, which trains technicians for the domestic telephone network operated by the Indonesian PTT, PERUMTEL.

9.1.1 Radio Frequencies and Broadcasting

The ITU administers the worldwide radio frequency spectrum, providing the primary forum for the subdivision of the spectrum for the various radio communication services that are in use today. The ITU's particular challenge is to allocate spectrum in such a way that the needs of all members are met and new technology can be accommodated. Traditionally, the area of greatest interest was in medium- and high-frequency channels (below 30 MHz) used for AM radio broadcasting, particularly because these "short wave" signals can propagate well beyond the borders of a country and may interfere with one another. *Radio frequency interference* (RFI), whether harmful or merely unacceptable, is something that the ITU seeks to avoid.

More recent technology like terrestrial microwave radio for line-of-sight transmission (between 1 and 30 GHz) and satellite communication for domestic and international services have challenged the ITU to adapt its ways and to produce workable regulations. It has been able to do this for the most part. On occasion, regional politics have materialized and some ITU actions unfortunately have not been in the best interest of the overall mission to facilitate the use of the spectrum. For example, South Africa was expelled from the ITU for political reasons, and now that country could choose to ignore the rules designed to prevent RFI (although it generally does not). Another case in point is the attempt by Colombia to claim that portion of the geostationary orbit which lies 22,000 miles above its territory along the equator. A much larger and more effective political force is the interest of developing countries, particularly those in Africa, to obtain special privileges. For the time being, such political moves are handled appropriately without subverting the basic mission of the ITU.

9.1.1.1 *The Radio Regulations*

The Radio Regulations of the ITU contain the rules and procedures for the planning and use of all radio frequencies by Administrations, which, in turn, assign individual frequencies to their respective government and private users. The Radio Regulations are contained in a set of two (or possibly three) loose-leaf volumes, called the *Red Books,* organized more or less topically. Unlike legal codes, the Radio Regulations are usually not drafted by lawyers, but rather by experts in frequency management and telecommunication. A key section shows the allocations of frequency bands to particular services, such as the Broadcasting Service, the Fixed Service (for communication between radio stations which are fixed on the surface of the earth), the Mobile Service (for ships at sea and land vehicles), and various satellite services for broadcasting, fixed communication, and mobile communication as well. In addition to being divided by service, there is a vertical division by geographical region. As mentioned in Chapter 1, Region 1 includes Europe and Africa, Region 2 comprises North and South America, and Region 3 consists of Asia and the Pacific. The Table of Frequency Allocations, as it is called, contains dozens of footnotes, which allow Administrations to make exceptions to the rules when assigning frequencies to particular services. For example, a footnote may permit a country in Region 1 to assign a terrestrial microwave radio user to a frequency in a microwave band allocated for satellite communication.

Key parts of the Radio Regulations define how an Administration can employ a particular allocation of the spectrum and assign a frequency to one of its users. The procedures are often quite complex, and may even lead to confusion and a lack of resolution. Theoretically, an Administration checks the allocation to see if the frequency fits a predefined category. If so, it follows the procedure for informing other Administrations whose radio communication services may be subjected to interference by the operation of the proposed radio station or stations. Space-based radio transmitters are more of a concern because of their greater potential to radiate into the territory of a larger number of countries. If the transmission can cause interference, the Administration must follow the convoluted process of *frequency coordination,* which is defined in the Radio Regulations. This gives the other Administrations a chance to decide if they want to allow this particular station to go on the air. A successfully coordinated frequency assignment is recorded in the *Master International Frequency Register* (the "Master Register") of the ITU and thereby gains international status.

The following is a brief overview of some of the parts of the Radio Regulations and ought not be used in an actual application. There are basically two types of frequency coordination: *terrestrial coordination,* for land-based microwave transmitters; and *space coordination,* for radio transmitters and receivers on satellites. Terrestrial coordination involves any land-based radio transmitter that potentially can radiate signal power across a border into a neighboring country. For example, the Radio Regulations and application-specific CCIR Recommendations provide technical analysis procedures to compute the *coordination contour,* which is a graphical depiction of the expected and worst-case power levels from a transmitting earth station after propagating through the atmosphere. A neighbor Administration analyzes the coordination contour to determine whether this level of power can interfere with the operation of domestic radio receivers that employ the same frequency band. If so, the two Administrations would, on a bilateral basis, make an agreement as to which frequencies would be used or how the transmit radiation pattern of the offending earth station should be altered. After coordination is complete, the Administration can register the new frequency assignment with the IFRB in Geneva.

The comparable process for a new communication satellite is similar but more lengthy because of the potentially wide radiation pattern across the earth and due to more frequency bandwidth usually being involved. Before launching a satellite and beginning operation, an Administration must take an extra step. It must inform all other Administrations of its plans through an Advance Publication of Information, which is published by the IFRB in its *Weekly Circular*. This first step is only a prerequisite for coordination, allowing any Administration to submit comments on how the new satellite network can interfere with the operation of its own existing or planned satellite networks. After a prescribed period, the Administration can initiate the coordination process for the satellite and earth stations in the network. Coordination is required with other satellite networks that fall into one of the following three categories:

1. Those that have been recorded in the Master Register of the IFRB;
2. Those that have already completed coordination;
3. Those that have entered coordination before the new network in question.

This system is referred to as "first come, first served" and has been the subject of controversy since the late 1970s. The fact that any satellite network which has at least entered coordination gains priority over newcomers is very significant to the entire process of launching a new satellite. Because of the importance of being early, the manager of a new satellite and earth station project should push the respective Administration to initiate the Advance Publication and Coordination process as soon as possible.

Coordination of satellite networks is a bilateral activity in which the newcomer must approach the incumbent and obtain its agreement regarding the potential for interference between systems. Such discussions and negotiations can take from months to even a year or more in particularly difficult or acrimonious situations. This difficult process could be helped by a new arrangement under the 1988 WARC Final Acts, wherein a Multilateral Planning Meeting (MPM) can be called by an Administration. The concept, which had not yet been tested in early 1990, is that the parties involved would work together more or less outside of the Radio Regulations to achieve a settlement.

The International Frequency Registration Board is the full-time body of the ITU that analyzes the proposed frequency assignments which Administrations submit for review. The IFRB "examines" the assignment to make sure that it fits the Table of Frequency Allocations and the coordination procedures have been followed. In particular, the new assignment should not cause unacceptable interference to an existing assignment that has already entered or completed review and coordination. While not conveyed with policing authority, the IFRB has power over Administrations because of the status given to frequencies recorded in the Master Register. When a frequency assignment is in coordination between two Administrations, the IFRB can assist in performing calculations of the expected level of interference and can make recommendations to the parties on how to prevent the interference.

Most of the time, coordination and registration are accomplished in a straightforward manner, taking anywhere from six months to three years, depending on the number of Administrations involved and the complexity of the technical analysis of potential interference. Occasionally, an Administration refuses to allow coordination of a new frequency assignment by another Administration. In one notorious case, a South American Administration refused to coordinate the assignment of radio frequencies to a U.S. domestic satellite, forcing the FCC to permit the satellite owner to proceed with operation. The Radio Regulations stipulate that the noncoordinated radio station (in this case, the satellite) can transmit only so long as it does not actually cause harmful interference to any legally operating receiver. Upon any complaint, the offending radio station must cease transmission. In this case, the South American satellite system does not actually exist, so the FCC bears little risk in its unilateral decision to allow transmission.

9.1.1.2 World Administrative Radio Conference

World Administrative Radio Conferences (WARCs) are held every several years to deal with the changing needs of Administrations for radio communication services. These conferences are also where the Radio Regulations are established and periodically rewritten. Even less frequently are Regional Administrative Radio Conferences (RARCs) held by one of the three ITU regions. The last RARC met in 1985 to develop a plan for Region 2 for the Broadcasting Satellite Service. Regions 1 and 3 already had developed their plans at the 1977 WARC, which was convened to create the planning process for this new service.

A typical WARC will last two months and be held at the conference center immediately adjacent to ITU headquarters in Geneva. The WARCs must consider the technical issues of interference between services as well as legal concerns for those who come first. Many weeks are taken to refine the two complex aspects until a general consensus can be found. Each country can express its position at general plenary meetings and may participate in smaller working groups that delve into critical issues. The Final Acts of the conference must be written, translated into the various languages, and approved by the entire body. Recommendations of the CCIR, discussed in the next section, are often used as the base for technical positions that the WARC takes and incorporates into future provisions of the

Radio Regulations. Some sections of the Final Acts become provisions of the Radio Regulations, and others become instructions to the IFRB and the CCIR. Following the WARC, the respective governments have the opportunity to confirm the Final Acts, incorporating them into domestic law. There have been instances where a particular government was not in agreement with the overall conference and so took exception to some provision or perhaps the Final Acts in their entirety.

A recent WARC concluded in October 1988, after three months of deliberations, and effectively rewrote the procedures for initiating new satellite communication systems. The 1988 WARC (WARC-88) was a difficult conference with a wide agenda, addressing the following [Taylor, 1989]:

1. Changes to the current frequency coordination rules for the Fixed Satellite Service (FSS), the allocation used by all domestic and international telecommunication satellites. The modified coordination procedures simplify matters for Administrations, but do not alter the basic approach, being the principle of "first come, first served."
2. Allotting new frequencies for the "expansion bands" to provide at least one dedicated orbit reservation for each Administration. This satisfies the political concerns of developing countries.
3. Allowing "additional uses" in the expansion bands, which can be introduced on a temporary basis before the more permanent allotments are applied in a particular region of the orbital arc. This opens the expansion bands for experimentation with new services such as mobile communications and private international networks.
4. Adding frequencies to the Broadcasting Satellite Service (BSS) plan which was first established by the WARC in 1977. The new frequencies are assigned to each country for the "feeder link" from the uplink earth station to the broadcast satellite.

WARC-88 had a difficult agenda and followed on the heels of a rather acrimonious conference in 1985, where developing nations presented resistance against serious progress. With three years between the two conferences, a more universal and conciliatory approach was developed. The conference was as long and arduous as any of its predecessors, but there was an atmosphere of goodwill and the need to work together toward the common goal of a viable set of regulations.

Under the new rules, a separate allocation of spectrum has been set aside for all Administrations, particularly those not having current launch plans. A satellite position is assigned to each country and held in reserve until the satellite is actually launched. This is designed to give the developing countries the confidence that a position in orbit will always be available for them. These countries had been concerned that the existing rules for coordination tended to favor the industrialized countries, which already had satellites in orbit and could afford to launch additional ones in the near future. The traditional approach to spectrum and orbit assignment is maintained for the bulk of applications. Even here, orbital space can become limited as satellites are squeezed ever closer together. In North America, spacing as close as 2° in longitude has been mandated by the FCC.

9.1.1.3 International Radio Consultative Committee

The allocation of radio frequencies by the ITU follows the principle of segregating services into different bands. This simplifies the job of evaluating the interference potential between services authorized by the Administrations on a worldwide basis. The technical characteristics of these services are studied and eventually defined by the CCIR, one of the two consultative committees of the ITU. In addition to supporting the WARCs and RARCs, the CCIR provides recommendations for the operation of radio communication systems, particularly those that convey signals and information across national boundaries. Broadcasting and satellite communication are two areas in which the CCIR is particularly active.

The actual work of the CCIR is conducted by technical experts from throughout the world. Every four years, there is a Plenary Assembly, where technical analysis procedures and standards are agreed. A key outcome is a set of technical volumes containing the well recognized CCIR Recommendations, which Administrations themselves use as guidelines for designing, operating, and coordinating radio communication systems. The volumes are organized according to the Study Groups, which are established at a previous Plenary Assembly, and which generate the reports and draft recommendations during each intervening four-year period. These Study Groups are listed in Table 9.1.

The specific composition of the Study Groups varies over time due to changes in technology and the interest of the Administrations. Until recently, Study Group IX on the Fixed Service using radio relay systems (i.e., terrestrial microwave) had

Table 9.1 Study Groups of the CCIR (1988)

Study Group	Title (Subject)
I	Spectrum utilization and monitoring
II	Space research and radio astronomy services
III	Fixed service at frequencies below about 30 MHz
IV	Fixed satellite service
V	Propagation in non-ionized media
VI	Propagation in ionized media
VII	Standard frequencies and time signals
VIII	Mobile, radio determination and amateur services
IX	Fixed service using radio relay systems
X	Broadcasting service (sound)
XI	Broadcasting service (television)

largely completed its work because interest in the technology reached a low. The introduction of wireless LANs and other packetized broadcasting techniques, however, could cause interest in this study group to increase. Study Group IV has been particularly important in recent years because of the focus of international meetings like WARC-88 on satellite communication. The most recent Plenary Assembly was held in 1988 prior to this WARC to ensure that the necessary technical supporting information and design tools for planning were in place. Another purpose is to allow representatives of the TAs and Administrations throughout the world to decide questions to be studied by the various Study Groups.

9.1.2 Interfacing Terrestrial Networks

The ITU is the premier world body dealing with issues associated with interfacing and interconnecting domestic and international telecommunication networks. The Administrations that constitute the ITU have a large stake in the efficient interoperation of their networks, as this is the only way that information can be effectively conveyed in electrical form between countries. Another important role is that of a standards-recommending body, providing the organization and discipline needed to propose technical and operational standards. Many nations of the world would follow these recommendations as if they were international law. These activities of the ITU are particularly important as the networks of the world become digitized and new applications and service providers appear. In fact, private and public international service providers, called Recognized Private Operating Agencies (RPOAs), are allowed to participate alongside government officials and PTTs. The old structure of analog networks and circuit switching is eroding, thus rendering obsolete the tried-and-true approaches that the ITU has previously supported.

In the next two subsections, we review the two key forums of ITU activity in the area of telephone, telegraph, and data networks. The International Telegraph and Telephone Consultative Committee (CCITT) and the World Administrative Telegraph and Telephone Conference (WATTC) are key world forums in which every nation participates. These are where the developing and developed worlds come together to resolve the interface problems that hamper the evolution of the information networks of the future. The cooperation of these groups, demonstrated at their respective meetings in Melbourne, Australia, in 1988, encourages us to have faith in the outcome.

9.1.2.1 *International Telegraph and Telephone Consultative Committee*

The CCITT has moved from a staid position as the housekeeper of international telephone and telegraph links to the forefront of discussion on data communication and ISDN standards. In line with the widespread acceptance of the CCITT among all Administrations, the participants in the various Study Groups and international meetings talk seriously about the importance of creating an efficient digital communication environment for public and private networking.

The Plenary Assembly in November 1988 was held in Melbourne, prior to the convening of the WATTC (WATTC-88) in that same city. Contained in the *Blue Books* are the results of the conference (including 368 new Recommendations) covering nearly all aspects of voice and data telecommunication network service [ITU, 1989]. This particular plenum made a strong statement about the future of the CCITT and the need for developing a more flexible structure. Under a resolution called the "Spirit of Melbourne," an accelerated procedure for the adoption of new technical standards was established. In particular, time-critical standards can be adopted before the next Plenary Assembly, which occurs four years after the previous one. The hope is that a new Recommendation can be created and approved as quickly as in nine months, which is nearly fast enough to be synchronized with the development of hardware and network capabilities that can employ the new standard. Some participants expressed the fear that the four-year adoption cycle was too long for the CCITT to remain competitive (in a standards-making sense) with other bodies such as ANSI in the United States and ETSI in Europe.

As with the CCIR, the CCITT has Study Groups that comprise delegates from Administrations, TAs, international carriers, and equipment manufacturers. The number of Study Groups and their respective responsibilities have varied over the years, but a total of 13 are in existence in 1990. These are listed in Table 9.2. Supporting these Study Groups are numerous Working Groups, which focus on particular issues, thus providing additional support and effort to the overall activities.

Table 9.2 Study Groups of the CCITT (1988)

Study Group	Title (Subject)
I	Services
II	Network operation
III	Tariff and accounting principles
IV	Maintenance
V	Protection against electromagnetic effects
VI	Outside plant
VII	Data communications networks
VIII	Terminals for telematic services
IX	Telegraphs networks and telegraph terminal equipment
X	Languages for telecommunication applications
XI	Switching and signaling
XII	Transmission performance and telecommunication networks and terminals
XV	Transmission systems and equipment
XVII	Data transmission over the telephone network
XVIII	ISDN

In the area of data communication, the work of Study Group VII has been key. This group produced the X-Series of Recommendations, including X.25 and X.400 for packet-switched data networks and electronic mail, respectively. Study Group VII is involved with implementing protocols and applications that follow the OSI reference model, requiring coordination with the ISO. Recommendation X.50 could also become extremely important as it sets standards for high-level multiplexing via public networks. This would effect the development of intelligent devices such as the T1 multiplexer and narrow-band multiplexers used on international data links (see Chapter 6). The benefit would be the introduction of standard interfaces and operating modes so that multiplexers built by different manufacturers could interoperate.

Gaining in importance is Study Group XVIII, on ISDN. Much of the focus of the CCITT and the ITU in general is on the promotion of modern digital networks for use by the Administrations. There is much activity in Europe, in particular, where the twelve members of the EC have agreed to interconnect their respective PSTNs by using ISDN technology and protocols. This is a precursor to the decidedly international trend, and Study Group XVIII will no doubt play a key role with facilitating recommendations that have worldwide acceptance. Some developing countries are already purchasing switching systems with the capability to introduce ISDN when the domestic backbone networks are digitized. This trend will strengthen the role of SS-7 as a standard pipeline to provide basic and enhanced telephone services. Importantly, this would be the first time that a single signaling system could become accepted throughout the world.

The 1988 Plenary Assembly recognized the role and importance of private telecommunication networks in the international environment. In the past, private networks were seen as a necessary evil, requiring PTTs to provide special services for companies engaged in such businesses as oil exploration and automobile manufacturing. According to John Ryan, head of Study Group XI, "Businesses now are demanding more control over their networks, and we will be looking closely at new switching and signaling technologies which enable them to do that more easily."

Study Group XVII, on data communication via the telephone network, has done an excellent job of defining workable standards for medium-speed modems. The V-Series of Recommendations, discussed in Chapter 6, are employed in modems working on analog telephone circuits, both private line and dial-up. These standards have resulted in a big market for manufacturers and have improved the performance of international data communication in the near term. At the Melbourne Plenary Assembly, Recommendation V.42 was adopted, providing standardized means of error correction for voice-band modems. Recommendation V.42 is an effective compromise, as it contains LAPM and MNP, two of the most popular error correcting protocols currently used in proprietary modems. This recommendation demonstrates the practical approach taken by the CCITT, encouraging the development of new technology that solves a real-world problem.

9.1.2.2 *World Administrative Telegraph and Telephone Conference*

Unlike the WARCs, which are routinely held to deal with operational problems of Administrations, the WATTCs are very infrequent. This irregularity is probably because the subject matter is incredibly complex and arriving at any kind of meaningful compromise is extremely difficult on matters of contemporary importance. As an exception, the Melbourne WATTC that was concluded in 1988 fulfilled its mission of resolving some key issues relating to the overall structure of international telecommunication networks. As stated previously, one important outcome was the creation of the International Telecommunication Regulations. Another was the agreement to sanction the continued development of specialized telecommunication services by public and private companies.

The Administrations actually are government agencies, and thus they support the principle that nations must retain their individual identity and sovereignty. An important issue at WATTC-88 was the balancing of national sovereignty with the birth and growth of independent telecommunication service providers. The RPOAs and other emerging business ventures need to have the freedom to experiment with new services like E-mail, teleconferencing, and document interchange; the public networks of the world, however, still need to work effectively and to interconnect with one another. Administrations are concerned that a special arrangement, through meeting a market need, could disrupt the public network and thereby work against the public trust.

WATTC-88 provided for this type of control of special service access in two ways. First, any arrangement should comply with relevant CCITT Recommendations. This places the burden on the provider and the CCITT, which must produce a new workable standard in response to an evolving need. The second approach is to give an Administration the power to allow a special arrangement to be introduced into its country (or, more importantly, *not* to allow introduction). These provisions give an Administration considerable control over any new and foreign concept, but represent the type of compromise needed to arrive at a consensus.

While the new regulations are still restrictive, they recognize the need for innovation. User groups that participated in the conference were generally pleased by the outcome. To quote Stephen Finch, head of the International Telecommunications Users Group (INTUG) [Pipe, 1989], "These regulations contain considerable freedom and flexibility within the boundaries of each nation's laws. Users participating here have a much better understanding of the telecommunications aspirations of all countries, and understand why they take different approaches to regulations. It is astonishing how many delegations sought our views during these deliberations.''

9.2 INTELSAT CONSULTATION FOR NEW SATELLITE SYSTEMS

The International Telecommunication Satellite Organization is a cooperative body established by international treaty. Because of this unique position of being both a commercial venture and an entity with treaty power, INTELSAT is particularly forceful in the universe of international telecommunication. Article XIV of the INTELSAT Agreement (which is the actual treaty) contains provisions whereby a member of INTELSAT must consult with that organization before implementing a separate commercial satellite network. This stipulation represents a second co-ordination process, beyond that mandated by the ITU Radio Regulations.

The language in Article XIV simply states that a member who intends to implement a satellite system that could either interfere with or compete with the INTELSAT system must "consult" with its Board of Governors (BG). The BG is the organ that effectively manages INTELSAT and the delegates to which have voting power proportionate to members' investment. After reviewing an evaluation of certain technical and financial information developed by the Executive Organ (the permanent staff of INTELSAT), the BG makes its "recommendations" as to technical compatibility and the possibility of "economic harm" that the new satellite network may cause to INTELSAT. All systems, whether domestic or international, must pass the technical compatibility criterion to be allowed by INTELSAT to proceed. Only those systems which could compete with INTELSAT, however, such as regional and private international satellite systems, are evaluated for the amount of revenue loss that INTELSAT may suffer.

INTELSAT consultation tends to be more difficult than ITU frequency co-ordination because economic as well as technical issues are involved. The BG typically reaches its conclusion by consensus and rarely totally blocks the creation of new satellite systems. In fact, political pressure from the largest members, the United States, the United Kingdom, France, Canada, and Japan, can counter the protective efforts of the Executive Organ or other members. A case in point is PanAmSat, discussed in Chapter 2, which was successfully consulted through IN-TELSAT, although the system was created to enter direct competition. These issues as well as the process are reviewed in the following paragraphs.

9.2.1 Technical Compatibility

Technical compatibility simply means that the proposed satellite network will not cause unacceptable interference to existing or planned INTELSAT satellite links anywhere in the world. Techniques for doing such analyses are nearly standardized and follow the basic approach taken by the ITU and the CCIR. INTELSAT, however, is almost always much more conservative than the ITU. This position is good for INTELSAT because it can make any changes that it wishes in the future without considering these separate satellite systems in detail.

In the case of RFI from a domestic satellite, the critical condition is always the case of an INTELSAT earth station located within the coverage area of the domestic satellite. This situation is common because nearly every country in the world has at least one INTELSAT earth station and potentially high interference cannot be ignored. The radiated power of the domestic satellite would be nearly maximum in the direction of the INTELSAT earth station, which should be pointed toward the closest INTELSAT spacecraft. The only way to control interference in this case is to spread apart the satellites so that the interference is minimized from the domestic satellite into the sidelobes of the INTELSAT earth station antenna. The other approach, and the one almost always taken, is for the new entrant to agree to restrictions on the frequency assignments and power levels that it employs on its domestic satellite.

Once INTELSAT's technical staff had determined that interference into the INTELSAT system would be acceptable, a report should be prepared for submission to the BG. If the technical data are in order, the BG typically approves the request for technical consultation, and that part of the evaluation is complete. Sometimes, the Executive Organ takes the position that the new system is incompatible, and makes recommendations on how the potential problem can be resolved. The requesting member may disagree with this conclusion, in which case an active debate ensues. Resolution of this dispute can last through several BG meetings for a year or more, but eventually a compromise is found.

9.2.2 Potential Economic Harm to INTELSAT

Regional and international satellite networks that can be used to convey telecommunication traffic between countries potentially may cause economic harm to INTELSAT. The basis for this concern is legitimate: INTELSAT is effectively the world's common carrier, offering satellite capacity to any member country and even countries (e.g., the Soviet Union) that are not members. The amount of investment in satellites and earth stations is in the billions of dollars, and so a reliable source of revenue is absolutely essential. Traditionally, traffic on the system has increased annually by several percentage points and revenue has increased accordingly. Entrepreneurs, telecommunication authorities, and regional cooperatives have been attracted to move into the field and to seize some of the existing thick route market or develop new ones.

As a treaty, the INTELSAT Agreement provides a mechanism for the system to protect itself to some extent from competition. In the 1970s, members wishing to launch domestic and regional satellites sought to avoid confrontation with INTELSAT. They constrained their satellite networks to supply only domestic services, forcing all transborder links to be carried by terrestrial systems. In the 1980s, however, new entrants such as Arabsat and PanAmSat successfully challenged the system. This produced a significant change in the environment, and the competing international satellite system became a reality.

9.2.3 Transborder Services on Domestic Satellites

Domestic satellites can provide transborder service because the coverage pattern usually has sufficient radiated power across a border to allow the full range of services. Any time that a border is to be crossed with telecommunications, however, an array of issues arises. In the current context, INTELSAT could consider transborder service as potentially causing economic harm to its system, in which case the entire consultation process would be required.

There are two basic arguments used to justify allowing transborder services to be rendered over domestic satellites:

1. The transborder link would be far less costly than the comparable link via the INTELSAT system and the service would be uneconomical if an INTELSAT satellite was required. Incidental radiation of television signals into the territory of a neighboring country can be received very inexpensively; having a dedicated INTELSAT link would be extremely expensive and would not be cost effective for reception by cable television systems, for example.
2. Transmitting to and receiving from an INTELSAT satellite would be very difficult from a technical standpoint as opposed to employing a domestic satellite system. There may not be a Standard A or B earth station in the vicinity of the point of service, or a double-hop through INTELSAT and a domestic satellite may provide unacceptable time delay for voice or data traffic.

To use a domestic satellite in the United States to serve points in Canada, the FCC would seek the approval of INTELSAT and the Canadian government. Economic harm to INTELSAT is unlikely in this case because there have never been links between these neighboring countries over INTELSAT satellites, nor would there likely be any in the future. Whether for one- or two-way transmission, INTELSAT consultation would still be required before links could be established. There is also the need to reach some kind of agreement on the nature of the services and the respective roles of the U.S. satellite operator and the Canadian operator, Telesat Canada.

Transborder links to Latin America and the Caribbean Ocean region are more challenging because INTELSAT satellites and earth stations are currently in use. The outcome will depend on the attitude of the foreign PTT toward the proposed service. There has been great difficulty to obtain the required authority to use a U.S. domestic satellite for two-way data communication between the mainland and an island in the Caribbean other than Puerto Rico. In another case, the Administration of Aruba, which is represented by the Netherlands in INTELSAT, sought and obtained full INTELSAT agreement for a transborder television link from Aruba to the U.S. Virgin Islands over the Westar 5 satellite. This easily went through the process, probably because there was no prospect of or interest in using INTELSAT for the service.

9.3 TRANSBORDER DATA FLOWS

A concern of many strategic units engaged in international business stems from difficulties in connecting computers so that conventional financial and business data can be automatically transmitted into and out of a country (e.g., on-line transaction processing). Multinational corporations, particularly those headquartered in the United States, wish to take advantage of information processing for such areas as employee records, financial reporting, and purchase order processing. Organizations, however, have encountered local restrictions on the transfer of such data out of a number of foreign countries. Without the use of data communication, a strategic unit would operate a separate data processing center in each country where restrictions were particularly onerous. This is costly and inefficient, as on-line records may not be available and transfer of key data can be delayed. In the worst case, selected data may have to be sent by messenger.

According to [Briner, 1987], governments and domestic industries in foreign countries pursue restrictions on transborder data flows because:

1. There is a concern for individual privacy; ironically, the United States was the leader in instituting privacy law, but only in connection with information held by the government. Other countries have extended this protection to private companies that hold data on its citizens. One problem is that a particular country may not allow a company to transmit private data to the United States because local law primarily protects data held by the government and not private organizations.
2. The domestic telecommunication and data processing industries could be bypassed by a strategic unit that connects to a system in the home country. As mentioned in Chapter 8, Brazil is a leader in such restrictions; unfortunately, for the telecommunication manager, many developing countries view Brazil as a pacesetter in this regard and have adopted similar policies.
3. National sovereignty should be preserved as well as cultural ties. This concern is more obscure than the previous two and mainly relates to information services brought into a country. A case in point is the importation of U.S. television services into Europe, an issue that continues to be debated in the EC.

The first concern applies when a strategic unit uses a computer network to manage the financial aspects of its business. Ideally, the network should operate as a single entity, with branch offices in foreign countries differing little from those at home. A restriction on particular information being transmitted outside of the country, however, could force a significant realignment of the data communication network architecture. A rule in Canada, for example, stipulates that all banks must maintain the records of domestic depositors in computers which are physically located in Canada, even if the bank is headquartered outside the country.

Fortunately, a majority of countries do not impose major restrictions on transborder data flow. There are, however, a few well known cases, principal among them being Brazil, Canada, West Germany, Mexico, and Austria. The United States has some restrictive regulations, particularly with respect to the exportation of information regarded as having national security significance. The concern is that sensitive reports, design information, and the actual products could slip into the hands of a present or future enemy to be used in developing war materiel. On a smaller scale, PTT regulations that force a telecommunication manager to use a less effective piece of equipment or network service can hamper the development of an effective information strategy.

Telecommunication managers need to evaluate the situation in each country for the type of facility to be installed. The in-country consultant mentioned in Chapter 8 should do the necessary research, possibly working with a local legal professional. In general, the business and government environment in the foreign countries of real and potential interest should be monitored for changes in policies on transborder data flows.

Chapter 10
TRENDS IN INTERNATIONAL
TELECOMMUNICATION

This final chapter is devoted to identifying trends that will continue to shape the international environment that the telecommunication manager will need to follow carefully. International telecommunication is fast becoming an everyday concern, as companies and governments realize that the future is necessarily global in nature. We have only to read the newspaper to see the dramatic changes occurring in the political and commercial worlds. These shifts are accompanied by hastened demand for improved infrastructures of telecommunication and information technology. Whether such improvements are made quickly enough remains to be seen. All told, the telecommunication manager will continue to be at the forefront, positioned to benefit from the trend by extending his or her understanding of how to operate in the still complex world of global networking.

Some of the primary factors which shape this trend toward a more unified global infrastructure are presented in Figure 10.1. At the center is shown the user, in a new position of leverage. Beginning at the bottom of the illustration, the technology of digital communication is at the foundation of network architecture. The newly commissioned networks are efficient in terms of the quantity and variety of telecommunication services that can be provided. More importantly, the facilities are said to be intelligent because they are processor and software controlled. Not only are intelligent networks more reliable and manageable than their predecessors, but capabilities easily can be subdivided to yield closed user groups that behave like private networks. As discussed later in this chapter, efforts are underway to create the intelligent networks of the future, which will carry all services, ranging from telephone and data to videophone and HDTV. A digital infrastructure can achieve such an end, but the need for this high degree of integration is yet to be demonstrated.

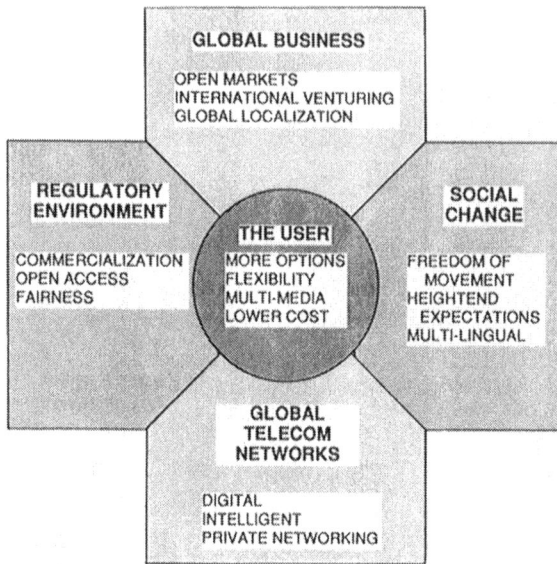

Figure 10.1 Primary factors which are shaping the future of international telecommunication.

If the recent experience of the European Community is an example, the old public telecommunication structures based on the PTT concept are being altered and even dismantled. The new environment (at the left of Figure 10.1) plunges the public telecommunication authority into the commercial world, where user needs are foremost and the profit motive is very strong. Governments are realizing that telecommunication is a business (and an attractive one, indeed). These new TAs are being allowed to finance their activities through private sources and to compete both domestically and internationally with other organizations. Governments also recognize the need to control the development of private telecommunication businesses within their borders, to prevent new monopolies, and to encourage some level of competition. As discussed later in this chapter, the principles of regulation by the central government, as demonstrated in the United States, are being examined with new interest and adapted to situations in other parts of the world.

There is significant social and political change in the world, as countries are discovering new freedoms. This is shown at the right of Figure 10.1. People have more freedom of movement between countries and are able to sample what others are enjoying, either by direct contact or, more conveniently, through electronic media such as television. Shown at the top of Figure 10.1, national businesses are combining with transborder partners to gain more market power and are becoming global in nature. In fact, as activities are spread over greater distances, companies need to develop procedures for more effectively managing local activities. Akio Morita, the chairman of Sony, has an expression for this, *global localization.* The concept is to integrate the overall operation through the effective use of information technology and to maintain degrees of local control by managers who have greater understanding of local markets.

Until 1990, the country where the most dramatic changes had occurred was the United States. With its natural resources, stable political system, and diverse population, the United States has prospered and developed as the economic and political leader of the world. What we are seeing in the 1990s is a shift of attention to other parts of the world. Already, Japan has taken the lead in manufacturing and exports; as a consequence, it is as strong financially as the United States.

The integration of the European Community may become a reality, coalescing the largest single market in the "free world." The Single European Act establishes December 31, 1992, as the deadline by which the EC is to become a single economic market, with fewer restrictions at the borders and a more integrated system of trade. How much integration actually occurs by that date remains to be seen, but many positive steps have already been taken as of 1990. There is the expectation that existing rivalries among individual telecommunication authorities will be subordinated to an overall mission of telecommunication integration. As a result, the fiber optic networks of the respective countries would be interconnected and open access to this infrastructure could be provided.

10.1 REDUCTION IN BARRIERS

Political and trade barriers have had a detrimental effect on economic development. This has manifested itself in Europe to such a degree as to produce the European Community, a powerful force for change. An elaborate governmentlike structure has been put in place, with permanent agencies dedicating themselves to outcomes of overall progress in a united Europe. In the years ahead, the EC could set the pace for international coordination of trade and transportation, telecommunication being one of the principal beneficiaries.

A particular epoch in EC development was marked by the passage of the Single European Act, which mandates a reduction in barriers by December 31, 1992. According to Article 8a of the Act,

The Community shall adopt measures with the aim of progressively establishing the internal market over a period expiring on December 31, 1992 . . . This internal market shall comprise an area without internal frontiers in which the free movement is ensured in accordance with the provisions of this Treaty.

This provision is still being interpreted by the European Commission, the Council of Europe, and the respective member governments. Although perfect integration and barrier reduction will not, in fact, occur, there are genuine efforts to improve the situation across twelve of the primary countries of Europe. Barriers will come down for many of the important aspects of physical trade; the same is potentially true for telecommunication and broadcasting.

Efforts toward integration are being conducted in earnest by the European Commission, but individual states still exert significant control over affairs. For example, member states may "take such measures as they consider necessary for the purpose of controlling immigration from third countries, and to combat terrorism, crime, the traffic in drugs and illicit trading in works of art." With regard to telecommunication and data processing, the respective countries are still at liberty to protect the private information of their citizens. As discussed in Chapter 9, this concern can be converted into restrictions on transborder data flows. Efficient network structure could be impeded by governments that use the principle as a means to force strategic units to use domestic computing facilities. Similarly, the European Commission is required to consider the possibility of economic harm to the existing TA structure throughout the EC. Although many TAs are concerned about the investments that they have made in terrestrial and satellite facilities, any steps taken to protect them from competition will tend to hamper the introduction of new services.

Companies in electronic media are concerned that Europe will raise high barriers to the importation of films and television programs produced in countries outside of the EC, notably the United States. Products of U.S. film producers are in great demand on the continent and sales represent a sizable source of revenue. Indigenous producers in Europe, however, currently do not have the financing ability of their U.S. counterparts, although efforts are underway in the United Kingdom, France, and Luxembourg to alleviate the situation. To counter this (taking a lesson from major Japanese manufacturers), U.S. film producers are establishing themselves in Europe through subsidiaries, joint ventures, and outright purchase of existing companies. After all, establishing a European production base to meet the political needs of the EC is not very different from the strategy recommended in Chapter 8 for telecommunication managers who must extend networks into foreign countries.

Barriers to telecommunication business development are also being reduced through the process of conversion of PTTs into more commercially oriented telecommunication service organizations. For the most part, the new telecommunication authorities are acquiring a profit motive, and therefore will better understand the relationship between serving customers and being profitable. In West Germany and the Netherlands, concrete steps were taken in 1989 to unleash the TAs so that they could be more businesslike. The final steps to privatization and competitive entry are still in the future for many countries, but the trend has already started in Japan and the United Kingdom.

On a worldwide basis, the trend toward the adoption of telecommunication standards based on ISDN is a very positive step. This will certainly reduce some of the technical barriers that still make global networking a challenge. As we have discussed in Chapter 9 and elsewhere, international standards agencies, such as the CCITT and ISO, provide forums for global telecommunication coordination. The agency in Europe that could become important is the European Telecommunications Standards Institute (ETSI); in contrast, CEPT, which is an organization of the former PTTs, has had problems coalescing the interests of members. ETSI has sought to simplify intra-European networking through the adoption of standards for terminal devices and ISDN features. This objective could motivate the TAs to open their networks to new applications and economically efficient services, which previously could not take hold. Experience in the United States has shown that such simplifications developed through industry groups such as ANSI and by the U.S. government have produced dramatic results for the overall quantity of telecommunication traffic carried on networks and the corresponding revenues that service providers enjoy.

10.2 ADOPTION OF AMERICAN REGULATORY PRINCIPLES

The United States has a history of employing government regulation of private industry, particularly for those which have established monopolies. In recent years, the government has sought to dissolve these old monopolies, but regulation still has a role in maintaining "rules of the game" that are reasonably fair. The newly rediscovered mission of regulatory bodies is to create a business environment where a company or individual is free to succeed or fail.

The U.S. federal agencies with a historical perspective on regulation are the Interstate Commerce Commission (ICC), the Civil Aeronautics Board (CAB) and, of course, the Federal Communications Commission (FCC). Both the ICC and the CAB no longer play a role in the United States but the FCC still serves several important purposes. Other U.S. government agencies that watch the operation of private companies and markets include the Federal Trade Commission (FTC), the Securities and Exchange Commission (SEC), and the Department of Justice. Bureaucracy has its disadvantages, as sometimes even simple transactions can take an inordinate amount of time.

The progressive nature of the U.S. regulatory structure is a relatively recent development. Much of the regulatory framework was set up during the 1920s and 1930s with an emphasis on controlling the private sector. As became recognized in the 1960s and 1970s, however, that regulation was serving its own masters and was not particularly positive for the overall economy. We can credit the administration of President Jimmy Carter for several of the key innovations, particularly in the deregulation of airline travel and shipping. President Ronald Reagan continued the pattern by appointing to these agencies people who had a free market bias. The FCC, under Chairman Mark Fowler, moved to reduce the controls on telecommunication companies to encourage innovation and competition. For example, satellite communication service providers were virtually deregulated in the 1980s. In addition, the Bell System breakup has both shaken and stimulated the telecommunication service and equipment markets in the United States.

Japan and Europe are examining the U.S. experience by using telescopes and microscopes, so to speak. The Japanese Ministry of Post and Telecommunications (MPT) has chosen to follow some aspects of the U.S. model as it institutes regulation of the newly opened domestic and international telecommunication markets. For example, a new entrant must file a tariff, which is reviewed in detail by MPT. Arbitrarily low prices are rejected, but the newcomer is allowed to offer service at a specified rate below that of the national carrier. MPT wishes to have competition but not when a new entrant in the Japanese domestic market engages in "irresponsible" business, causing all participants in the industry to lose money. The European Commission, conversely, is still studying the situation, taking a long-term strategic view of the future competitiveness of European industry. The EC considers information technology to be an important facilitator in business as it is applied across nearly the entire value chain.

10.3 MODERN DIGITAL NETWORKS

In our book on private networks [Elbert, 1989], we expound on how modern digital networks have become the common denominator in U.S. public and private telecommunication. The same can be said about the situation with respect to international telecommunication, although we must add several qualifiers. First, digital terrestrial facilities are generally available in the industrialized countries of North America, Western Europe, and Asia. INTELSAT has been in the position of providing digital links for more than a decade, but the user must be able to reach the national gateway or IBS earth station from locations within cities. A suitable local digital link must be available, which is often not the case. As discussed in Chapter 2, fiber optic cables are moving into operation across the Atlantic, Pacific, and Indian Oceans, delivering wideband digital capacity for use in private networks. In some cases, this capacity has already been claimed. Another concern is that a break in cable operation may lead to interruption of service unless a backup arrangement on another cable or by a satellite link has been set up.

Such short-term dislocations will be resolved during the 1990s so that users will be able to depend on having at least the backbone of facilities. There remains the challenge of interfacing and integrating the networks on an international basis. We have emphasized the importance of ISDN as a nearly universal standard architecture for international voice and data services. Terminal devices and switching systems will be standardized, allowing users to port their applications nearly anywhere in the industrialized world. Standardization is not the present situation as of 1990 because many of the details of the protocols (e.g., SS-7) are being debated by the public telecommunication operators.

As with any networking technology, once users have become accustomed to the operation of global ISDN, the situation will start changing again. There is a strong movement in Europe, the United States, and Japan to develop a broadband infrastructure with the ability to integrate voice, data, and broadcast-quality video services. This was discussed as a technology called *asynchronous transfer mode* (ATM). In Europe, the project is called *integrated broadband communications* (IBC). The timetable as of 1990 was to have IBC technology fully developed and in place to a significant degree by the year 2000 [Ungerer, 1988]. IBC is the cornerstone of the Research and Development in Advance Communications Technologies in Europe (RACE) program of the EC. Ideally, IBC would be a single solution for all communication services. The network would be programmed for mixed-mode multimedia communication for flexible adaptation to service needs and future extensions. There is even a place for HDTV. One fiber optic cable to the user's premises (home or business) would carry all services. The telecommunication authorities would then be responsible for carrying and transferring all services, which should provide a substantial economy of scale. Whether this project will be converted from research into reality remains to be seen, but the prospects are enticing.

The global digital infrastructure is useless if users cannot attach their devices and employ the networks in an effective manner. There exist today a wide variety of interface standards and network architectures, making the potential of telecommunication on a global basis difficult for users to realize. There are exceptions, of course. Telephone and facsimile service is truly global in scope, and users enjoy the effectiveness of these media. The challenge is to invent and promulgate the standards that will give other applications the same type of utility. This process is taking a long time; many feel that it is taking longer than it should. There are useful forums, however, such as the CCITT and OSI. Interestingly, vendors are making significant contributions by introducing technology and applications almost more quickly than the standards-makers can deal with them. As stated in Chapter 9, one of the outcomes of the 1988 CCITT Plenary Assembly in Melbourne was to introduce a "short cut" process for publishing new standards.

10.4 IMPROVED SERVICE QUALITY AND RELIABILITY

Modern digital telecommunication networks, particularly those based on fiber optic systems and two-way interactive satellite networks, have raised the quality of public and private telecommunication services. Today, one expects that a telephone call from New York to Los Angeles will have the clarity of a local call placed within a community. A similar trend is underway for transatlantic and transpacific telephone service. Satellite telephone circuits also have been improved through the use of digital links and echo cancellers. Underlying the transmission media are the switching systems, which have also been upgraded. Today, common channel signaling greatly speeds call setup and increases the probability of call completion on the first attempt. CCS also allows international carriers and TAs to offer new calling services for businesses, such as credit card calling and software-defined networks.

The reliability of these systems has increased markedly over the years, but this area is where future improvements will be the most dramatic. Users of private line services are accustomed to interruptions in service, which, in many cases, is unacceptable for business transactions. This problem is why private and public packet switching is gaining a strong foothold, particularly in the international area. Telecommunication managers, however, appreciate the flexibility and versatility of digital private lines because they are able to do their own aggregation of voice and data traffic. As the bandwidth on these links becomes more attractively priced, we can expect to see more international T1 and E1 networking. Reliability will then become paramount to users. The international carriers are responding to these demands by adding more digital routes, using a combination of fiber optic cable, digital microwave, and satellite links. As more private line traffic is applied to these networks, loading and profitability will justify the investments.

The current frustration of users with the reliability of international private lines can also be addressed with technology. Network management systems have become quite sophisticated due to the introduction of computer workstations and automated testing systems. The networks of the future will have more of the automatic routing and dynamic bandwidth allocation features now found in privately owned systems like the STAT MUX, packet switch, and high-level multiplexer used on T1 links. While already a facet of some of the PDNs, the network infrastructure of the future could be "self-healing," a buzz word that refers to the use of intelligence throughout the network to alter the actual routing of circuits without changing the apparent topology. Obviously, this entails a combination of diverse paths and automatic routing features. The technology exits in 1990, but the organizations will need to cooperate in creating this type of capability.

On a country-by-country basis, domestic and international carriers are feeling the effects of competition. Users are gaining leverage in dealing with service suppliers, often generating meaningful alternatives. This principle was addressed in Chapter 8. The domestic and international carriers, however, are behaving more like businesses. Profit motive is strong for two reasons. First, the owners of these enterprises want to see a return on their investment. Even a government will demand financial performance from the TA. Second, profit is needed to remain competitive, allowing the organization to add and improve facilities. For a business to thrive and make profits, however, there is a strong need for customers (this should be obvious, but some TAs have acted as if customers were a relatively low priority).

Telecommunication carriers of the public and private varieties will feel the pressure in two ways. First, they must attract customers to their service offerings. Second, and more importantly, they must hold onto those customers by maintaining satisfactory service quality and reliability. Binding the customer with a long-term contract may work temporarily, but smart users have a way of compensating. Service providers will quickly discover that the best way to build a long-term international telecommunication business is to continue to give customers value for their money. Many surveys have shown that, although first decisions are often made on price, customers remain loyal more because of superior service.

10.5 INFUSION OF INFORMATION TECHNOLOGY

The infusion of information technology into the international business environment is both obvious and important. As we have stated, information technology is pervasive in the value chain of businesses and even industries. Most companies in the United States, Japan, and leading European countries are avid users of centralized computer systems for a multitude of data processing applications. Networking in the information technology field is quite advanced for many of the specific applications, including reservations systems, banking, and point-of-sale inventory control. Information technology executives and telecommunication managers are accustomed to building networks using proprietary architectures such as SNA or DECnet; alternatively, the services of a PDN or value-added network provider are available.

Telecommunication authorities in some countries view this trend toward multiple vendors as a threat because it gives all the appearances of being uncontrollable. Users, however, enjoy innovating through the use of technology and services from a variety of sources. Managers of telecommunication and information systems face the complexity of integrating and managing the new multivendor environments. Conversely, without multiple vendors, information technology could not have become such a powerful business tool.

Extending these applications on a global basis is not a simple endeavor, as discussed under the topic of international data flows in Chapter 9. Some countries wish to protect their domestic manufacturing and service industries. For Western Europe, in particular, fulfillment of the aims of the Single European Act in terms of a removal of many of the petty border restrictions is very encouraging. Among these are the restrictions on transborder data flows, but also there are issues relating to connecting devices to public networks. Western Europe can be seen as a pacesetter for much of the Third World to follow.

10.6 WORLDWIDE INFORMATION AVAILABILITY

Efficient global telecommunication networks are not enough to bring worldwide information access to the point of critical mass. Convergence of information technology and telecommunication media has been predicted, but a crucial element is missing. This is the technology that allows users to access the system on their own terms, rather than those of the network or the service provider. For example, there are hundreds of different languages in use throughout the world. English is currently the *lingua franca* of international business, but the number of nonnative speakers is certainly small compared to the overall working population. The system should allow users to engage in their transactions in their own language, and not be forced to translate in their heads. Technology is under development in Japan that promises to give translation between languages in natural form. After this infant technology is converted to practical and inexpensive applications, there is no reason that language translation cannot be built into the information technology environments of the coming century.

Information is the underlying commodity of telecommunication. Whether in verbal, textual, graphic, or visual form, information is the *raison d'être* for billions of dollars of investment around the world. Telecommunication has evolved over the past century, and is realizing its potential as the world's fastest and most sophisticated highway of communication (broadly defined). Telecommunication represents an attractive business for service providers and manufacturers, a situation which will continue well into the next century. We hope that this volume will aid managers and users to position themselves properly for these developments in the international telecommunication environment.

BIBLIOGRAPHY

[Bell, 1983] R.F. Ray, ed., *Engineering and Operations in the Bell System,* prepared by Members of the Technical Staff, AT&T Bell Laboratories, Murray Hill, NJ, 1983.

[Borthick, 1989] Sandra L. Borthick, "Prime Time for PictureTel," *Business Communications Review,* June 1989, p. 62.

[Brabner, 1989] Steven Brabner, "X.500: A Global Directory Standard," *Telecommunications,* February 1989, p. 71.

[Briere, 1990] Daniel D. Briere, *Virtual Networks — A Buyer's Guide,* Artech House, Norwood, MA, 1990.

[Bullington, 1959] K. Bullington and J.M. Fraser, "Engineering Aspects of t.a.s.i.," *Bell System Technical Journal,* 1959, Vol. 38, p. 353.

[Bush, 1989] John Bush, "Frame-relay Services Promise WAN Bandwidth on Demand," *Data Communications,* April 1989, p. 102.

[Canright, 1988] Collin Canright, "EDI: Business Standards," *DATAPRO Research Reports on Data Communication Management,* Report CS93-220, Datapro Research, McGraw-Hill, Delran, NJ, October 1988.

[Carter, 1989] W.T. Carter, "An Evolutionary Path to Integrated Communications Using Pure ATM Techniques," *International Conference on Communications — Proceedings,* Boston, MA, June 1989, p. 0695.

[Caswell, 1988] Stephen A. Caswell, *E-Mail,* Artech House, Norwood, MA, 1988.

[Cerf, 1974] Vinton G. Cerf and Robert E. Kahn, "A Protocol for Packet Network Intercommunication," *IEEE Transactions on Communications,* Vol. COM-22, No. 5, May 1974, p. 637.

[CSI, 1989] *Network Management Perspectives,* Vol. 2, No. 8, Communications Solutions and Information Group, 3Com Corporation, Santa Clara, CA, August 1989.

[Curran, 1989] John R. Curran, "Using Testing to Pick the Right Multiplexer," *Data Communications,* July 1989, p. 113.

[DATAPRO, 1987] "An Overview of Value Added Networks," Datapro Research, McGraw-Hill, Delran, NJ, 1987.

[Elbert, 1987] Bruce R. Elbert, *Introduction to Satellite Communication,* Artech House, Norwood, MA, 1987.

[Elbert, 1989] Bruce R. Elbert, *Private Telecommunication Networks,* Artech House, Norwood, MA, 1989.

[Elliott, 1989] Dennis E. Elliott, "Global Communications: the Need for Connectivity," *Proceedings of the Eleventh Annual Conference,* Pacific Telecommunications Council, Honolulu, Hawaii, January 15–18, 1989, p. 131.

[Emmett, 1989] Arielle Emmette, "The Dual-Protocol Modem Duel," *Datamation*, February 15, 1989, p. 81.

[Foley, 1989] John Foley and Jennifer Samuel, "GE Goes Global with Net," *Communications Week*, May 29, 1989, p. 1.

[Hills, 1989] J. Hills, "Telecommunication Policy: The Movement Towards Liberalization and Privatization — Japan and Australia Compared," *ITU Telecommunication Journal*, Vol. 56-III/1989, p. 163.

[Humphrey, 1988] John H. Humphrey and Gary S. Smock, "High Speed Modems," *Byte*, June 1988, p. 102.

[Humphrey, 1989] John H. Humphrey and Gary S. Smock, "Whither the Modem," *Byte*, January 1989, p. 281.

[INTELSAT, 1988] *INTELSAT Earth Station Standards (IESS)*, Introduction and Approved IESS Document List, Doc. IESS-101 (Rev. 10), December 14, 1988.

[INTELSAT, 1989] INTELSAT, *Annual Report*, Washington, D.C., 1989.

[ISD, 1988] *International Satellite Directory*, 1988.

[ITU, 1989] "CCITT to Enter a New Era," *Telecommunication Journal*, Vol. 56-I/1989, p. 7.

[KJH, 1987], *IBS in Europe — A Guide to Services and Facilities*, KJH Communications, Atlanta, GA, 1987.

[KJH, 1988] *Wideband Digital Services in the Atlantic Ocean Region, Volume 1: Offerings of the European Carriers*, KJH Communications, Atlanta, GA, 1988.

[Korostoff, 1988] Katheryn C. Korostoff, "Packet Switching Equipment Revenues Down but Not Out," *Business Communications Review*, July-August 1988, p. 90.

[Kwok, 1989] Conrad K. Kwok and Biswanath Mukherjee, "Cut-through Bridging for CSMA/CD Local Area Networks," *International Conference on Communications — Proceedings*, Boston, MA, June 1989, p. 614.

[Lippis, 1989] Nicholas John Lippis, III, "Coping with the Cost Realities of Multipoint Networks," *Data Communications*, July 1989, p. 125.

[Lockwood, 1987] Russell Lockwood, "High Speed Modems," *Personal Computing*, December 1987, p. 233.

[Luhan, 1989] P. Luhan, "The Yearbook of Common Carrier Telecommunication Statistics, Published by the ITU," *Telecommunication Journal*, Vol. 56-IV/1989.

[Marney, 1986] Victoria C. Marney-Petix, *Networking and Data Communications*, Reston Publishing Company, Reston, VA, 1986.

[Matsumoto, 1989] Mitsuji Matsumoto and Norman D. Kenyon, "Standardization Trends in Audiovisual Service," *Pacific Telecommunications Council Conference — Proceedings*, Honolulu, Hawaii, January 1989, p. 147.

[McConnell, 1989] Kenneth McConnell, Dennis Bodson, and Richard Schaphorst, *FAX: Digital Facsimile Technology and Applications*, Artech House, Norwood, MA, 1989.

[McEntee, 1988] J. A. McEntee, "Rewiring the World — Fiber Optics in the Global Telecommunications Network," *Fiber Optic Sourcebook*, 1988.

[McQuillan, 1989] John McQuillan, "Interworking on the Frontier: Bridges, Routers and Gateways," *Business Communications Review*, July-August 1988, p. 19.

[Minoli, 1988] Daniel Minoli, "An Overview of Digital Cross Connect Systems," Report CA30-020, *DATAPRO Reports on Communications Alternatives*, Datapro Research, McGraw-Hill, Delran, NJ, October 1988.

[Morgan, 1988] Walter L. Morgan and Denis Rouffet, *Business Earth Stations for Telecommunications*, John Wiley & Sons, New York, 1988.

[Morita, 1989] Maseo Morita and Akira Tashio, "Financial VAN Service by FENICS," *Proceedings of the Eleventh Annual Conference*, Pacific Telecommunications Council, Honolulu, Hawaii, January 15–18, 1989, p. 484.

[Muller, 1989] Nathan J. Muller, *Minimum Risk Strategy for Acquiring Communications Equipment and Services,* Artech House, Norwood, MA, 1989.

[Nakamaki, 1989] Kyoici Nakamaki *et al.,* "Traffic Control for ATM Networks," *International Conference on Communications — Proceedings,* Boston, MA, June 1989, p. 713.

[NEC, 1987] Nobuhiko Shimasaki, ed., *NEC Research and Development — Special Issue on "ISDN" 1987,* NEC Corporation, Tokyo, Japan, 1987.

[Noll, 1986] A. Michael Noll, *Introduction to Telephones and Telephone Systems,* Artech House, Norwood, MA, 1986.

[Noll, 1988]. A. Mitchell Noll, *Television Technology: Fundamentals and Future Prospects,* Artech House, Norwood, MA, 1988.

[Nosaka, 1989] Kunishi Nosaka, "Digital Video Transmission by Satellite," KDD, Tokyo, Japan, 1989.

[Novy, 1987] Henry J. Novy, *World Communications — Ways and Means to Global Integration,* Gaston Lionel Publications, Istituto Geografico de Agostini Officine Grafiche Novara, Italy, October 1987.

[NTIA, 1983] *Telecommunication Policies in Seventeen Countries: Prospects for Competitive Entry,* National Telecommunication and Information Agency, NTIA CR 83-24, U.S. Department of Commerce, Washington, D.C., May 1983.

[NTIA, 1985a] *Telecommunication Policies in Ten Countries: Prospects for Competitive Entry,* National Telecommunication and Information Agency, NTIA CR 85-33, U.S. Department of Commerce, Washington, D.C., March 1985.

[NTIA, 1985b] *U.S. International Information Services,* National Telecommunication and Information Agency, NTIA CR 85-32, U.S. Department of Commerce, Washington, D.C., March 1985, p. W-9.

[NTIA, 1988] *NTIA Telecom 2000 — Charting the Course for a New Century,* National Telecommunication and Information Agency, NTIA Special Publication 88-21, U.S. Department of Commerce, Washington, D.C., October, 1988.

[O'Reilly, 1988] Cormac O'Reilly, "How to grow a world-class X.25 network," *Data Communications,* May 1988, p. 193.

[Owen, 1989] Richard Owen and Michael Dynes, *The Times Guide to 1992 — Britain in a Europe Without Frontiers,* Times Books Ltd, London, 1989.

[Pipe, 1989] G. Russell Pipe, "WATTC Agrees on New Telecom Rules," *Telecommunications,* January 1989, p. 19.

[Schwartz, 1987] Mischa Schwartz, *Telecommunications Networks — Protocols, Modeling and Analysis,* Addison-Wesley, Reading, MA, 1987.

[Seymour, 1988] Michael Seymour, "It's Got to be DCME — Says BTI," *Via INTELSAT,* December 1988, p. 3.

[Steinbruck, 1989] "Use of Satellite Trunks in a Packet Switched Public Data Network [PSPDN] and their Impacts on Performance," *IEEE International Conference on Communications — Proceedings,* June 1989, Boston, MA, p. 77.

[Sulkin, 1989] Allan Sulkin, "Hidden Costs of PBX Options," *Business Communications Review,* June 1989, p. 40.

[Sundstrom, 1987] R.J. Sundstrom, *et al.,* "SNA: Current Requirements and Direction," *IBM Systems Journal,* Vol. 26, No. 1, 1987.

[Taylor, 1989] Leslie A. Taylor, "Depoliticizing Space WARC," *Satellite Communications,* January 1989, p. 28.

[Thomas, 1989] Michael Thomas, "Technical Design Issues in International Private Networks," *Telecommunications,* March 1989, p. 39.

[Toyoshima, 1989] Kan Toyoshima, Misa-aki Sasagawa, and Ikuo Tokizawa, "Flexible Surveillance Capabilities for ATM-Based Transmission Systems," *International Conference on Communications — Proceedings,* Boston, MA, June 1989, p. 699.

[UN, 1987] *The Telecommunication Industry — Growth and Structural Change,* Economic Commission of Europe, United Nations, New York, 1987.

[Ungerer, 1988] Herbert Ungerer, *Telecommunications in Europe — Free Choice for the User in Europe's 1992 Market,* Office for Official Publications of the European Communities, Luxembourg, 1988.

[Unsoy, 1980] Mehmet S. Unsoy and Theresa A. Shanahan, "X.75 Internetworking of Datapac and Telenet," *IEEE Seventh Data Communications Symposium,* 1981, p. 232.

[Vignault, 1987] Walter L. Vignault, *Worldwide Telecommunications Guide for the Business Manager,* John Wiley & Sons, New York, 1987.

[Voros, 1989] Gregory L. Voros, "Forecasting the Future for Facsimile," *Business Communications Review,* January 1989, p. 44.

[Welch, 1979] S. Welch, *Signalling in Telecommunications Networks,* Institution of Electrical Engineers (UK), Peter Peragrinus, Stevenage, U.K., 1979.

[Yurow, 1983] Jane H. Yurow, ed., *Issues in International Telecommunications Policy,* The George Washington University, Jane Yurow Associates, Chevy Chase, MD, 1983.

INDEX

ABOUT THE AUTHOR

Bruce R. Elbert, a graduate of Pepperdine University (MBA), the University of Maryland (MSEE) and the City College of New York (BSEE), is a Member of the IEEE and the AIAA. As a senior telecommunication manager at Hughes Aircraft Company, Mr. Elbert planned and implemented networks and services in North America, South America, Asia and Western Europe. His experience in international telecommunications also includes work on the Technical Staff of COMSAT, and he has served as Assistant Vice President at the Western Union Corp. He is the author of numerous articles and two Artech House books, *Private Telecommunication Networks* and *Introduction to Satellite Communication* (1987). He is an experienced manager of telecommunication systems, not only having worked on COMSAT and in other private sector positions but also as a member of the United States Army Signal Corps in both the United States and Southeast Asia.

The Artech House Telecommunications Library

Vinton G. Cerf, *Series Editor*